D1154033

DEPTHS OF THE EARTH

The lure of caves. *Charlie and Jo Larson*

DEPTHS OF

REVISED AND ENLARGED EDITION

William R. Halliday, M.D.

DIRECTOR, WESTERN SPELEOLOGICAL SURVEY

THE EARTH

Caves and Cavers of the United States

HARPER & ROW, PUBLISHERS

NEW YORK, HAGERSTOWN, SAN FRANCISCO, LONDON

1817

Designed by Sidney Feinberg

Library of Congress Cataloging in Publication Data

Halliday, William R, date
  Depths of the earth.
  Bibliography: p.
  Includes index.
  1. Caves—United States.   2. Speleology.
I. Title.
GB604.H32   1976   551.4′4   75-6338
ISBN 0-06-011748-6

76 77 78 79 10 9 8 7 6 5 4 3 2 1

This book is dedicated gratefully to the American cavers whose names do not appear in its text: the unsung spelunker, the patient speleologist, and the expert caver whose name chanced not to be a part of this particular narrative. For their part in advancing our knowledge of our netherworld, all merit far greater recognition than I or anyone else can ever provide.

Contents

Introduction to the Revised Edition

Ten years ago, American speleology was appropriately proud of its achievements, chronicled in the first edition of this book. Yet, in that short decade, the face of American caving has changed radically. The Flint Ridge cave system now is part of Mammoth Cave. Jewel and Wind caves, and West Virginia's Organ or Greenbrier cave system have vaulted high on the list of the world's longest caves, shooting far past the Butler-Sinking Creek system, where progress has been maddeningly slow. New techniques have brought mastery—for some—of such superlative vertical caves as Fern and Ellison's, and hundreds more. Burgeoning hordes of expert cave hunters have unrolled entire new caving areas in the Rocky Mountains, Alaska, the Southwest, and elsewhere. The entire field of glaciospeleology has come from nowhere to become a major subscience in its own right. Under the wing of the new American Spelean History Association, scholarly research has restructured much traditional history of Mammoth Cave, and considerably altered that of many others, including Carlsbad Cavern. I personally have come to doubt so much of even respected historian Otto A. Rothert's old re-creation of the bloody history of Cave-in-Rock that I deleted that entire sequence from this edition.

In the course of the extensive revisions for this edition I again had the warm assistance of so many cavers and speleologists that the long list of acknowledgments merely begins to express my obligations and thanks. Inevitably the problem of selection was even worse than ten years ago. Alert readers will note many exchanges, replacing accounts of earlier triumphs with the new. I beg the indulgence of those whose favorite cave or favorite tale had to be shortened, replaced, or omitted. Omission of Minnesota's Mystery Cave and Florida's Warren Cave was painful, and that of the unparalleled months of Michel Siffre inside Texas. Yet everything has a limit. Only through the understanding indulgence of Harper & Row was I able to wheedle its editors out of much more drastic cuts.

Not all the changes of this past decade are splendid advances. Nickajack Cave is lost, and Marmes. Some of the finest in California's Mother Lode country have followed its once-celebrated Hawver Cave into the insatiable maw of the cement industry. But Hellhole was saved from a similar quarry in West Virginia, and as recounted in Chapter 16 preservation of caves from flooding and from pollution is suddenly making great strides. Conservation through development, too, is protecting many of our newfound wonders, especially in Arkansas, where both the National Park Service and the U.S. Forest Service have undertaken admirable projects.

Our next decade will bring equal or greater change, though no man can confidently predict their nature or location. The increasingly scientific nature of today's caving, however, strongly suggests that what is past is mere prologue. The grand Kentucky junction of Chapter 1 may well be announcing the golden age of American speleology. Only time will tell.

Yet the greatest possible single achievement of the next decade would be recognition by every caver of his responsibility—responsibility to each cave he enters and to his fellow cavers: his responsibility to leave each cave and each caver a little better for his passing.

Increasing population pressure is probably the greatest single present threat to our beloved caves. For every reader of this book, every concerned American caver has a heartfelt plea: help us save the irreplaceable glories of which I write. Join us and speak out persuasively in our cause.

If you yourself are or become a caver, act responsibly and bring credit upon all of us. As upon yourself.

To those who share in this, the best of our beloved netherworld—whether vicariously or as one in the forefront of our most formidable discoveries—good caving always!

WILLIAM R. HALLIDAY, M.D.

Seattle, 1975

Introduction to the First Edition

In nearly two decades of caving, it has been my privilege to watch American speleology come of age. When I joined the National Speleological Society in 1947, we thought we were doing well to locate a new cave or single virgin corridor. Now, from coast to coast, caves are being integrated into well-comprehended systems of remarkable complexity and size. State-wide surveys are increasingly advancing our systematic knowledge. We have come far, but much is still to be accomplished—in discerning the history of our caves as in exploration and underground study.

Though only a little of it intentionally so, much of what has been written about American caves has been erroneous. Underground, truth is more exciting than fiction. Yet it is sometimes difficult to reconcile the divergent viewpoints of cavers on opposite ends of the same rope. Fact often is irrevocably intertwined with cavern folklore. The first explorations of many of our most important caverns were not set down until much later—and then in clearly distorted form. Basic source material is terse, inaccurate, scattered, yet so voluminous that a team of historians would need a lifetime to locate and sift it all. Great gaps appear at crucial points, requiring the chronicler to re-create the scene if he is to bring the dramatic story to life.

To minimize these problems, I asked the help of leading cavers throughout the United States. That help was warmly given, often despite considerable difficulty. Not even such basic terms as "caver," "spelunker," and "speleologist" have the nationwide identity of context I necessarily give them here. Not all the experts agree and certain key references elude us all. If you note something that is not in accord with knowledge you may have, I ask your indulgence. It would be remarkable if the research for this one book successfully weeded out every error that has long been incorporated into traditional accounts. By and large, however, I think we have achieved considerable success. The very coherence of the narrative which has emerged is reassuring.

Inevitably, this book is a personal view of our caves and their physical and intellectual exploration. The names of some of our best-known cavers are absent from these pages simply because they are not a part of the particular caves selected to unroll this narrative. Someone else recounting his view of the enthralling story might choose different caves—and their explorers—for half his chapters. I have been underground in each of the major cave areas portrayed here, obviously more in some than in others. Yet if the reader seeks a first-person account of hair-raising exploits leading to ever greater records, he has the wrong book. Record breaking has its place. We all want to know which is the deepest or longest cave. But most cavers of my acquaintance find record breaking only a small part of our spelean satisfaction. There is glory underground, and excitement, and there are moments of awe in comparatively small caves that appear drab and unimportant to those uninitiated in their entrancing lore. We need no sensationalism, no overdramatization, no individual stars in our close-knit teamwork. It is our fervent hope that each reader will come to share our deeply rewarding comprehension.

In order to satisfy many who know little of caves, I ask the patience of my expert caver friends for what they may consider oversimplification —and for omitting many of their favorite caves and cave tales. Of less expert readers I ask equal tolerance of confusing cave names: Wind Cave and Cave of the Winds; Blowing Cave and Breathing Cave—and Overholt Blowing Cave. At last count there were eight Crystal Caves in California alone. Sometimes it seems that half the caves in the eastern United States are named Saltpeter Cave.

On first encounter, the language of cavers might seem an equal barrier. Nevertheless, most of the technical terms are standard, most cavers' cant self-descriptive. For those who have not previously differentiated "glacieres" from "glaciers," or have considered "chimneys" man-made objects for smoke, I have appended a glossary. Even advanced cavers may find it useful, for the caves of our different regions—and thus the language necessary to describe them—vary more than I would have thought possible a decade ago.

The frequency with which superlatives appear in this book might seem to negate what I have just written about overdramatization. Yet even the least of our caves is somehow exciting, entrancing, and these are extraordinary caves of which I write: our most magnificent, most historic, most intriguing, most challenging.

Here is the best of our beloved netherworld for you to share.

W.R.H.
1965

DEPTHS OF THE EARTH

1

The Eternal Standard

The Story of Mammoth Cave

Up to his quivering underlip in chill black water, John Wilcox momentarily halted, incredulous. Intoxicating triumph somehow sang through
his utter fatigue: "I see a tourist trail!"

Strange words these were for instant history. Yet, to thousands of avid
American cave explorers they spoke volumes. American cavers long had
insisted that this grand Kentucky junction someday would occur deep
beneath the earth: the greatest speleological achievement of all time. For
eighteen years, skillfully recruited teams of incredibly self-disciplined
cavers had sought to connect Mammoth Cave to their own beloved Flint
Ridge cave system. Yet these finest of American cavers had failed year
after year. When I wrote the first edition of this book in 1965, the grand
Kentucky junction seemed remoter than eleven years earlier, at the start
of the famous week-long C-3 Expedition into Flint Ridge's Floyd Collins
Crystal Cave. Now a rare combination of scientific exploration, dogged
determination, and just plain luck had seen the wistful dream come true.

To the older generation of American speleologists, newspaper headlines of the 1972 breakthrough brought a special satisfaction. Happily
they recalled bold 1955 headlines in the tradition of the endless controversies of the Mammoth Cave region: EXPLORERS DISCOVER WORLD'S
LARGEST CAVE IN KENTUCKY, and so on. Throughout America, they had
nodded enthusiastic approval. The forum then was the yuletide meeting
of the august American Association for the Advancement of Science. The
spokesman was lanky Brother G. Nicholas Sullivan, Roman Catholic

This view of the tourist trail in Mammoth Cave's Cascade Hall told John Wilcox that eighteen extraordinary years of systematic exploration finally had linked the Flint Ridge cave system to Mammoth Cave. *Cave Research Foundation photo taken on the first photographic trip through the connecting passage, courtesy Roger Brucker*

biology professor, equally at home in clerical and spelunking garb. Already he was an obvious choice for future president of the National Speleological Society, the world's most noted organization of cavers and cave scientists. His fellows were among America's most respected speleologists and the discovery locale was where everyone expected: Flint Ridge, the sprawling, sandstone-capped limestone plateau just north of Mammoth Cave Ridge itself.

The facts had seemed unassailable. A special National Speleological Society project had devoted eighteen months to systematic exploration, mapping, and study. Knowledge of the Flint Ridge netherworld had advanced far beyond the much-publicized C-3 Expedition in Floyd Collins Crystal Cave. Now hardy explorers had proven that intricate cavern "the nucleus of a great system of interconnected caves." Twenty-three miles had been surveyed, with another 9 miles explored. The total of 32 miles exceeded that of Switzerland's Hellhole Cave (Hölloch), where 30 miles was on record. Even more important, "the explorations showed two connections between Crystal Cave and other nearby caves." Much more cave might be confidently expected.

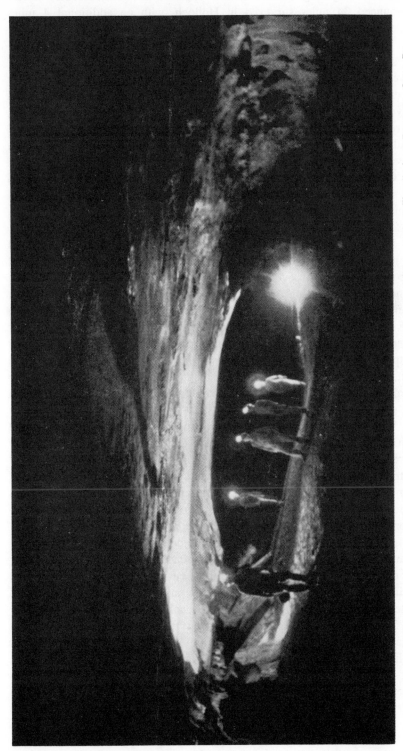

Deep inside Flint Ridge, members of the National Speleological Society's C-3 Expedition explore Floyd's Lost Passage. *Cave Research Foundation photo, courtesy Roger Brucker*

Then everything seemed to fall apart. Within days, cables tersely announced 34 miles surveyed in Hölloch, with an additional 4 miles under survey.

The Louisville *Courier-Journal,* too, scratched its collective head. The Crystal Cave property didn't seem large enough to hold 32 miles of cavern: "Controversies of the Kentucky cave country just don't seem ever to end. . . . If they're not all within the 302 acres of private Crystal Cave property, Mammoth Cave Superintendent Perry Brown reasons, there's no place for them to go except under national park property." And a local policy change had prohibited spelunking within Mammoth Cave National Park.

Congratulating Swiss speleologists, the Flint Ridge cavers attempted no self-vindication. Tersely they declined to state what caves they had connected or to release maps of their discoveries.

It seems almost as if the entire history of Mammoth Cave is keynoted by just such dreams, controversy, and imperfect knowledge. Traditional accounts place its discovery on one of several dates after 1800. Kentucky Land Office Certificate No. 2428, dated September 14, 1798, however, mentioned "two saltpeter caves": Mammoth Cave and nearby Dixon's Cave, which geologically is a detached fragment of Mammoth. No prior reference to Mammoth Cave is known, but its "discovery" by pioneer saltpeter seekers undoubtedly occurred several years earlier. Aboriginal American spelunkers preceded them by at least 2,000 years. Some of the hardiest mined gypsum and medicinal minerals almost halfway through the sprawling cave.

With the new nation's European sources of powder cut off by the British blockade of 1812, the pioneers' demands for saltpeter suddenly were eclipsed. The pace of the mining skyrocketed. Some 200 tons of saltpeter are said to have been extracted from its peter dirt. A gang of some eighty Negro miners worked with teams of oxen hundreds of yards inside Mammoth Cave. Without this 200 tons of saltpeter, the gunpowder-short War of 1812 might well have had a different outcome.

At the end of that war, the saltpeter boom collapsed. Following the death of a co-owner, Hyman Gratz purchased that half of the property for only $200. Ten years later he sold the cave to Franklin Gorin, the first white man born in Glasgow, Kentucky, and great-great-great-uncle of noted Nashville speleologist Standiford (Tank) Gorin.

Tank Gorin's affinity for caves seems to have been hereditary. His great-great-great-uncle owned Mammoth Cave for less than two years, but with him began the great period of exploration carried on by his successor, Dr. John Croghan. Gorin's nephew, C. F. Harvey, became lost somewhere in the complex cavern and was not found for 39 terror-ridden hours. Disturbed, Gorin "determined to have further explorations." A young slave named Stephen assumed increasing responsibility during these "further explorations." Soon he found leadership thrust upon his willing shoulders.

Already, Mammoth Cave had been widely proclaimed "the most extensive and stupendous vault in the world," the eternal standard to which all other American caves traditionally were compared. To those who saw the vast volume of the Historic Route illuminated only by flickering torchlight, the boast seemed appropriate. Each visitor who has halted spellbound on first glimpse of the immensity of the Rotunda, only a minute's stroll into the cave, will pardon that early exaggeration. Today's tourists cannot comprehend that Echo River and the Bottomless Pit then ended Mammoth Cave. Today, they are its mere threshold.

Electrically illuminated and bridged by a sturdy span, the Bottomless Pit irrevocably has lost its ferocious image. A widely printed 1837 account depicted its intimidating aspect in the dull glow of Stephen Bishop's lard lamps:

> The Bottomless Pit in the Mammoth Cave of Kentucky is suspected by many to run nearly through the whole of the earth. The branch terminates in it, and the explorer suddenly finds himself brought up on a projecting platform, surrounded on three sides by darkness and terror, a gulf on the left, and before him what seems an interminable void. He looks aloft; but no eye has yet reached the top of the great overreaching dome, nothing is there seen but the flashing of the water dropping from above, smiling as it shoots by, in the unwonted gleam of the lamps.—He looks below, and nothing there meets his glance, save darkness thick as lamp black, but he hears a wild mournful melody of waters, a wailing of the brook for the green channel left in the upper world never more to be revisited. Down goes a rock, tumbled over the cliff by the guide, who is of the opinion that folks come hither to see and hear, not to muse and be melancholy. There it goes—crash; it has reached the bottom. No—hark, it strikes once again; once more and again still falling. Will it never stop! One's hair begins to bristle, as he hears the sound repeated, growing less and less, until the ear can follow it no longer . . . for two minutes, at least, we can hear the stone descending.

None could see what lay beyond. Yet the key to Stephen Bishop's discoveries was his crossing of this dread pit on a rude cedar ladder, together with an unsung Kentucky gentleman-explorer named Stephenson. In 1963 Tom Barr and I stood on its rounded brink and picked out an easy-looking route along the pit wall. In the days of Stephen Bishop, however, powerful headlamps and belay ropes did not exist. Cave explorers were hardly accustomed to climbing along tiny chinks on the walls of Bottomless Pits. Those who halted, daunted by its monstrously shadowed depths, deserve no condemnation. Indeed, despite all the glories beyond,

I would not care to repeat Stephen's unbelayed crossing of that jagged void.

Some have deprecated Stephen Bishop's achievements on the ground that, as a slave, he had no choice. But Stephen went far beyond the grudging minimum of slavery. Famed traveler-author Bayard Taylor left us a vivid contemporary description of this extraordinary person:

Stephen, who has had a share in all the principal explorations and discoveries, is almost as widely known as the Cave itself. He is a slight, graceful and very handsome mulatto of about thirty-five years of age, with perfectly regular and clearly chiselled features, a keen, dark eye, and glossy hair and moustache. He is the model of a guide—quick, daring, enthusiastic, persevering, with a lively appreciation of the wonders he shows, and a degree of intelligence unusual in one of his class. He has a smattering of Greek mythology, a good idea of geography, history and a limited range of literature, and a familiarity with geological technology which astonished me. He will discourse upon the various formations as fluently as Professor Silliman himself. His memory is wonderfully retentive, and he never hears a telling expression without treasuring it up for later use. In this way, his mind has become a repository of a great variety of opinions and comparisons, which he has sagacity enough to collate and arrange, and he rarely confuses or misplaces his material. I think no one can travel under his guidance without being interested in the man, and associating him in memory with the realm over which he is chief ruler. . . .

Stephen and Alfred belonged to Dr. Croghan, the late owner of the cave, and are to be manumitted in another year, with a number of other slaves. They are now receiving wages, in order to enable them to begin freedom with a little capital in Liberia, their destined home. Stephen, I hear, has commenced the perusal of Blackstone, with a view to practicing law there, but from his questions concerning the geography of the country, I foresee that his tastes will lead him to become one of its explorers. He will find room and verge enough in the Kong Mountains and about the sources of the Niger, and if I desired to undertake an exploration of those regions, I know of few aids whom I would sooner choose.

It would indeed be heart-warming could I add that Stephen Bishop achieved his long-merited goal. It was not to be; manumission, yes, but not Liberia, for the illness which caused his death in July 1857, was soon apparent. For a score of years his body lay in an unmarked slave grave near his beloved cave until an indignant but frugal philanthropist purchased a second-hand tombstone in his memory. Somehow it ended up with the wrong date: 1859. Surely Stephen Bishop, to whom we owe so much of Mammoth Cave, deserves a better memorial.

After Stephen Bishop's far-flung explorations, few questioned the supremacy of "150-mile-long" Mammoth Cave. Intensive explorations continued intermittently for many years. Only an occasional iconoclast noted that the magic "150 miles of passages" never grew with the widening discoveries.

Stephen Bishop's secondhand tombstone. *Photo by T. C. Barr, Jr.*

In 1921 came the thunderclap of George Morrison's New Entrance to Mammoth Cave. As related in the next chapter, it opened up the entire east end of the cave and fanned the flames of the Great Cave War of Kentucky. The commercial caves in nearby Flint Ridge, mostly operated on a hopeful shoestring or fervent prayer, were among the fiercest com-

Two local recruits of the 1927 Neville expedition holding a cloth banner commemorating the event before leaving the cave. The same names appear on the fragment of breakdown behind the thinner figure. *Photo by Russell Trall Neville, courtesy Mrs. Burton Faust*

petitors for the tourist dollar. Perhaps sincerely, some of their touts—locally known as cappers—claimed lustily that *their* cave was part of Mammoth Cave. "That's what the perfessers from the city said!"

As at nearby Mammoth Cave Ridge, the beginnings of the story of Flint Ridge and its great underground network are lost in the shifting mists of time. Unlike Mammoth Cave, these were not saltpeter caves. Tens of centuries before the pioneers, however, long-vanished tribesmen mined Salts Cave minerals. In the early 1920s Russell Trall Neville used Salts Cave for much of the world's first underground movies. His 51½-hour expedition proved to forgetful skeptics that prolonged underground stays do not cause insanity or other dread maladies. Transcontinental caver George Jackson still shivers in recollection of Neville's sketchy sleeping arrangements during that bold venture: each explorer carried but a single blanket. "While there were plenty of soft, sandy spots we found few of them that were confined enough so that the warmth from all of our bodies would heat up the 'room.' "

Perhaps next discovered on the Flint Ridge was Unknown Cave, northwest of Salts Cave in an arm of Three Sisters Hollow. Unknown Cave seemed rather small and unimportant to most visitors. Found here, however, are marks smoked on the walls at obvious survey points, plus the name of Edmund Turner, the leader of the first speleological venture within Flint Ridge. Modern speleologists at first believed that neither Turner nor other early cavers made any effort to force a way through the great shaft and breakdown areas which terminate the entrance section of Unknown Cave. A little-known composite sketch map made in 1903, however, showed Unknown Cave as an entrance to Salts Cave. Apparently based on then-current local belief, other portions of this map are laughably incorrect, but the indication of this interconnection is unequivocal. Yet, through the decades, knowledge of the linkage—and of the 1903 map—faded from memory.

Shortly before 1900 someone found Colossal Cave, a major cavern located on the south side of Flint Ridge. Twelve hundred feet inside, explorers encountered a breathtaking 135-foot dome-pit—Colossal Dome. Below lay a broad, spacious corridor some 10,000 feet long with a stalagmite seemingly 75 feet high. Elsewhere its walls were marvelously encrusted with gleaming gypsum.

The Colossal Cavern Company surveyed its cave in detail and excavated a new entrance at the north end of this Grand Avenue. For many years intensive exploration and tourist promotion waxed simultaneously. Some 14,000 feet of passage were mapped. Another three miles seem to have been explored, including an underground river which proved elusive in later years. In time it became clear that Colossal Cave should have

The Neville expedition's uninspired sleeping arrangements in Salts Cave in 1927.
Photo by Russell Trall Neville, courtesy Mrs. Burton Faust

Using flash powder and glass plates or nitrate film, Neville accomplished some excellent photography in Mammoth Cave in the 1920s. This photo shows the famous Tiger Lily, an especially beautiful gypsum flower in the New Entrance section.

been discovered twenty-five years earlier. One narrow, sinuous side corridor of a nearby little cave, previously noted only for the discovery of an Indian bedquilt a quarter century before, proved to lead into Grand Avenue.

The great cavers of that fading era spoke enthusiastically of this splendid cavern. In 1903 French speleologist Max le Couppey de la Forest predicted the eventual connection of Colossal Cave to Mammoth Cave or "the nearby caves." Despite its spectacular vistas and glorious mineralizations, however, its commercial operation became a casualty of the Great Cave War.

Two names stand out in the early twentieth-century story of Flint Ridge: Floyd Collins and Edmund Turner. In 1925 Floyd Collins became a household word as a result of the Sand Cave tragedy described in Chapter 2. The name of Ed Turner is virtually forgotten, although his contributions were more soundly based and may well have surpassed those of his ill-fated fellow. But in 1917 Floyd Collins dug his way down a sinkhole just three hundred feet from his home on the far north rim of Flint Ridge. Inside was tremendous Crystal Cave. In eight years Floyd explored perhaps five miles of corridors, canyons, chambers, and crawlways. With his death, however, exploration and casual visits alike became unsystematic. Many remote corridors lapsed into legend for almost a generation.

With the exception of Colossal Cave, accurate mapping in Flint Ridge was as unpopular as in nearby Mammoth Cave Ridge, where the managers of the Croghan estate rightly feared the opening of new entrances beyond their property lines.

In 1904 E.-A. Martel, "father of modern speleology," crossed the Atlantic to emphasize to a distinguished American audience the need for accurate surveys of these caves. By coincidence, in that lecture he first brought to American attention the newly discovered Hölloch "with nearly six miles [of passages] which may prove the [largest] European cave when the present explorations have been finished. . . ." The explorations of which he spoke are still incomplete sixty years later, with more than twelve times that length of passage surveyed in Hölloch.

In the halcyon days of early Flint Ridge discovery, local explorers and visiting speleologists alike predicted linkage of the area's caverns into a single vast network. Through the years, however, the long-sought interconnections seemed ever more elusive. A contrasting concept developed: each cave seemed limited to its own arm or section of the ridge. Separating them were deep sinkhole valleys—perhaps the collapsed and eroded remnants of connecting caves. Passages might exist beneath narrow spurs which link the arms of the ridge. But these highland necks were disturbingly small. Furthermore, most of the major passages seemed to follow the edge of the ridge top rather than pointing toward the intervening sinkhole valleys. Morrison's New Entrance to Mammoth Cave led to renewed speculation, but skeptics pointed out that the entirety of Mammoth Cave lay within a single long ridge. Folklore insisted that Floyd Collins sometimes had popped out of holes miles away from the entrance of Crystal Cave. Such tales were so manifestly untrue, however, that by 1940 the entire concept of interconnections was largely discredited. From the whole era of purely local exploration, the only knowledge of linkage that survived was the unimportant one between Colossal and Bedquilt caves. In the years before World War II, exploration was increasingly relegated to the guides of the various caves which had been developed as tourist attractions.

At Mammoth Cave, the traditions of the guides were particularly notable. For some, guiding and exploration were part of a century-old pattern. For 150 years, theirs was at least a share of every major discovery. Thus it was scarcely unusual when Leo Hunt and Carl Hanson in 1938 pushed far up Roaring River, where even Stephen Bishop had turned back. Heavy rains cause Roaring River to rise as much as six feet an hour. At the notorious Keyhole, the ceiling is only four feet above the normal level of the black channel.

Streams of Mammoth Cave contain blind fish which rarely exceed three inches in length. *Photo by T. C. Barr, Jr.*

Hunt and Hanson alternately paddled and dragged a flat-bottomed scow up Roaring River. In a series of lantern-lit explorations they encountered a worthwhile succession of virgin side avenues and crawlways. Where digging was necessary, they recruited Carl's son Pete and Claude Hunt, Leo's cousin.

The going was tough. After many hours' struggle with primordial nature, the explorers were near exhaustion. Pete Hanson was in the lead, resting on his back just past a particularly nasty squeezeway. Idly he glanced upward, then sprang to his feet, fatigue forgotten. At his shout, the others rushed to behold a gleaming paradise.

Here, in this pristine New Discovery, glistening crystals of fibrous white gypsum bedecked an enlarging corridor which had never seen light. Hundreds of feet of snowy deposits covered walls, ceiling, even the floor. Here and there, larger clumps sprouted into crystalline flowers. In size they far surpassed any previously discovered in the far-flung cavern. But in this fabulous new natural palace, each was—and is—unsmoked, unvandalized. Foot-long petals of giant lily-like forms are rivaled only by strange gypsum corkscrews and huge transparent needles. Here is gypsum as delicate as cotton; nearby it is broad, firm, and glistening. And, as the marveling explorers brought their lanterns close for a better look, great clear needles began to wave in convection currents of heated air.

While news spread of this New Discovery, the guides returned again and again. More than four miles of virgin passage soon had been explored. The National Park Service concentrated its local research program in this unmarred area. The principal route was mapped and a new entrance blasted to bypass the tortuous Roaring River approach. Perhaps someday you and I will be privileged to see what guides Hanson and Hunt found.

In both Flint Ridge and Mammoth Cave Ridge, the death of Pete Hanson and other World War II tragedies divide the old from the new. One last prewar effort, however, laid the foundation for much that was to come. In 1941 Harry Dennison and Ewing Hood came upon rusting cans of Floyd Collins' supplies in the legendary Floyd's Lost Passage of Crystal Cave. As news of the rediscovery seeped around the war-torn globe, cavers of the newly formed National Speleological Society dreamed new dreams. At the end of the war a new breed of Flint Ridge cavers began to emerge—active participants in the society.

In 1947 Crystal Cave manager Jim Dyer sparked the first systematic explorations in Flint Ridge in two decades. Bill Austin, Jack Lehrberger, Luther Miller, and others soon joined him.

Except for Crystal and Great Onyx, all the major Flint Ridge caverns had been incorporated into Mammoth Cave National Park before World

Paddling such an unwieldy scow, Mammoth Cave guides advanced up Roaring River to the New Discovery. The paddler nearest the camera may be Pete Hanson but the identification is not confirmed. *United States Department of Interior National Park Service photo*

War II. Diurnal duties at first channeled the new efforts into nightly ventures deep into the complexities of Crystal Cave. With the consent of National Park Service personnel, however, the teams began to probe Salts and other caves. Soon the endurance barrier was a factor everywhere. In Crystal Cave, six struggling hours led only to the end of Floyd Collins' footsteps.

In 1951, Floyd Collins' old acquaintance Ellis Jones showed the modern cavemen a hidden passage Floyd had thought might lead to Mammoth Cave. It didn't—at least not exactly, or right away. Beyond lay a bewildering maze of great and small corridors, pits, chambers, and canyons. Time and energy always ran out—not the possibilities of the cave. Increasing numbers of "outside" cavers began to join the explorations, but they were equally defeated by the intangible barrier. A new approach was urgent.

In the summer of 1951 Roy Charlton proposed a three-day expedition. With the particular assistance of Joe Lawrence, it was attempted the following December. A fantastic outlay of energy bulldozed bulky equipment through crawlways as low as ten inches and a quarter mile long. Sleeping bags and other impedimenta had to be dragged laboriously across the lips of black "bottomless" pits. The back-breaking load precluded any significant achievement, but perhaps a full-scale effort. . . .

So, in 1954, the C-3 Expedition came to Flint Ridge, its leaders bursting with confidence that modern speleological techniques would readily yield them the long-dreamed link to Mammoth Cave. After all, were there not two lengthy caverns in between, placed exactly as if they had been planned for the purpose of eliminating the blank spaces on the underground map of the ridge?

Yet, a frustrating week later, hard-bitten Roger Brucker wept as he penned the final defiant entry in the expedition log. Despite herculean effort that yielded 2,000 feet of tangled new passage, no caver had even broken out of the little arm of Flint Ridge that contained Floyd Collins' Crystal Cave. Not one inch of the new-plotted passage extended into the huge white gaps on the expedition's large-scale base map.

But Roger Brucker is a stubborn man. He and many another C-3 participant vowed to return. More importantly, they vowed to learn how to conquer the infuriating "bowl of spaghetti" that is Flint Ridge's underworld.

Wherever cavers assembled in the eastern United States, restless discussions debated the future of Flint Ridge. The base-camp supply-team concept had been proven less useful here than most had hoped. The leaders of the Flint Ridge project began a series of experimental expeditions. Parties of various sizes entered Crystal Cave at every possible hour, under every possible condition. They remained underground briefly, or as long as twenty-four hours. Through the months, new techniques of rapid assault began to evolve. Crucial was the concept that every participant must be out of the cave before he became a burden to others.

Rather than mass assaults, the Crystal Cave optimum proved to be four to six experienced, conditioned cavers. Moving rapidly, such teams could arrive fresh at the start of their assigned tasks deep in the cave. Operating at maximum speed, halting only for snacks—some as odd as Red Watson's beloved canned oysters and cranberry sauce—these attack teams were able to push onward for fifteen to twenty hours. When the need for warmth, food, and rest became imperative, they could pass on their findings to a second team coming in to carry on their work for an equal period. Or half a team might map perhaps thirty stations, then trade tasks in surprisingly effective leapfrog advances. In certain key areas, far beyond the C-3 camps, caches of food and supplies slowly accumulated for additional support. In the following three and one half years, such teams spent more than 12,000 man-hours in Crystal Cave.

Crystal Cave was not the only target of these systematic advances. In November 1953 Jack Lehrberger and Jack Reccius probed a hole in Salts Cave that was seemingly just like a thousand other unimportant holes. This one continued, though the explorers had to move aside rocks to advance. Beyond almost half a mile of crawlway and an equal stoopway was an amazingly spacious avenue. Undisturbed artifacts revealed that Indians had mined here, entering through some long-sealed portal. Indian Avenue led excitingly toward the forbidding shafts of Unknown Cave.

Unfortunately, new restrictive policies then gravely affected explorations in the National Park caverns. Acting under specific orders, the park staff began patrolling the caves under their administration.

Normally, responsible American cavers honor any governmental regulation, then work to improve those which are unsound. But in Flint Ridge the situation was unique. Convinced that they were on the verge of a great breakthrough, participating cavers reacted in ways understandable only in this curious land of eternal controversies. Some gave up the project in disgust. Some sought redress through political channels. Others, less patient, found ways of eluding the patrols.

Especially well kept was the secret of Unknown Cave, officially nothing but a jagged pit and crawlway beyond a short entranceway. In November 1954 explorers who must still be nameless followed the footsteps of a select few Flint Ridge pioneers. Below Unknown Cave's pit lay a 200-foot pit complex—and the underground heart of Flint Ridge.

Probably unexcelled in the annals of American caving was this first modern venture past the shafts of Unknown Cave. It also was one of the most tiring, for in one day's venture, popeyed explorers penetrated five miles—real, honest-to-goodness 5,280-foot miles—of virgin passage. Most was traversable by ordinary walking.

From this extraordinary discovery, interlacing corridors beckoned everywhere. Salts Cave's Indian Avenue was almost within shouting distance—somewhere. Colossal Cave lay not far to the south. The remote southwest edge of Flint Ridge seemed within reach. Other corridors led northeast toward the arm of Three Sisters Hollow separating Unknown Cave from Crystal Cave.

Not all this fantastic new netherworld could be assimilated at once, but the Crystal Cave explorations were immediately redirected toward Unknown Cave.

Soon Bill Austin and Phil Smith encountered the huge, smooth-walled cylinder of Overlook Pit, 40 feet wide. From the natural balcony from which they gaped, it reached 80 feet downward and 70 feet dimly upward. Rocks heaved into the black abyss kerplunked into deep water. Was this the long-sought underground river which many thought bone-dry Crystal Cave should have? Guess-reckoning suggested that the passage which it interrupted was heading roughly toward cavernous Pike Spring. If such a river existed, it ought to be somewhere hereabouts.

Overlook Pit offered no easy route to the reverberating water below, but a muddy little passage continued opposite the balcony. Traversing the wall of the spectacular pit with particular caution, Bill and Phil entered a mud-floored canyon they began to call the Storm Sewer. Obviously this whole section of the cave sometimes flooded deeply.

A low section forced the explorers onto muddy hands and knees. Beyond, a broad black passage led down to another corridor, where deep pools in the rocky floor at last revealed the long-sought river. Eyeless Fish Trail, they called it. Upstream they sloshed toward Unknown Cave for half a mile in shin-deep water, past openings of passages at water level. Soon they jubilantly crossed beneath the deepest valley in Flint Ridge, without any special difficulty.

By January 1955 the secret of Unknown Cave was out. Officially, frequent surface patrols ended ventures there by even the most daring. Now it can be told that it wasn't quite 100 percent that way. By October, grubby-looking men had pushed the unknown so far up Eyeless Fish Trail that certain risks were unavoidable. One midnight Roger Brucker and Red Watson were to listen at a certain likely spot for pounding that somehow was scheduled to happen in Unknown Cave. Perhaps a trifle tardy, Red and Roger came charging up the river passage. Sure enough they heard a faint pounding somewhere ahead. Onward they raced, seeking its source. But the pounding had ceased. Hours later they learned that the Unknown Cave explorers unexpectedly had discovered that their predetermined spot was beneath a loud waterfall. They were lucky to have been able to hammer at all.

But Roger and Red found a telltale natural shaft with acorns and other near-surface debris. So Bill Austin and Jack Lehrberger plunged back into Flint Ridge—never mind where—and came out via the Eyeless

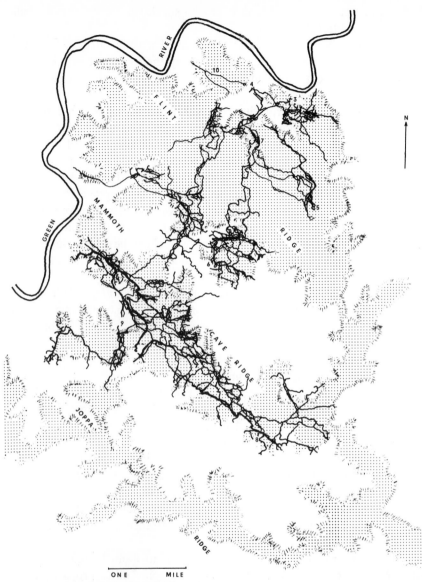

Sketch of Mammoth Cave and the three flat-topped ridges beneath which it extends. (1) Cascade Hall, reached from Flint Ridge by C.R.F. explorers after passing beneath the intervening valley and all but the farther slopes of Mammoth Cave Ridge itself; (2) Historic Entrance of Mammoth Cave; (3) Floyd Collins Crystal Cave entrance; (4) Austin Entrance; (5) Entrance to Salts Cave; (6) Colossal Cave section; (7) Unknown Cave section; (8) George Morrison's New Entrance; (9) New Discovery entrance (artificial); (10) Great Onyx Cave (not part of Mammoth Cave). Shaded ridge-top outlines based on U.S. Geological Survey topographic maps, cave passages simplified from maps by National Park Service, Max Kaemper, U.S. Geological Survey, Cave Research Foundation, and others

Fish Trail, wearing smug, Mona Lisa–like smirks that were already well known as the Flint Ridge smile.

No doubt about it. Flint Ridge contained a truly integrated cavern complex. And, as the speleologists totaled up their miles of explorations, the new cave system exceeded any other recorded anywhere else in the world.

The situation was perplexing, incredible. The Flint Ridge explorers had obtained information of exceptional scientific and popular significance. Yet incautious release of data might mean admission of violation of government regulations. Mapping of certain key sections was carefully avoided. Elated explorers might name one corridor Turner Avenue because they rightly suspected it ran beneath or near the grave of Edmund Turner. Yet pending completed surveys, all was technically mere surmise. "Are you trespassing?" Bill Austin was asked formally. "I don't know," he replied—correctly. Perhaps with tongue in cheek he added—again accurately—that he hadn't seen any NO TRESPASSING signs down there. "If I do, I'll respect them."

The quandary, however, was no joking matter. Independent cavers caught in Unknown Cave the weekend after the Crystal-Unknown breakthrough had been fined $200 and sentenced to jail—sentence suspended. Yet the great discovery had to be announced to the scientific world, for all Flint Ridge research had to be reoriented. Little wonder that the famous 1955 announcement was disappointingly terse.

Much more than halfway across Flint Ridge already, the outer limits of Unknown Cave, too, were on the verge of the endurance barrier. An easier ingress was badly needed. Thoughts turned to a new entrance on the Crystal Cave property, only minutes from the heart of the ridge via newly found Pohl Avenue. A careful resurvey through five miles of difficult cave located the selected point within eight feet. Bill Austin and a group of volunteer assistants devoted five months, 3,600 man-hours, and a ton of dynamite to what is now the Austin Entrance. Months later an ecstatic Phil Smith could still report that "from the entrance it is merely minutes to unexplored cave."

With this problem solved, another began to loom increasingly: the well-meaning red tape which has hampered some other projects of the National Speleological Society. In response, several leaders of its Flint Ridge project organized an independent Cave Research Foundation in 1957. Its impressively expanding program of research and systematic exploration produced a partial relaxation of restrictions. When Crystal Cave and Great Onyx Cave were acquired by the federal government the new group was granted permission to continue and expand its studies.

Officially paramount in the program of the Cave Research Foundation was the concept that a cave is not properly studied without a careful examination of even the least accessible regions. The interrelationships of all penetrable passages and their relationship to the surface topography must be included.

Even though to some the breakthrough to Mammoth Cave was a greater target, the Cave Research Foundation painstakingly set out to fulfill its tedious systematic goal. Some forty assault teams pushed deeper into the unknown each year. Many obtained no important information except that certain passages need not be re-explored. Others yielded undramatic but accumulating data. A few produced spectacular discoveries. Great advances extending the Crystal-Unknown complex to the south edge of Flint Ridge west of Colossal Cave. Exciting archeological data. Curious gypsum deposits. New minerals whose nature changed almost the moment they were removed for study. Mile upon mile of passage that never before had known light.

Key to this new effort was an unusual subgeneration of extra-tough cavers, long immersed in the new techniques Flint Ridge was teaching American speleology. One of their brand-new approaches was especially unexpected. In much of the world cavers follow underground winds as clues to vast unknown areas. In Flint Ridge, however, the success of Eyeless Fish Trail was repeated over and over. The newly self-styled SOBs of Flint Ridge quickly learned to follow the water even more than the wind. Almost automatically they turned from great upper-level trunk passages near their collapsed terminations. Down jagged shafts they scurried, seeking nasty, humid natural drains and miserable sewer passages. Often these led to tributary drains of other shafts, which in turn led back up to other throughway corridors that ate up sinuous miles. Old misunderstandings increasingly resolved, they re-scoured all the caves with permits granted by the National Park Service.

By the end of the decade, a breakthrough from Crystal-Unknown into Colossal Cave seemed imminent. Thunderstruck Flint Ridge SOBs instead connected Colossal to Salts Cave quite accidentally, following strictly normal routine. In August 1960 Jack Lehrberger returned to a river passage of Colossal Cave with David Deamer and "Spike" Werner. Climbing up into a pleasanter, drier area, Jack suddenly realized that his surroundings looked strangely familiar. With a sense of shock, he realized that they were in a lengthy corridor of Salts Cave leading southward from Indian Avenue. Many hours later the trio dragged themselves from the entrance of Salts Cave. Now there were two great cave systems within Flint Ridge: Crystal-Unknown and Salts-Colossal.

Almost exactly a year later Dave Deamer surfaced with another Flint Ridge smile. With Bob Keller and Judy Werner he reattacked the Crouchway area near the point where Bill Austin and Jack Lehrberger had accomplished the final breakthrough between Unknown Cave and Crystal Cave. At the extreme end of Lower Crouchway, Deamer paused to rest. Seemingly emerging from the pores of the rock on which he sat, a cold breeze blasted his posterior. Wrestling aside the forty-pound rock, he opened a hole that blew out his headlamp.

Keller led onward, surveying through a tight, muddy crawlway. Beyond was a canyon two feet wide. On its walls were hints of human passage. Climbing exuberantly to its end the trio soon found itself in Salts Cave, near the west end of Indian Avenue. Man could now travel underground between Crystal, Salts, Colossal, Bedquilt, and Unknown caves.

From Colossal Cave and the farflung southwest ramifications of Unknown Cave, explorers crept farther and farther beneath Houchins Valley toward Mammoth Cave. Soon they had passed beneath the basement fringes of Mammoth Cave Ridge itself. Through the years the surveyed and resurveyed miles accumulated impressively: 8.6 miles in 1960, 11.1 miles in 1961, 7.9 miles in 1962. When I wrote in 1965, 40.52 miles painstakingly had been plotted on master maps of the integrated system.

But by 1965, hopes for the grand Kentucky junction had ebbed badly. A miserable low natural sewer dubbed Candlelight River had brought the Cave Research Foundation despair. Across the full breadth of Flint Ridge, then beneath the huge compound sink called Houchins Valley, its teams had probed branch after upstream branch. Some extended well into the northeastern fringes of Mammoth Cave Ridge. The last ended at a choke of huge sandstone boulders, less than 300 feet from Mammoth Cave: survey point Q87. For a decade, it has resisted all efforts.

As president of the C.R.F., Joe Davidson told St. Louis philosophy professor Red Watson to take a surveying party into the nearby part of Mammoth Cave and see what he could find. That academician complied, more or less. His "survey party" ended up with forty cavers—enough for ten or more survey teams—that mapped the whole northeastern flank of that part of Mammoth Cave. Fortunately they included Cascade Hall, of which more later. For the moment, it was all for naught.

So were all the other 10-mile, 26-hour round trips back to that miserable boulder plug at Q87.

Then one August day, along the way to Q87 petite Patricia Crowther looked down a tiny hole nobody else had bothered with. After a tight 15-foot crawl, she found herself looking further down into a small room where an attractive little canyon passage carried running water.

Bearded Dick Zopf, Roger Brucker, and Roger's son Tom set out to map Pat's lead. To move a big rock where even 115-pound Patricia had barely squeezed by, they carried a jack. Even with that, Roger just didn't fit. In the darkness, he sat and thought and thought while the others pushed on. Hours later they returned, excitedly telling him of better going beyond, but tricky climbs and wet squeezes intermixed. At a fork in the passage, each had taken one branch. Tom's penetration had produced more than a thousand feet of corridor, and a brand-new river with white crayfish and blind fish.

"Which way does it go?"

"I don't know!" Tom's admission was sheepish. As it happened, they had been carrying only one compass, and Zopf had had it.

The river passage, however, had to go somewhere important. On August 30, back went Zopf and Tom Brucker, surveying with Pat Crowther, babbling in their eagerness for her to share their excitement over the new river. Fourteen hours in, they brought the survey to the point where Tom had turned around—precisely on schedule.

Lacking any pressure to surface immediately, the trio decided to spend 15 minutes in the pure joy of exploring—on hands and knees— along the sloshway where no human had gone before. Pat lagged a trifle.

Suddenly Tom and Dick began to howl like madmen. Joining them at full speed, Pat understood without need for a word. On a mudbank were inscriptions common in the New Discovery section of Mammoth Cave: the initials P.H. and an arrow pointing downstream—out! On a wall nearby was the name Pete H.: Pete Hanson of Mammoth Cave!

Tremendously charged with unsuspected energy, the trio raced on and on, wondering and hoping. As they advanced, the river passage enlarged, expanded.

This time, the team had a compass. At first it told them they were headed northwest, parallel to the axis of Mammoth Cave Ridge, mostly toward nothing but the gorge of the Green River. Then the streamway abruptly turned left, southwest, under the ridge itself, seemingly without end.

Seemingly, too, this new river was all by itself in the basement of Mammoth Cave Ridge, without any intersecting shafts that might lead up to the famous avenues somewhere above. Could it somehow miss Mammoth Cave entirely? More and more the question came to three minds: How had Pete Hanson reached that remote spot, nowhere near his famed New Discovery section?

C.R.F. parties simply do not overstay the four-hour grace period each is allowed before rescue teams are alerted. Even by stretching their limit to the utmost, only one hour could be allotted to this unplanned venture. The cavers' energy now sinking lower and lower with their dwindling hopes, reality intervened. With the end nowhere in sight, the trio inscribed the date 8-30-72 where it could not be overlooked and sadly turned back. Utterly drained, they surfaced at dawn, haggard from their twenty-six-hour marathon, the cave spinning around them, their minds "on automatic." For all they knew, the great connection was only a few yards farther on.

Or nowhere.

Had Pete Hanson merely explored and left his marks in one of the many unimportant little caves opening in the rocky flanks of Houchins Valley? It seemed less likely than his coming from somewhere inside

Mammoth Cave, but even a new entrance to the Flint Ridge system would be welcome here.

If Hanson had really reached this point from somewhere in Mammoth Cave, perhaps he had branched off from Roaring River on his way to the New Discovery. When Pat recovered her energy, she received the dubious honor of leading a wet-suited crew that searched each foot of the chill, black Roaring River route. A connecting passage just wasn't there.

Meanwhile, another team was trying from the Flint Ridge side, but somebody goofed. Among those assigned to John Wilcox's breakthrough-determined team was another caver who—like Roger Brucker—just didn't fit through Pat's Hole. The mission had to be aborted.

And so on September 9, 1972, Patricia Crowther had the chance to lead from Flint Ridge again. This time the group was not only superstrong but superskinny: Richard Zopf, geology graduate student Steve Wells, Gary Eller, a research chemist, National Park Service ranger Cleve Pinnix. And John Wilcox.

Although this was correctly billed as a routine mapping trip, C.R.F. leaders saw the situation as a bit beyond the ordinary. Quietly they set up a nationwide alert for a possible all-out rescue effort, in case irresistible temptation should—just once—overcome the stern Flint Ridge self-discipline. Across the continent, cavers slept fitfully, one ear cocked toward the telephone.

Leapfrogging two teams, the cavers mapped onward, routinely, stolidly, hour after hour, through the chill black sewer. Strength and endurance drained away, but this was equally routine, expected. Huddling 600 feet short of the previous turnaround, the tired teams decided the mapping was over. Soon they too would have to retreat, their time exhausted. Perhaps another try from the Mammoth Cave side . . .

John Wilcox retained a contagious spark of energy. Sloshing through deepening water, he led onward, though the mud-covered ceiling lowered ominously. A hundred feet more, 200, 400, now through thigh-deep water in a lowering passage, each caver bent almost double. Increasingly heads tilted sidewise to keep mouths out of water. If anything in the eighteen-year struggle had seemed hopeless, this was it.

Totally exhausted, five of the sextet dropped blankly on a mudbank. Even Wilcox could manage little more. Yet—

"The ceiling's getting higher!"

John's sudden call sang clear despite the watery echoes. Suddenly he burst into a great black vault, staring incredulously. Across Mammoth Cave's long-famous Cascade Hall a man-made railing reflected carbide light. The Grand Kentucky Junction was a reality.

For long moments, John's Flint Ridge comrades could not comprehend his shouted triumph. Then weariness vanished. Chest-deep in chill water yet incredibly savoring the incomparable moment, all slithered beneath the last duckunder, cheering, shouting, hugging each other, jumping up and down with the broadest, most triumphant Flint Ridge smiles that ever existed.

Pat Crowther slipped, ducked completely under, totally soaked. The elated group sobered, recalled its fatigue and its obligations. Already it was doubtful that the sextet could regain the Austin Entrance short of thirty-five hours' elapsed time—fearsomely close to human limits. In the other direction, Mammoth Cave is well locked at night, and standing around in the chilly cave for hours is less than healthy in wet clothes. The nationwide rescue alert suddenly seemed no theoretical practice.

But ranger Cleve Pinnix flashed an extra-special Flint Ridge smile. Trimuphantly he fished a key from his pack: the key to the service elevator at the Snowball Dining Room, only a twenty-minute walk ahead. At 4 A.M. in Ohio, Roger Brucker's family phone rang insistently: "Get Brucker out of bed!"

Roger admits that for a moment icy fear stiffened his body. Then came the telephone version of the Flint Ridge smile: "This is Pete Hanson!"

Cascade Hall, with a caver igniting a flashbulb at the low entrance of Hansen's Lost River whence John Wilcox first looked into Mammoth Cave from Flint Ridge. *Cave Research Foundation photo, courtesy Roger Brucker*

An incredible smile transformed Roger's face, broader and broader: an eighteen-year smile indeed.

But self-discipline triumphed. Even before packing up their gear, the junctioneers plunged back through the icy duckunder for the thousand-foot survey that tied all of underground Flint Ridge into Mammoth Cave. The map of Mammoth Cave suddenly showed more than 144 miles of surveyed passages: some 58 miles in the main section and 86 inside Flint Ridge. Soon the traditional 150 miles lay far behind. As I write, the total stands at 169.2 miles, more than twice the length of Hölloch, still number two in the world. Today no man can even guess the range in which the total will eventually level off. With the full cooperation of the National Park Service, C.R.F. surveyors are finding that the "old" part of Mammoth Cave is "going" just like the Flint Ridge section. Countless years of effort and triumph clearly lie ahead.

A few weeks after the grand Kentucky junction, I sat in Roger Brucker's comfortable Ohio home, toasting with champagne each new carload that assembled for the C.R.F.'s unheralded private celebration. Each bore a new tale, hilarious or wry. Biologist Tom Poulson had been part way up Hanson's Lost River studying blind fish. Harold Meloy pointed it out on Stephen Bishop's 1845 map of Mammoth Cave, and on later maps by the Reverend Horace C. Hovey and R. Ellsworth Call, M.D. Dozens had seen its entrance; I had, myself. Roger's own tale topped all the rest. "It's a silly little hole," he insisted grandly. "When I was eight, I asked the guide where it went. Boy, did I feel little when he answered back: 'It doesn't go anywhere, it stays right there. Haw!' " How many other kids, he mused aloud, had wondered the same thing. . . . Then he joshingly tried to make us believe he'd vowed to find out, then and there, the answer to his own childish question.

But the news that emphasized the grim reality was the report about the current water level in Mammoth Cave. At that moment the grand Kentucky junction was several feet under water. Hidden by water is its usual condition, and nobody seemed to have realized it earlier.

The nostalgic party continued throughout the weekend. Quietly but deeply, exultation reigned. Yet more than a hint of poignancy tempered the celebration. Suddenly their beloved Flint Ridge cave system was no more.

For a time, a determined effort urged formalization of the name "Flint Mammoth Cave System." C.R.F.'s Joppa Ridge specialist Gordon Smith retaliated by archly urging "Flimmouth Cave" or even "Mammint Cave."

But the weight of historical momentum and public desire was an overwhelming obstacle anyway. American science may find value and validity in "Flint Mammoth Cave System," but for a century the experts had insisted that it was all part of Mammoth Cave. Today's public agrees: It's all Mammoth Cave now.

Yet the fantastic underground labyrinth of Flint Ridge will never become just another part of Mammoth Cave. The tremendous outpouring of human energy here altered the entire course of American speleology. In this unique part of the world's largest cave the Cave Research Foundation discovered and seized the first opportunity to study a truly vast American cave in its wilderness entirety.

Just as Mammoth Cave has long been the standard of comparison for all other caves, the Cave Research Foundation here set the standard for all future triumphs of speleology.

2

In a Lonely Sandstone Cave

*The Story of Floyd Collins and the Great Cave War
of Kentucky*

It was such an innocuous little rock, something like a flattened, gray leg of lamb. Could this irregularly rounded slab of water-pitted limestone have been the focus of half the world for seventeen dramatic days in February 1925? Practically all the writers insisted that solo spelunker Floyd Collins had been trapped by a gargantuan boulder weighing six or seven tons. As I hefted this oblong piece of limestone, I guessed its weight at not much more than twenty pounds.

"Twenty-seven pounds." Seated on the front porch of his attractive Cave City home, genial Arthur Doyle was reading my mind with polite amusement. He pointed to a peculiar gooseneck at the smaller end of the bulgy rock. It bore a curious oval flat about three inches long. "That's where it broke off from the wall when Floyd's foot pushed on it. It just slipped down a little and got wedged so he couldn't move. That was all it took."

Arthur Doyle knows every detail of that fateful rock. As a boy he was present when Floyd Collins was boarding at his father's home in January 1925. Doyle's father owned the site of the tragedy—Sand Cave. Young Doyle himself saw miners tenderly and cautiously haul the corpse of Collins to the surface—and the fatally stubborn little rock which I held in my hands. I turned it round and round, seeking clues to its story. The larger end—I could hardly put my hand around its bluntness—was smoothed as if by the unintentional polishing of countless hands. Surely many had hefted this keystone of the dramatic 1925 epic of the hill country of Kentucky. Some firsthand descriptions of it should have reached the world long ago. I queried Mr. Doyle.

"No, most everybody around here's seen the rock, but not many outsiders," he explained. "Except a few of you cavers. I figure that end got so polished from their poking, trying to work it off Floyd's leg."

Thanking Mr. Doyle profusely, I drove a few miles, parked, and hiked the furlong to Sand Cave. In the peaceful twilight of its broad, arching alcove, I could not immediately visualize its mob scenes of epochal heroism and turbulent roistering. Only a few feet below a deserted section of the flat top of Mammoth Cave Ridge, I was alone in a rounded, spacious sandstone overhang. All the world seemed at peace.

What I could see wasn't much of a cave. The broad, inviting entrance promised much, but immediately changed its mind and melted into solid rock walls. Just at the outer edge of the ten-foot ceiling line was a funnel-like pit twenty feet across: the slumped remnant of the famous shaft. Beyond, broken rock and earth sloped into a little valley-head ravine.

Near the back wall, new-looking masonry attracted my attention to a locked gate. It barred a vertical opening so small I wondered whether the gate was really necessary. I could have slithered to the right-angle bend visible below, but it would not have been easy.

As I looked, I noted solid limestone below the sandstone strata visible elsewhere. That was curious. The old accounts of the long race against death spoke of the cave as a mere "sand hole" or "sandstone cave." Perhaps the old tales might not be so wild!

I climbed the twenty-foot cliff and scouted the lightly wooded environs where gawking thousands had milled. As I mused, the puzzle pieces of Sand Cave began to fall into a pattern quite unlike its oft-repeated, ever-changing "authentic" story.

Probably we will never know all the details. To the self-reliant people of Kentucky cave country, this never was a proper concern of outsiders. Gingery, graying Lee Collins greeted a friend with a Biblical quotation, expecting one in return. Yet the chunky, God-fearing farmer under oath refused to give the name of bootleggers whose actions contributed to his son's death. In the tradition of the hills, the court seems not to have expected him to do so. Not even the basic records agree. All that is wholly beyond question is the date. The epic which brought out the best and worst in mankind began Friday, January 30, 1925.

The story of Floyd Collins really began years earlier, perhaps as early as the Great Cave War of Kentucky itself—and today no man knows that date. Already by 1893 famed spelunking pastor Horace C. Hovey was moved to protest:

Several (lesser caves) are lauded by their owners as rivals to Mammoth Cave. This petty jealousy cropped out . . . where we had to change [railroad] cars, to the effect that Green River had broken into Mammoth Cave so as to make its avenues impassable; that visitors were not admitted at this season; that the hotel was literally dropping to pieces and had been closed.

Nor did anyone really begin the Great Cave War. Rooted in the bone-grinding poverty of the clannish back country around the caves, it just grew. Perspicaciously, the *New York Times* recounted in 1925: "The land in this region is worth little for any other purpose than the exploitation of caves. Rugged, rocky, broken by hills and streams, it is a hard land to cultivate." To many of the impoverished natives, caves seemed hardly more than potential gold mines of tourist dollars—a view still all too prevalent. Mammoth Cave was the standard by which other caves were measured. Dreamers searched for "another Mammoth Cave." Others sought a private entrance to Mammoth Cave itself. It seemed as if every property owner for miles around had dreamed of fortune—a cave "bigger than Mammoth" beneath his lands. Many a sinkhole in three counties was hopefully excavated. Quite a few led into caves, and lanterns were cheap. In later years some fortunates found caverns good enough to persuade spelunking bankers to loan them a few dollars. On shoestring budgets they built curio shops and strung lightbulbs underground, often on bare wires. Newly commercialized caverns appeared out of nowhere. So did ever-improving ruses for snaffling tourists and their dollars.

Whether or not they were beauteous, most of the more accessible caves did reasonably well. Hard pressed, however, were splendid caves located some miles from the none-too-good highways. In this region where brawls traditionally were settled without bothering the law, the remoter caves' advertising campaigns soon degenerated. Directional signs at strategic road forks began to disappear.

Accustomed to working out their differences within their naturally evolved code, the back-country folk at first had little truck with law or lawyers. The devoutly Baptist community background soon established an unwritten but well-comprehended code for the Great Cave War. Outright lying was banned, but it was fair enough if the outsider jumped to well-planned conclusions. Killing was taboo, and crippling, but just about anything else went.

A variety of advertising gimmicks multiplied. Cappers—the experts at rerouting tourists headed for Mammoth Cave or some other private cave—became increasingly bold. "Central Cave Offices" and "official" information booths were plentiful. Military-appearing cappers pretended to write down the license numbers of cars ignoring their police whistles and red flags—a device which usually brought the "offender" back, all excuses. Never telling a direct lie, they succeeded by implication in diverting perhaps a third of the Mammoth-bound tourists to other caves. Each afternoon, they were careful to point out that the all-day trip had gone—"the one that most people think is the good one." Early-morning visitors instead were brainwashed with the great length and arduousness of the all-day trip. The elderly were regaled with accounts of the 500 steps in Mammoth, the young with the alleged beauties of some competing cave. To the naïve listener, Mammoth Cave assumed a semblance of a New York subway, dirty and smoked up by kerosene fumes: "They might make you sick, but that's all right. They keep plenty of stretchers handy. . . . You figure on going anyway? O.K. The first cave entrance is just up the road here. . . . Yes, probably it's part of Mammoth. Geologists say all the caves around here are probably connected."

Not every cave enterprise could afford to build information booths— or to replace those which mysteriously caught fire on moonless nights. The better-financed caves found their trade nibbled away by nonuniformed cappers who brainwashed tourists at stoplights and railroad stations as much as fifty miles away. Some Kentuckians aver that the cappers contributed mightily to the disappearance of running boards.

Rock fights often broke out where cappers battled for choice locations. As the Great Cave War settled down to a long campaign of stratagem and attrition, however, car pools began to carry competing cappers to their daily locations. By midday, the struggle for customers still might lead to fisticuffs, but by evening the battlers usually had settled their differences and they rode home in amity.

As time passed, ruses were evoked which would have shamed carnival pitchmen or circus barkers. In retrospect, much of the Great Cave War was high comedy, though grim enough to the participants. Innocent-looking pseudo-tourists appeared among unsuspecting throngs, relating how much better they had liked some other cave. At least one cave retaliated by offering a "money-back guarantee of satisfaction." Somehow dissidents seemed to have trouble collecting their "money-back guarantee"— until a rival cave owner sent a renowned local prize fighter in disguise.

For those who avoided the roadside capper, billboards found ever-new devices. "This is NOT the road to Great Onyx Cave" screamed one on the road to Mammoth Onyx Cave. Mammoth Onyx Cave and Diamond Cavern both claimed the title of "Kentucky's Most Beautiful Cave." Crystal Cave advertised as "The Grandest Cave of All." Great Onyx Cave topped everyone with "It's Better than the Best." An Edmonson County grand jury felt it necessary to admonish even quasi-public organizations then operating the competing ends of Mammoth Cave. At least as late as 1940, a judge felt it necessary to lecture some of the others. A byword sprang up: "When God made Mammoth Cave, he should have stopped right there."

In 1921, George D. Morrison hit the jackpot: the New Entrance to Mammoth Cave itself.

To Morrison, the triumph was doubly sweet. Five years earlier, he had gained a pyrrhic success where many had failed. The shrewd proprie-

Capper and anti-capper billboards of the 1950s.

tors of Mammoth Cave had long forbidden publication of any accurate map of the cave. Those of Stephen Bishop, Horace Hovey, and R. E. Call alike were curious contrivances, showing interrelationships of certain important passages without acknowledging that far-flung ramifications of the tremendous cavern extended beyond property lines. Many had attempted to dig into Mammoth Cave from other properties—unsuccessfully. But, by bribing the right people, Morrison sneaked in a mapping party, which secretively worked overnight at high speed. The resulting map lacked pinpoint accuracy, but careful reconnaissance, local inquiry, and sixty days' work by a hired gang led Morrison into Mammoth Cave through what is now known as the Cox Entrance. All for naught. The owners of the Cox property were friends of the Mammoth Cave management and enjoined Morrison from using "their" entrance.

The setback was only temporary. In the next five years, he was able to purchase or lease "cave rights" where his map showed him Mammoth Cave extended. Success was sweet indeed. His Mammoth Cave Development Company promptly set about commercializing "his" end of Mammoth Cave. Soon he had discovered the Frozen Niagara section, far more beautiful than anything in the historic section. Publicity releases went out all over the world. Billboards blazed with announcements of the New Entrance to Mammoth Cave, practically at the back door of Cave City. Why bother going all the way to the heavily smoked Old Entrance? Adding insult to injury, his guides began to appear out of the shadows of the netherworld. With courtly aplomb, they invited tours entering via the historic entrance to extend their trip and emerge through the New Entrance —free!

The original Mammoth Cave management reacted like an elephant with a trunkful of hornets. They denounced Morrison, claiming his New Entrance was a completely separate cave. Then they sought another injunction.

Neither side had an ideal position; Morrison's maps were crude by modern standards and of little legal value. He made no attempt to halt parties guided from the Old Entrance at what he thought was his underground boundary line. Though they may have suspected that much of their tours ran beneath Morrison's leased land, the Croghan trustees were unwilling to concede his maps *any* accuracy.

When the court convened, Morrison was armed with affidavits from tourists he had enveigled into leaving the "regular" tours and emerging through the New Entrance. They turned the tide, as recounted in the *Proceedings of the Kentucky Bar Association:* "The court held on September 4, 1926, that the word 'Mammoth Cave' was fairly applicable to all this general system or labyrinth of caverns and possessed no special or secondary meaning which could be appropriated by the trustees of the Mammoth Cave estate."

Floyd Collins grew up in the thick of the Great Cave War. Legend recounts that, as a boy at the turn of the century, he saw a mule break through to its knees where no cave had been known. Immediately he felt called to explore it. Mule Cave never amounted to anything, but Floyd soon became one of the young mountain folk who neglected their chores for caving. Castner Browder was to write perceptively in the *New York Times* of this curious subtribe:

Strong in physique, slow of speech, shambling of gait, bright of eye and as shy as deer, these explorers live to themselves in little farm houses along the creeks. Life is simple, their wants are few and it requires little effort above ground to obtain food and the little clothing they wear. . . . With little education in the schools but much in the way of wild things they are impatient of restraint or discipline. They make cavecraft a science, like woodcraft as practiced by the Indians who formerly roved this region.

Although only occasionally did he drop in at the little country school on nearby Joppa Ridge, Floyd Collins soon stood head and shoulders above his contemporaries. Seven years before the Sand Cave tragedy, his efforts were crowned by discovery of "Great" Crystal Cave, just a stone's throw from his family's ridgetop home.

The Collins family promptly went into the commercial cave business. Floyd concentrated his explorations in Crystal Cave with considerable success. But few tourists could be enticed over the deep ruts which served the remote Flint Ridge homestead.

As Floyd Collins lost his post-adolescent pudginess, he studied the sinks and ridges of the land—today we would term it karstic geomorphology. More than ordinarily taciturn, he came alive picking the brains of each visiting geologist. All too often he found that the book-learnin' of the experts contrasted pitifully with his self-taught lore of twenty years under the earth. His few smiles largely were reserved for the pioneer spelunkers who spoke his language: Indiana's George Jackson, Russell Trall Neville, Ed Turner. But an inevitable cultural barrier divided Floyd Collins from even these few peers who came to know his ways. Neville once told the late Clay Perry:

Floyd was an expert caver, but an uneducated man with some superstitions, among them the belief that he had some sort of magnetism in his body that enabled him to tell directions in a cave without a compass.

We tested him once, with my sister holding a compass without Floyd knowing it. He would stop and stand rigid and give a little shiver and point to where he believed was north.

He was wrong, of course, but we never let him know. It made no difference which way was north, anyway, in a cave, unless you were travelling by map.

I sat on that compass by accident and broke it. I gave the needle to Floyd and he kept it. It was on his body when it was finally brought out. I think he believed it would help his "body magnetism."

One additional mental quirk set Floyd Collins apart from fellow pioneer spelunkers. Despite their efforts, that stocky young explorer saw no harm in solo expeditions into the unknown with pitifully limited equipment. Naïvely he boasted of ability to get out of trouble. Once he had had to pile rocks high to climb out of a smooth-walled pit into which he had jumped, overconfidently hanging free from a not-so-reachable ledge. His trust was in God, his brothers, and his own ability. Had not his brothers saved him when he was trapped, lightless, for twenty hours in another cave?

Almost any modern caver would have predicted tragedy ahead.

To Floyd Collins, the narrow neck that connects Flint Ridge with Mammoth Cave Ridge seemed a particularly promising area. On a fifty-fifty basis (with the property owners providing room and board) the thirty-four-year-old spelunker began to investigate Sand Cave early in January 1925. First penetrating downward along the limestone buttress of the overhang, he soon encountered a tight stream-course crawlway twisting downward amid cliffside jumble. Occasionally resorting to dynamite, he moved much broken rock in pursuit of an elusive current of

Floyd Collins (front) and Homer Collins (rear) with Russell Trall Neville party in Crystal Cave. Neville (at left) apparently took the photograph with a self-timer. *Courtesy Burton Faust*

warm cavern air. A few days' work brought him a half hour into the netherworld.

But on Friday, January 30, Floyd Collins did not return for supper. Nor by bedtime. Nor by the wet, chill dawn of the 31st.

Alarmed neighbors found his coat and yesterday's lunch on a ledge near the black tunnel mouth. Hesitantly squirming down the squishy black crawlway, their shouts at last produced weak answers. Somewhere in the blackness ahead a dull-eyed Floyd Collins lay trapped. His gasped story held seeds of conflict.

Almost twenty-four hours earlier, Floyd's left foot had pushed on an out-jutting bit of limestone as he squirmed downward, feet-first through a particularly difficult spot. With a loud, metallic snap, the limestone had broken away from the wall. Falling across his extended leg, it had locked his foot neatly into a narrow slot without even bruising it.

Turned half on his side, more upright than prone, Collins lay with his

Bee Doyle in Sand Cave at actual point where Floyd Collins was trapped. *Photo by Russell Trall Neville, courtesy Burton Faust*

other leg flexed awkwardly beneath him. An overhanging ledge held his right arm immobile. A trickle of storm water found its annoying way to his unprotected face.

Floyd awkwardly strained every muscle to free himself. His panting struggles dislodged small rocks and packed gravel ever tighter around his tortured body. Soon he could no longer use even his left arm. The chill hours sapped his endurance. Eventually wakening from a troubled doze, he found his lantern had flickered out, leaving him in black, heart-shrinking aloneness.

But it was just a little rock. "Go tell my folks," he begged. "With a crowbar my brothers'll be able to get me out of here without much trouble."

His friends earthwormed to the surface and haltered two mules—the fastest available transportation. The Collins' skew home was eight miles away. As it turned out, Homer Collins, Floyd's closest brother and an able spelunker, was in Louisville. A messenger was dispatched to the nearest telephone. The oft-deplored rural party liners spread the word like wildfire. Twenty-five neighbors were waiting to help when Lee and Marshall Collins reached the site. Leading a small group into Sand Cave, Marshall encouraged his shivering brother, freed his upper body, and diverted dripping water from Floyd's face. Warmed by a gasoline lantern, the trapped explorer felt some vitality seep through his chilled body.

But even his most knowledgeable neighbors and kin were novices by Floyd's standards. No one could decide what to do next until Homer was rushed from the station in a battered Model T. In eight fierce hours, he hand-picked two bushels of loose rock and gravel from Floyd's body, scorning ragged, bleeding fingers as he scorned cold and fatigue. Each rock tugged from its socket and passed upward along the human chain should have allowed a crowbar to slip beneath the pesky rock. Should have but did not.

Early on Sunday, mental and physical collapse slowed progress to a halt. Friends finally persuaded Homer Collins to seek momentary rest. Exhausted and encased in soppy mud, he left the scene with grave misgivings. Even by dawn, ominous events portended a coming debacle. Moonshine had appeared mysteriously to resuscitate the bone-chilled rescue workers, and the bootleggers had found other customers. In the hills, it was inhospitable for a man with a jug not to offer his friends a swig, or for a man to refuse it unless he had taken The Pledge. Fighting waves of fatigue, Homer Collins offered a reward of $500 for the rescue of his brother, then was led away to seek the rest of exhaustion. A shocking Saturnalia followed. Some sincere helpers appeared among the drunken crowd. Yet, hour after hour, nothing seemed to change.

By Monday morning, Floyd Collins had been in his rocky tomb for

The rock which trapped Floyd Collins, shown with his kerosene lantern and boots. The imprint of the rock is visible on one boot. *Photo by Russell Trall Neville, courtesy Burton Faust*

three days. His living death "300 feet underground" (actually less than 70 feet) had become of more than local concern. Even with the first garbled reports, the effort to free the trapped man seized the imagination of the nation. The public vicariously began to keep vigil at the mouth of Floyd's tunnel. That new plaything called the radio found its first great human interest drama.

Unable to visualize the fearful conditions of the rescue, the nation still shuddered collectively with changing columns of newsprint. Never before had such suspense been recounted so widely, so rapidly: a man trapped deep in the earth under appalling conditions, yet where anyone slender enough could crawl down to him, talk with him, feed him and work to free him. When the responsible people of the region appealed for help, warm-hearted citizens responded overwhelmingly. Tons of equipment, thousands of messages, and hundreds of volunteers began to pour into Cave City: the vanguard of the Army of Sand Cave.

Two reporters reached the scene early Monday. The first merely crawled in far enough to shout to Floyd and to convey a message of hope to a hopeful world. But after him came skinny red-haired cub reporter William Burke (Skeets) Miller of the Louisville *Courier-Journal.* Crawling head-down in semidarkness, 110-pound Skeets Miller met Floyd Collins by sliding head-on onto him, seeking the story he could not obtain on the surface.

"Don't go crawling around in any cave," Skeets had been told. Touched by the pathos of the unearthly scene, however, his first curious trip led to another and another and another. Danger, hunger, fatigue, and threats paled before the obvious need that compelled this slight youngster. Within a week he had become an integral part of a great American legend—a bizarre situation which today obscures his near-total inexperience with both caves and journalism.

In retrospect, the young cub reporter was a better spelunker than journalist. Naïvely he telephoned his first lead article to a competitor newspaper, which was delighted to print it. The *Courier-Journal,* however, quickly recouped. Its city editor Neil Dalton knew and understood the drama in Skeets' subsequent terse calls. Dalton was a seasoned rewrite man. His skillful accounts—carrying Miller's byline—immediately brought glory and prosperity to the *Courier-Journal.* As a by-product, they quickly made Skeets Miller chief hero of Sand Cave to all the world except Dalton and cynical fellow veterans of journalism.

Lieutenant Robert Burdon, rescue expert of the Louisville Fire Department, repeatedly entered the cave with Skeets and Homer Collins. Burdon was skeptical that Floyd could be freed via the tight crawlway. He suggested digging through the cliffside rubble, but the idea was generally unpopular. Someone suggested pulling the trapped explorer loose. The trio were willing to try.

Fed three times Monday, Floyd Collins had rallied. Homer slipped a hastily fashioned leather harness around his brother's chest. The three men began to pull, then harder.

A low moan burst from Floyd's clenched teeth. Instantly they stopped, solicitously questioning the trapped man.

"Go ahead. Pull me out even if you tear off my foot," Floyd urged tautly. They responded until sweat stood out on all four brows. Something slipped an inch, three inches, two more. Floyd's body straightened from its cruelly cramped position, but nothing more. Renewed tugging caused Floyd too much pain to continue.

Returning to the surface to confer with relatives and friends, the brave trio found pandemonium rampant. Moonshine was flowing freely. Drunken crowds reeled wildly about the rescue scene. Fistfights sprang up faster than the responsible minority could break them up. Each swig of fiery corn likker suggested a new way to rescue Floyd—providing someone else did it, of course. Shouted obscenity and profanity dinned so loudly that at times rescue workers had to bellow vital messages into each other's ears.

Monday evening saw the first attempt to fill the leadership vacuum. A party emerged announcing that Floyd had asked John Gerald to take charge of the aimless operation. Outspoken and more than a little indiscreet, Gerald soon became a hero to the Collins family. To some other people, he seemed a villain. There is no doubt of his immediate grasp of leadership, however. Regardless of the enemies he knew he was making among his own people, he acted ably to exclude quarreling, hampering factions. With a college president and a professor of mathematics, he worked underground for seven bitter hours Monday night. With both hands now free and resuscitated by hot food, Floyd was able to help his rescuers appreciably.

"I can live in here two weeks if you'll just feed me," he told Gerald. Then, wrapped in a quilt and head pillowed, he dozed off into a sound, healthy sleep.

But Floyd Collins' strength had been sapped, and for a particularly tragic reason. Behind many a rock along the constricted crawlway, Gerald had found food cached by those who had volunteered to feed Floyd but turned back unnerved. Nearly all of it was in the outer half of the crawlway.

Working frenziedly with only two hours' sleep, Gerald drove himself far too far. A turning point faced him when he emerged stiffly at dawn Tuesday. With Homer and Marshall collapsed, Lee Collins referred to Gerald a momentous decision. Skilled stonecutters from Louisville had spent five hours fighting their way through the uncontrolled mob to offer their peculiar skills. Perhaps they were a trifle curt to one unaccustomed to outsiders, perhaps Gerald was overencouraged by the night's progress, perhaps he was utterly tired of the thousands of well-meant suggestions. Lying on a cot, shaking with exhaustion, perhaps he was merely overly irritable. In any event, the stonecutters caught the next train home.

For the first time, the press hinted at a fundamental conflict among the workers: a culture-sharpened conflict inherent in many a cavern rescue. Though little educated and disorganized, some of the natives considered themselves comparatively skilled spelunkers. Floyd was one of them and they'd get him out. But the stonecutters were merely a vanguard of city people with very different skills. Outsiders soon inundated the country people. Authority and the press spoke their language and automatically swung to their side. To the cave-wise natives, the city folk were intruders with newfangled ideas who didn't know the first thing about caves. To the educated newcomers, the hill folk much resembled the long-eared mules they bestrode.

As more and more rock was passed upward, however, success seemed moments away. The rescue gangs now could see the fateful rock itself. Crews led by Johnny Gerald and Skeets Miller competed hotly to be in on the finish. Rumor after happy rumor swept the now-buoyant crowd. Some long-forgotten newsmen slightly overdid the cherished journalistic tradition: Get the news fast, catch the early editions, and get the facts straight later on. Even the *New York Times* found itself embarrassed with a premature heading: PULL CAVE PRISONER FROM GRIP OF ROCK: TAKEN OUT ALIVE. It didn't quite happen.

Wednesday morning saw Skeets Miller and fireman Burdon leading a human chain of thirteen men. Down the sinuous burrow they passed wires and affixed a series of old-fashioned light globes. One against Floyd's chest provided a little wonderful warmth and a harsh light, equally wonderful in the black hole. Next came a crowbar, a tiny jack, and a series of wooden wedge blocks. Delicate teamwork maneuvered the crowbar past the trapped victim and under the rock which was indenting his left boot. Just two inches and the epochal struggle would be only an enduring memory.

Skeets Miller assembled the Rube Goldberg unit. Lying full-length atop Collins, he began to exert pressure on the crowbar. The rock wiggled, shifted, rocked, rose perceptibly—and slipped back.

Hours had raced by, again exhausting the team. With victory in plain sight, Miller's gang was forced to leave the struggle to the next crew.

But as Skeets Miller was driven to Cave City for a Turkish bath and a short rest, a car overtook him with shocking news. The next gang had encountered an alarming slide about fifteen feet from the trapped man.

At first it didn't seem too ominous. "A few bushels of earth and one heavy stone," a news service termed it. Too, one worker revealed that Collins had weakly called that shifting rock had freed his foot. No one asked if it was delirium or a frantic fear of being abandoned. It made the day's headline, though the press did point out that because of the slide "Collins is not much better off."

As time rolled on, however, a cloud of despair settled upon the rescuers. The roof of the hand-enlarged tunnel continued to crumble. As Floyd's friend Norman Parker studied the tragic rockfall, his head was grazed by a rock which would have crushed it like an eggshell. Contagious reports began to circulate that the whole lower crawlway was squeezing shut. By 4:30 Thursday morning, expert miners swore hysterically that they'd never again venture into the burrow: "I'd never come out alive!" Many a weary worker now forgot his determination when next offered a jug of white lightin'. Inebriation reached another climax. Appeals went out for the National Guard to control the reeling, brawling mob.

But, while some sought relief in whiskey, other grim-faced rescuers began their efforts anew.

Led by the state's lieutenant governor, the Kentucky National Guard arrived Thursday. Promptly the rescue was organized army-style. The area was roped off and the journalists restricted to bull pens, where communiqués were issued four times daily. The bootleggers vanished magically, but folk talk ever after blamed "the martinet army officer" for everything that went wrong. In truth, Lieutenant Governor Brigadier-General Henry H. Denhardt was not ideally suited to delicate intercultural interactions. A decade later his stormy career came to an abrupt end, "fittin' " indeed, according to the hill folk. On the eve of a retrial on the charge of murdering a voluptuous but naïve Kentucky beauty, he interacted permanently with seven slugs from her feudin' brothers' handguns.

Even before the arrival of the bayonet-equipped troops, many valiant workers were turning increasingly to an unobtrusive outsider with the look of a professor of divinity. In reality, Henry T. Carmichael was a tough, experienced tunnel expert and manager of the Kentucky Rock Asphalt Company. Under the authority of the National Guard, he moved into leadership as surveys hastened the shaft long proposed by fireman Burdon. Bitter stories later charged that this shaft was through hard sandstone and thus hopelessly slow from the start. This was not so. The cliffside rubble was fairly easy digging, and the crews averaged ten feet daily. Despite timbering, however, it was dangerously loose and collapsing— "like digging in a sack of peanuts," to paraphrase Carmichael. Constant rockfall and flooding were no deterrent to these shock troops of the Army of Sand Cave.

Across the continent, the enthralling suspense kept the front pages. Buried alive, yet so near . . . "Pray for my boy," Lee Collins begged, and the heart of America went out to him. One New York newspaper is said to have sold an extra 100,000 copies daily. Each edition sold out still damp from the presses. Radio bulletins kept a constant din. So did telephones in radio and newspaper offices, swamped by anxious queries. Eager reporters at the scene were pressured for new material to assuage the incredible hunger for news.

As though to a county fair, gawking hordes of perhaps 20,000 responded to the thrilling newspaper accounts. Parking was nonexistent within four miles. Except for mules. Poke-bonneted or bewhiskered country folk and their long-eared transportation delighted the reporters, soon 150 strong: "Parson Goodblower, feet nearly to the ground on each side of his drowsy mount . . ."

As the carnival atmosphere peaked, hastily organized lunch wagons ran out of food. Profiteering of hot dogs, apples, and soda pop was rampant. Pitchmen hawked patent medicines. Jugglers and sleight-of-hand artists reaped a macabre harvest. So did shell-game operators and lesser swindlers. At least one itinerant preacher muscled in on the local pastors, loudly praying for the deliverance of Floyd Collins and the salvation of his hearers. During solemn prayer services in the little cove, volunteer soloists unmelodiously bellowed different hymns. Lee Collins too prayed aloud for his son. Perhaps dazed by fatigue, he went about loudly offering rewards to anyone who could rescue Floyd—and distributing Crystal Cave handbills.

Not all those who came to Sand Cave were parasites or sensation seekers. A steady stream of men and machines came to help. A United States Bureau of Mines rescue team arrived from Indiana. The Louisville and Nashville Railroad gladly loaded oily equipment on its crack passenger trains.

A stout wooden railway was assembled from tons of supplies piling up at the cave. Hundreds from every walk of life volunteered to work with pick, shovel, or bare hands. Quickly they were organized into effective teams. College football teams competed for the toughest jobs. A crude hand winch began hoisting rocks and dirt as the shaft deepened. Human mules dragged handcars to the dump. "Dig, dump, and pray" became a catchword.

The country people bitterly opposed "the outsiders' shaft." It was in the wrong place, they said, and anyhow their Floyd would surely die before it reached him. Against combined authority and civilization they did their impotent best. Until barred at gunpoint, they continued to claw at the fatal "squeeze." Then they vainly attempted to drum up overwhelming popular support for their unpopular position. Homer Collins once eluded the troops and squirmed past the "squeeze" to dig farther into the final little slide. So high were feelings and so great the confusion that the

Sand Cave a few months after the Floyd Collins tragedy. The group is peering down the shaft re-excavated for the removal of Collins' body. Note the boom and cable still in place. *Photo by Russell Trall Neville, courtesy Mrs. Burton Faust*

welter of conflicting motivations may never be clarified. Sensational journalism helped not at all. Denhardt had Governor Fields demand the recall of an Associated Press reporter for a controversial dispatch headed: DOUBT COLLINS IN CAVE SPREAD BY NEIGHBORS: MANY ASSERT MAN KNOWS EXITS OR HAS CACHE OF FOOD HIDDEN. The Associated Press formally refused, but the reporter soon disappeared from the scene.

Controversy was brewing even before the A.P. dispatch flashed across America. Two days earlier, a high-voltage wire somehow fell across the ordinary wire leading to Floyd's vital light bulb. Engineers opined in print that Collins had been electrocuted—and wondered if the contact was accidental. The press hinted grave charges against Lee Collins and Johnny Gerald. Some even mentioned a "murder theory"—perhaps enemies of Floyd had caused the walls of the cave to collapse. Facts belying the charges were buried deeply and obscurely in the reports or omitted entirely. Still other newsprint hinted the whole operation was a giant scheme promoting the commercial caves of the region. One widely published account stopped just short of accusing Johnny Gerald of murder by sidetracking food and water. Three different news services recounted a death threat supposedly thrown over a newsman's transom. The three quotations were barely recognizable as the same note. Two of them were ideally designed to inflame public opinion against Johnny Gerald.

In this land of the Great Cave War, almost anything seemed possible. It was obvious that not all the rivalries had been suspended. Quite without publicity, George Morrison and his New Entrance employes were in the thick of the rescue operations. Names of the staff of the Mammoth Cave Hotel and Historic Entrance were conspicuous by their absence.

County Prosecutor J. Lewis Williams announced an investigation of the charges. Governor Fields, however, took the matter out of local hands "for the honor of Kentucky." General Denhardt convened a "court martial of inquiry," which labored mightily and produced little.

As the second week slipped by, progress on the deep, narrow shaft was agonizingly slow. Staticy noises from a radio amplifier attached to Collins' light wire were encouragingly interpreted as his heartbeat and breathing. As the shaft deepened, the diggers began to encounter cave crickets within the rubble. Amid constant showers of rockfall, the miners grimly dug on. Fast-thinning crowds tacitly revealed the expected verdict, but miners who quietly re-entered the "collapsing" cave on Friday, February 13, still detected signs of life. "DIG ON" was the watchword.

Early Monday afternoon, February 16, miner Edward Brenner broke through a thin layer of limestone and found himself in the little chamber at Floyd's head. The explorer's eyes were deeply sunken and his jaw rigid. No sign of life remained.

The six-foot miscalculation necessitated a prolonged conference before any announcement could be made because amputation would be necessary in order to recover the body.

Now reconciled with authority if not the press, John Gerald descended to make the identification. Dr. William H. Hazlett also made the perilous descent. Then came his official announcement: "Floyd Collins died of exposure and exhaustion." Four decades later, he still recalled the pose of a cave cricket—lengthwise on the victim's nose, facing the nibbled tip.

Sorrowfully the Collins family signed an authorization for the amputation. But, before the body could be freed, the knowledge of Floyd's death collapsed the tremendous drive of the rescue gangs. Dozens were willing to risk their lives if any chance remained for Floyd. Not one would again descend the tight, rock-strewn shaft to remove a corpse. As the Army of Sand Cave melted away, Carmichael and Lee Collins bowed to the inevitable and ordered the shaft filled. Memorial services were held, ballads sprang into being, and Floyd Collins took his place in the folklore of America:

> . . . now his body lies a-sleepin'
> in a lonely sandstone cave.

Perhaps it was the grave he would have preferred.

And in a deserted Pennsylvania mine a little boy was crushed to death playing "Floyd Collins in the cave." Few paid any attention.

The Collins family soon reversed its agonizing decision. They sought help to re-excavate the shaft properly and safely so that their boy could be laid to final rest. But the once-generous help had evaporated. Homer was forced to take to the vaudeville circuit, and senile old Lee Collins was talked into doing the same. Lee's act soon collapsed, and Homer was a reluctant star. Yet, retelling his personal version of the tragic days, he raised the necessary $2,800. A contractor's crew redug and timbered the shaft, blasted through a limestone ledge, and emerged behind the corpse. Tenderly they lifted the fatal little rock from the imprisoned foot. The emaciated body was laid to rest in a sunny grave marked by a splendid stalagmite and a pink granite headstone: GREATEST CAVE EXPLORER EVER LIVED.

For a time, curious tourists were guided to Sand Cave for a fee, shown the mouths of the crawlway and shaft, and a gasoline lantern lowered to the bottom of the fearsome shaft. With the coming of the national park, all indications of its location were removed. With amazing swiftness, the Kentucky wilderness reclaimed its own.

Cub reporter Skeets Miller received a Pulitzer Prize. Unmentioned was the city editor responsible for his fame.

Inevitably, a tremendous upsurge in cave consciousness accompanied the tragedy of Sand Cave. Seemingly every cave owner in central Kentucky tried to profit from the opportunity, and as the Reverend Hovey had recounted a generation earlier, "There are said to be 500 caves in Edmonson County [alone]." Counting both ends of Mammoth Cave, at

Central portion of the Grand Canyon of Crystal Cave, with Floyd Collins' casket and headstone in foreground. *Cave Research Foundation photo by William T. Austin*

least twenty local caves were displayed simultaneously. Some were outstanding. Others were tourist traps.

The war even spread to Missouri, where half of Onondaga Cave was operated as Missouri Cavern in fierce competition with Onondaga per se. Rivalry was terrific. Signs advertising one cave kept appearing six inches in front of those of the other. Overnight they were ripped apart, seemingly by miraculously localized tornadoes. Deep holes appeared mysteriously in strategic roads. Missouri Cavern cappers hopped on running-boards: "Hey, mister! Don't go to Onondaga. They jist found out the water's p'ison!" (The same stream courses through both halves of the now reunited cavern.)

Underground, a barbed-wire fence separated tours of the two enterprises—but it kept getting moved back and forth. Heightening rivalry led to underground rock fights between the guides. Tourists cheered their team and sometimes joined the melees. As lawsuits raged, two fences came into being with a no man's land between. An epochal 1932 state supreme court decision established the principle of mandatory underground surveys. Nevertheless, for a long time the surveyors' stakes somehow wouldn't stay put.

"When Harry Truman was running for Senator in 1934," Lyman Riley recalls, grinning, "he visited Missouri Cavern. They were about the only Democrats in Crawford County then. It was the first time any Democrat candidate'd ever come here. Maybe that's why he won. Anyway,

Sand Cave ticket office during brief commercialization in 1925–26. One member of the party holds the rock which trapped Floyd Collins. *Photo by Russell Trall Neville, courtesy Mrs. Burton Faust*

Missouri Cavern had gone electric by then—the first in Missouri. The Onondaga Republicans met him full force at the fence with gas and kerosene lanterns. Quite a few words got tossed back and forth, but no rocks that day." Until Blanchard Spring Cavern got into the 1964 Arkansas gubernatorial campaign (of which more later), Harry Truman was considered America's only underground politician.

Aging veterans of the Great Cave War now talk as if they were no more than glorified practical jokesters. As they "set an' rock," the eavesdropping outsider sometimes still may catch mirthful snatches of sophomoric exploits long considered top secret:

"Remember the time I dammed up the creek across your road and you thought it was my flivver that got mired in the middle? And you put the shoats in it? That jake sure was mad as a pig under a gate when he came back to pull it out!"

"Yep, and I recollect the time it *was* your old car, and we h'isted it up on top of a big snag!"

But there is no doubt of the bitterness which once stalked this pleasant land. Over the years, the flaring strife and the swarming nuisance alike turned a once-amused public against the cappers. Unintentionally they gave a great impetus to the creation of Mammoth Cave National Park.

Over the years, modernity came to the Mammoth Cave country, and business practices changed. Yet, even today, stand after rickety stand hopefully offers to sell the gutted remains of many a once-fine cave to the unthinking tourist. Nor are the old ruses wholly dormant. The culturally attuned visitor need not seek far to find latter-day cappers on at least one road to Mammoth Cave. The ghosts of the Great Cave War still stalk this pleasant land.

Even in death, Floyd Collins continued to be chief victim of the cave war. Under bizarre circumstances which spawned several lawsuits, his body was exhumed and placed in an expensive casket in his beloved Crystal Cave—by now renamed Floyd Collins Crystal Cave and operated by new ownership. Under even more mysterious circumstances, the new proprietor one morning found that handsome coffin broken open and the embalmed body missing.

Bloodhounds were quickly assembled for a grim, heavily armed posse. Had they not promptly found the ill-treated corpse, thrown over a nearby cliff, Floyd Collins might have made more headlines. Half a century after his death, controversies still swirl, lawsuits still emphasize the tragedy of Sand Cave. Even today, the story of Floyd Collins is far from told.

Was there foul play in those overwrought seventeen days? Under pressure of exhaustion, hysteria, and cultural misunderstanding, was the wrong decision made? Could Floyd Collins have been saved by renewed excavation of the rockfall in the cave?

We know that reliable witnesses who entered Sand Cave late in the rescue found it quite different from hysterical early reports. We know that, before the National Park Service barred Sand Cave three decades later, expert caver Larry D. Matthews crawled to the final "squeeze" and found an opening still continuing. We can only wonder.

We can only wonder, too, what Floyd Collins found. Today, guesses vary enormously. Reliable native-born insiders in the momentous drama have no doubt that Floyd discovered a magnificent cave. The contractors' crew which freed Collins' body found him "on the edge" of an impressively deep pit. I myself have seen, less than a hundred paces down the cove, a very active little sinkhole swallowing the hillside at an avid rate. Some sort of cavity is down there.

On the other hand, the "outsiders'" rescue gangs generally believed that Floyd got stuck going in, not coming back from any great discovery. In those seventeen days in February, wishful thinkers could choose from seventeen hundred rumors.

The key to the complex enigmas of Sand Cave still lies far down a nasty little "sand hole." But it now appears that its gripping climax may come sooner than any of us had thought possible. Mapping their way eastward from George Morrison's end of Mammoth Cave, unsung teams of the Cave Research Foundation have been surveying at least two routes that lead in the general direction of Sand Cave. In the very basement of this fascinating junction of the two cavernous ridges, one grubby team found a side lead heading south, directly toward the site of Floyd's death.

These new-found passages are far below the level where Floyd was trapped, probably far deeper than he ever penetrated in Sand Cave. Yet the map of Mammoth Cave now has reached a point closer to Sand Cave than the C.R.F. was to Mammoth Cave when I began this book. A single strategically located dome-pit here could make an enormous difference.

In these classic Kentucky karstlands, dome-pits characteristically occur along the edges of sandstone ridge caps: precisely the situation at Sand Cave. That little rock may have set history back more than anyone but Floyd Collins dreamed.

3

Beginnings Old and New

The Story of Famous Caves of the Virginias

The flickering candle cast enormous, wavering shadows on the pocketed cavern walls. Determinedly, the young Virginian slowly deepened his signature into the pitted limestone. This was true adventure—as adventure had followed new adventure in this wonderful wilderness between the great mountain ridges.

Just above the floodplain of marshy Evitt's Run, the black mouth of the cave yawned in a shallow, asymmetrical sink in the rolling Virginia bluegrass country. Inevitably, its air of mystery had drawn the attention of the eager youth during the noon halt. Munching venison, he approached the cave warily, fascinated, yet a trifle fearful of the shadowy unknown.

Cautiously the young surveyor descended a natural path among locusts, elm, hackberry, and red oaks. Soon he could see that most of the broad cavernous alcove was merely a shadowed recess beneath an overhanging limestone wall. But at the south end a dark orifice contrasted with the brown-streaked gray limestone. He swerved, picking his steepening way around bare mounds of broken limestone, avoiding thick patches of budding poison ivy. With the thrill of discovery, he stooped and peered into the hillside.

In the midday bright, a spacious gloom was dimly visible. Curving walls and dim piles of rock hinted at hidden alcoves leading to endless corridors. As he knelt irresolutely, a faraway drip . . . drip . . . drip called the youth like the song of some subterranean Lorelei. An occasional PLOP! told of water dripping into some hidden pond or silent river.

Lured by the age-old call of the unknown, the youth inched into the portal. No hissing serpents or sulphurous vents—instead, the air tasted cool and fresh and curiously invigorating, exciting. He found himself sidling farther and farther into a commodious, shedlike chamber of rock. To his right, the dim outlines of arched grottoes tempted him. To his left, stray rays of light showed a low extension leading north—how far?

As his pupils dilated, he edged into the darkness. Step by cautious step, he avoided shin-barking limestone fragments and slick little mud slopes. Ahead and to his right, plenteous dripping splashed amid rocks, then trickled onto a strange waist-high cascade of stony terraces. They seemed to flow across his path, then into a widening cavern lost in the gloom.

A sudden need to know what might lie concealed beyond the darkness gripped the sixteen-year-old surveyor. He picked his way to the sun-drenched entrance and dug into the saddlebags for a tallow candle. Parrying good-natured banter of his frontiersmen companions, he hurried back to the pleasant, now-familiar freshness of the cave. Lighting the candle with flint and tinder, he began to retrace his steps. Now he could admire details of ornate works of nature half imagined before—though the cave did seem to shrink most remarkably.

Past the stone terraces and around to the right he advanced, to a blank wall. But off to the left a widening room seemed to funnel down to a low orifice half hidden by rocky slabs, some as large as a pianoforte.

A glimmer of blue daylight still accompanied the lone adventurer. Hitching up his pouchy hunting shirt with a deep breath, he resolutely turned from the friendly outer chamber, inspected the low opening, and approached it on his stomach. Candle in hand, he winced more than once, crawling on the sharp-edged rocks.

The hole which had looked so small proved amply spacious for the lithe youngster. Swinging his feet wide, he sprang up triumphantly, inspecting the chamber at hand. To his left, a steep slope seemed covered by a petrified waterfall. Ahead, strange stony hangings reached down to stubby rock mounds. For a moment, a crevice looked promising, but a second glance proved it too small for even his flexible torso.

To the right, the little room sloped down to a dripping pipe organ of stone perched over a tiny pool of clear, tasty water. The precocious youth drank deeply, thankfully. As he stooped, his candle showed more cavity past hanging draperies, but no way through. He had reached the end.

Maybe it wasn't much of a cave, he mused, but he had explored it despite the joshing of the others. It was *his* cave now, by right of conquest. Setting his guttering candle on a convenient ledge, he unsheathed his ever-present knife. Laboriously he began to carve boyish block letters into an alcove at chest level: G. WASHINGTON—1748: the earliest known date of any spelunker in the mainstream of American caving.

Some might find overstrained any claim that the Father of His Country is also the father of cave exploration in his country. George Washington was not the first American spelunker. Still earlier in the eighteenth century, back-country pioneers manufactured gunpowder from saltpeter found in caves in what was then western Virginia. Legends link Johnny Appleseed to Devil's Den near Webster Springs, West Virginia. Cave-in-Rock on the Ohio River was shown on river charts by 1729, and French voyageurs knew a saltpeter cave in Minnesota by 1701. As early as 1674, Friar Rodrigo de la Barreda visited an impressive Florida cavern "with three apertures buttressed by stonework of unusual natural architecture."

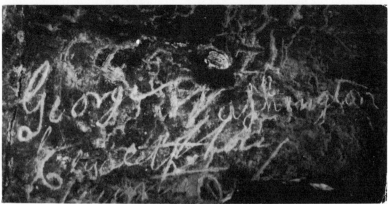

Signatures of George Washington in George Washington's Cave, West Virginia (the 1748 date is barely visible below the name); and Madison's Cave, Virginia. *The lower photo from the Clay Perry Collection by G. Alexander Robertson, courtesy Paul Perry*

Early New Englanders ventured into crevices near their settlements. These isolated visits, however, were far from the mainstream of American caving, perhaps comparable to pre-Columbian Viking expeditions to Vinland. Just as pseudo-discoverer Columbus opened up the New World, so began American caving with George Washington in March 1748.

Eleven years younger than first spelunker George Washington, the first American speleologist was five years old in 1748: Thomas Jefferson. In 1784 the President-to-be published the first edition of his celebrated *Notes on the State of Virginia*. In it Jefferson discussed details of the use of spelean saltpeter in the gunpowder industry of Revolutionary times. He also included the first known map of any cave in the United States— Madison's Cave near Grottoes, Virginia.

Had this great statesman and patriot no other laurels, his name would have been perpetuated through his interest in caves. In 1796 saltpeter miners in a cave in West Virginia came upon the skeleton of a huge-clawed quadruped as big as a bull. The skull was missing, but scientist Jefferson was able to obtain most of the other bones.

Jefferson was perplexed by the bizarre remains. Puzzling over the odd feet and huge claws, he concluded they had belonged to some great cat with three times the bulk of an African lion, "as formidable an antagonist to the mammoth as the lion to the elephant," he told the venerable American Philosophical Society. Since the skull and telltale teeth were missing, it was not a bad guess.

Jefferson's deduction about the great bulk "of the Great Claw, or Megalonyx," was accurate. His speculations about the nature of the beast were not. But, even before Jefferson's memoir appeared in print, other early scientists had caught the exciting scent. Caspar Wistar, M.D., vice-president of the American Philosophical Society, thumbed through precious new reference books at the little University of Pennsylvania library, where he was adjunct professor of medicine. Wistar noted and recorded resemblances between the foot of Jefferson's animal and that of the sloth. The first ground sloth ever found by science bears the name of the great statesman and speleologist: *Megalonyx jeffersoni*.

Jefferson, like Washington, was a trained surveyor. It appears that he himself prepared the historic sketch of Madison's Cave. His descriptions of that and other caves are those of one who has spent considerable time and thought underground. According to some accounts, Jefferson himself first chanced upon this pretty little cave and named it for his young friend James Madison—the Madison who succeeded him as President and became "the Father of the Constitution."

John Randolph was another early spelunker, and Benjamin Franklin was a member of the Saltpeter Committee of the Continental Congress.

Probably a thousand caves were found during the first century of spelunking in the Virginias—more than twice that number are recorded today. At first glance, few of them appeared remarkable to the matter-of-fact pioneers except when their gunpowder supply ran low. In 1804 or 1806, however, young trapper Bernard Weyer inadvertently provided a hint of things to come. Two years later, the respected John Edwards Caldwell of New York wrote:

Madison's Cave, mentioned in Mr. Jefferson's Notes, is now abandoned as an object of curiosity, and is about a quarter of a mile from Weir's [sic] Cave, which was discovered in February, 1806 [*Bernard Weyer 1804* inscriptions in the cave are a still-unexplained mystery—W.R.H.] by a pole cat's being caught in a trap, and retreating for shelter to the cave, to which a dog pursued her. The owner of the dog enlarged the hole by which the animal entered, and discovered the place from which I now write to you. It is certainly the most remarkable subterraneous curiosity on this continent, or perhaps in the world, and is well worth the attention of an observing traveller.

Reports of the glorious splendor of the newly discovered cave spread rapidly. Hordes of the curious clamored to see Weyer's cave, today known as Grand Caverns. It wasn't quite as easy as it is today. In 1837 a well-traveled if grandiloquent author graphically described these long-metamorphosed commercial tours:

The preparation for exploring one of these cyclopean caves consists of a supply of pitchpine sticks, faith in your guides and folly in yourself.

The sticks are about two feet long, and each one as thick as a little finger; 15 or 20 of which being held together in the hand and fired at the opposite end, make the best of torches, will burn bright for two hours, and distinctly show the floor, sides and roof of the cave through the palpable obscure.

Little magazines of sticks are judiciously left at intervals of a quarter of a mile, as you penetrate deeper and deeper into the bowels of the land, to replenish from time to time the moribund luminaries.

In few places is America's progress so evident as in today's splendid tourist facilities. Today the fame of Bernard Weyer's Grand Caverns is little dimmed by the flamboyant advertising of later cavern discoveries. Even our most skeptical spelunkers nod concurrence with the proud boast: "Universally recognized as one of the very few great caves of the world."

As Civil War clouds gathered, a bearded maniac with a righteous cause made use of a convenient cavern on the outskirts of Harper's Ferry. There John Brown stored arms and ammunition for his seizure of the nearby United States arsenal.

Or so the story goes. Cavers are inclined to agree with the local historian who heatedly termed the tradition "a come-on for suckers," designed to enrich local cabmen. John Brown's Cave is small and wet. Until recently commercialized, its entrance was so low that pack mules would have had to crawl in on hooves and knees. "Anybody that would store powder and arms there ought to have his head examined," snorted the late Burton Faust. Yet we cannot wholly exclude John Brown and "his" cave from the story of the caves of America. Federal troops under Colonel Robert E. Lee soon reversed the abolitionist's temporary suc-

The Angel's Wing of Grand Caverns. *Photo by Flournoy, courtesy Gladys Kellow*

cess, but his soul went marching on. The youthful United States and its spelunking never again were the same.

The first southward thrust of Union forces along the Shenandoah Valley brought young, blue-clad Ohio Volunteers to Melrose Cavern. Almost within rifleshot of the front lines, the pleasant cave offered welcome relief from remarkably bitter spring weather. Though some accounts attribute its discovery to David Harrison around 1818, local tradition recounts that Melrose Cavern sheltered settlers during the French and Indian War of 1754. The boys in blue followed their legendary example in considerable comfort. Garrison stoves were packed in, and holes drilled here and there for support of torches. Whole troops bedded down inside, comfortably out of the wet. Until some unpopular officer intervened in time to save most of the beauty of the cave, thoughtless soldiers amused themselves by shooting down stalactites and gracefully fluted draperies; scars and neatly aligned bullet holes are still visible today.

In Melrose Cavern, the troops soon spotted inscribed names of members of the family of their hero President: Old Abe's ancestral home still stands nearby. A JOHN LINCOLN inscription may have been carved by the President's great-grandfather.

In those remote days of infrequent cavern visitors, no one realized the long-term effects of such inscriptions. Many a young soldier paused to smoke, scratch, or pencil his name on a conveniently overhanging ledge or low ceiling. Those more patient methodically carved their names into the rock walls or the softer coating of some stalagmitic column. A few added their dates and regiments. One hero worshiper smoked on a wall a crude portrait of his President. In addition to his name on the massive registry column, artistic W. P. Hugus of Company C, 8th Regiment, Ohio Volunteer Infantry, carved his regimental coat of arms on another stalagmite. It did not yet seem that war was entirely hell, that quiet April of 1862.

But officialdom had blundered. This time it had underestimated Stonewall Jackson. His brilliant maneuvering soon forced General Banks's bluecoats northward, completely out of the great valley. The Union forces enjoyed Melrose Cavern less than three weeks.

Suspiciously, the advancing Confederate forces reconnoitered the scene of so much activity. Entering the deserted cave with pistols cocked, they found it filled with maddening inscriptions. To these they devoted part of their fury at the destruction abroad in their verdant native land before turning northward to more important matters.

The role of Melrose Cavern in the Civil War then grows dim in the shadowland between fact and tradition. Some 1864 dates with Union inscriptions may be mementos of Sheridan's infamous scorched-earth raid. Few Confederate inscriptions can be seen; the late Clay Perry noted one by Captain George Koontz, who is said to have fired the first artillery shot at Gettysburg. Most Confederate activity was limited to hacking away at earlier Union signatures.

Some inscriptions of Melrose Cavern appear hardly a day old. To the Civil War enthusiast, they seem remarkably fresh links to the great conflict. In this cool, moist cave, however, smoked names will never survive as perpetual mementos like the prehistoric cave paintings of Europe.

Melrose Cavern's Registry Column is a subterranean history book. *Photo courtesy Endless Cavern*

Especially in its eastern section, something inherent in the cave already has rendered many of them fading and illegible. Many a caver, entranced by the attractive, smooth-walled corridors, feels it better so. To inscribe one's name today means virtual ostracism ("What have *you* done that makes *your* signature history?"). Yet something valuable will be lost when these little sidelights of history are washed away or hidden by thickening traces of calcite. Consider the names on the stalwart stalagmite dubbed The Torpedo: E. P. Shepard, Oberlin, Ohio; L. G. Wilder, May 4, 1862, C. 7th O.V.I. Below is another name: L. W. Shepard, Washington, D.C., with the date 6–9–22 recording a painstaking search for the name of his deceased father. Such humanness in the story of Melrose Cavern offsets much of the bitterness of the War Between the States.

With Sheridan's terrible raid and the ebb and flow of the Shenandoah Valley battle lines, this once-rich granary of the Confederacy was soon desolated. By the end of the war, cave hunting lagged elsewhere in the now-divided Virginias. In the pauperized Shenandoah Valley, however, the meager but steady tourist income of Weyer's Cave was widely envied. The decimated populace avidly sought similar sources of hard cash—but warily. Superstition was rife, and at least as many feared caves as sought them.

Not long after the conflict, former New York State photographer Benton P. Stebbins fell in love with the great valley and established himself in the valley at Luray. During the summer of 1878 he became intrigued by nearby Cave Hill. On that modest ridge 300-foot Ruffner's Cave had been known since 1795.

With a friend, Stebbins visited Ruffner's Cave and pondered. Broken rock at the lower end seemed to block openings that might lead to a deeper, grander cave. He enlisted the help of a neighbor, William C. Campbell. Leaving no movable stone unturned, they investigated Ruffner's Cave intimately, without success.

On August 13 they had planned another visit to Ruffner's Cave with Campbell's brother Andrew and cousin Quentin (Billie) to provide additional manpower.

Andrew Campbell, however, expressed a decided preference for sinkhole digging despite the jeers of townspeople who had taken to calling the cave hunters "cave rats" and less complimentary names. Five hours' hot work led the cave hunters down ten feet to an open space: low, flat, and about fifteen feet wide.

Stubby Andrew Campbell was the smallest of the group. With his brother and Stebbins steadying a rope, he lit a candle and wriggled into the tight orifice. As his eyes adjusted to the gloom, a gaping black chasm opened below him. Balancing down a slippery slope with the aid of the rope, he cautiously made his way downward around successive stalagmitic ledges. Soon he reached the floor of the startlingly beautiful antechamber of Luray Cavern, spellbound as its gloriously fluted Washington Column came into view. Covering the hole with rocks, the party adjourned for a well-earned dinner, then returned to explore. They quickly found Campbell's beautiful chamber bounded by subterranean lakes that seemed endless, bottomless in the light of their flickering candles. With a group of helpers, they enlarged the opening and built a small boat underground. Beyond the lakes were enormous chambers and great corridors where almost every trace of the walls was richly adorned by splendid stone draperies. Age-old stalagmites of surpassing beauty often barred the way. The three cave hunters envisioned themselves as men of wealth—and more than a million visitors have since viewed their great discovery and come away entranced.

Luray Cavern in 1878. *Painting by Mrs. Benton Stebbins, courtesy Arthur Stebbins*

The local newspapers, then the nation's press, took up the discovery of "the most beautiful cave in America." It was no great problem to build stairways and bridges, or to lay out pleasant paths, or to drain the lakes. From the beginning, visitors were entranced. In the magnificence of its splendid draperies, in the symmetry of its hanging cascades, Luray stands unchallenged among American caves.

The great Luray discovery was the talk of the Shenandoah Valley. Thousands of rabbit holes were hopefully excavated. A year later two boys and their dog flushed a bunny on the farm of Reuben Zirkle, near New Market. Automatically they pulled apart its boulder-pile refuge. Almost at once they encountered a wide shaft slanting steeply downward. Below was an extensive cave which many rank with Weyer's Cave and Luray Cavern. This was promptly commercialized as Endless Cavern.

On August 14, 1920, new owners reopened Endless Cavern to the public. With electric lighting, it became an immediate success, especially after the inspired management invited the Explorers Club to come spelunking. That club responded enthusiastically and lightheartedly. With cooperative publicity, several expeditions left bottled notes in far recesses of the intricate cavern. As one of the Explorer spelunkers remarked, "If anyone ever finds those bottles and carries them farther, they deserve more than an empty bottle." Endless Cavern became a household word.

Today Endless Cavern is still endless, but it must be admitted that no truly expert group of modern spelunkers has ever sought its end. Members of the Virginia Cave Survey have studied and sketched about one mile of this delightful commercial cave, and traced nearly all of the explorers' routes. Yet even that methodical crew left one region uncrawled. In print they smiled benignly: "Its official length is not publicly known." Unspoken is our need for an Endless Cavern.

After the rapid-fire discovery of Luray and Endless caverns, the hunt for other tourist caverns was redoubled in the Valley of Virginia. Though not really explored until 1884, magnificent Shenandoah Cavern was encountered in 1868. Blasting for reconstruction of a branch line of the Southern Railroad opened a narrow crevice, long supplanted by a spacious entry. Today its sparkling rose-pink, faceted flowstone walls and its giant "bacon" draperies elicit the praise of connoisseurs of caves. In 1937 the story of Luray Cavern was almost duplicated with a scientific touch. Walter S. Amos of Winchester had studied a sinkhole-strewn cove near the northern end of the lofty Skyline Drive. Nearby is Allen's Cave, to which the late Clay Perry traced Civil War legends of Mosby's raiders and the glamorous spy Belle Boyd.

Geologist Amos deduced the hidden presence of a more important cave. On December 17, 1937, excavation opened extensive Skyline Cavern and its matchless anthodites. With something of the appearance

The Explorers Club Endless Cavern team roping up. Mountaineers now consider this technique obsolete and dangerous. *Clay Perry Collection photo, courtesy Paul Perry*

of petrified, pure white bunch grass, the anthodites and other features of this unique cavern have delighted thousands annually.

A little less than two years later, the beginnings of modern American speleology stirred fitfully beneath a different valley: Germany Valley in West Virginia's Pendleton County. Here are Seneca Cavern, Schoolhouse Cave, Hellhole, and more than a dozen others known only to cavers.

The name of Schoolhouse Cave is not as ridiculous as it sounds. Generations of children have played in its entrance room, only a hundred yards from the old Cave School on the back-country Harper Gap road. As in many another West Virginia cave, saltpeter was mined here a hundred years ago and more. Dozens of miners and hordes of Tom Sawyer–minded schoolchildren had gawked from a Jumping-off Place five hundred feet inside the steep, airy entrance. Crude searchlights dimly revealed shadowy walls stretching far ahead in the eternal midnight. From the slippery ledge, the cave seemed to implode downward into a vast, unknowable abyss. Rocks tossed over the rim took too many heartbeats to crash resoundingly in the blackness. Their clattering echoes seemed to fade away in infinity.

Thus, on the sunny, cool Armistice Day of 1939, the country folk of Pendleton County were certain that fifteen weirdly clad city people were at least slightly mad. Enormous piles of ropes, wire, telephones, timbers, and unidentifiable packs poured out of beaten cars. An overloaded human line slogged down into the cave ravine. Odd characters who termed themselves "spelunkers" were going to tempt fate at the Jumping-Off Place.

At a turn to the left, the wide maw of the cave yawned, condescendingly, 40 feet wide and a third as high. Gray in the half light, a wide sloping chamber came into view. After the cool autumn crispness, it seemed a bit muggy.

In the abrupt twilight, Bill Stephenson, president of the newly formed Speleological Society of the District of Columbia, raised his hand for a brief halt. "Better load up your carbide here. Is everybody O.K.?"

The origin of the anthodites of Skyline Cavern has long baffled speleologists.
Photo courtesy Skyline Cavern

It was a good question. The official log of the society included just eight relatively unimportant caves. Only a few months separated it from its curious origin as a Sunday school class. Those few months and a few cautious ventures had greatly advanced the skill and confidence of its enthusiastic adherents. Nevertheless, this was the first major venture of the young society into a really major cave. A touch of jitters was permissible.

From an ancient hand-hewn trough beneath a dripping crevice, the milling cavers filled the upper compartments of their miners' lamps. Shaking rough, grayish pebbles of carbide into the lower chamber, they screwed the units together and adjusted the control lever. Each took his lamp in hand, cupped his palm momentarily over the curved reflector, then whirled the striker with the heel of his hand.

Usually the gas jet ignited on the first few tries. Here and there among the group, repeated failures and decreasing nonchalance revealed inexperience. For those unable to "pop" their uncooperative lamps, neighbors obliged with "the speleologist's kiss"—a touch of the acetylene flame to the other's unkindled gas jet. With fifteen lights throwing warm circles or long beams of light deep into the receding gloom, the trek continued.

Beyond the gaping maw, the throat-like entrance room sloped downward steeply. At the bottom of the narrowing slope, the high, curving chamber seemingly ended at a steep wall of clay and rock. From rock to slippery slope to old wooden pole, the long line slowly worked its cautious way almost to the roof.

Ahead was the expected low passage. Deepened by saltpeter miners, it led the cavers farther and farther south. At times, all could stand comfortably. More often, stooped backs ached to throw off the burdensome loads. The eager chatter of the entrance room gave way to determined silence. More than one robust caver panted unashamedly.

Another curve, another . . . duck down, straighten up—no, not quite. Duck down. THUNK! Someone's helmet hit the roof. A hundred paces, two hundred . . . The party strung out wretchedly, the girls and some of the men already staggering. Then all at once there was nothing in front of lead scout Don Bloch.

A flimsy ladder was hung into nothingness. As the most experienced of the group, Don tied a safety rope around his waist. Leaning far out with a battery-powered searchlight, he shook his head.

"It's still going. I can't see the bottom."

Handing the light to Dwight Vorkoeper, Don nimbly started down the twisting 75-foot ladder. More and more of the safety rope snaked over the edge. A dozen feet below he halted, vainly seeking a fleeting glimpse of something—anything.

Somewhere far off, a slick wetness reflected a faint glimmer from some distant pinnacle or wall. Off to the left in the black depths, a spattering cascade echoed distantly, thinly. His headlamp revealed a second ledge sloping into all-engulfing darkness. Ahead the broad, level ceiling receded into unreachable shadows. Only a few dozen feet of solid limestone separated him from the crisp, colorful Appalachian autumn. But in this light-devouring void, it might as well have been a hundred times as much.

This was what the team was seeking. Soon Don's attenuated voice rose from the blackness: "I've got a good ledge. Come on down!"

The safety rope came up. In turn, three other cavers joined Don Bloch in a slippery little grotto where man had never been. Twirling downward rung by rung, Don again took the lead. Again he found a ledge, so tiny that only Dwight could join him. Again he began the descent, this time to the full extent of the 75-foot ladder. There his feet rested lightly on a queer sort of shelving floor.

To one side, a great rounded shaft disappeared into the silent darkness. On the other, silvery rain of a little waterfall glistened vaguely in other half-glimpsed depths. Ahead? Maybe there was a way between the latter-day Scylla and Charybdis to the rock jumble visible beyond their yawning depths. But that was not for today, for this first scouting trip.

The electric lantern was lowered. Don scanned the far cavern, calling his findings upward in ecstatic staccato shouts which broke through the rolling echoes.

"Huge cave beyond! Hundred feet high! Pit to the side! Keeps going!"

Then the wicked, arm-wearying ascent, far more difficult than the descent. With tremendous exertion, Don inched up the jerking ladder, now seemingly fighting him like a disjointed serpent. Far above, the unskilled crew took in slack rope on a hand winch. As Don scrambled onto Dwight's ledge, trembling with exhaustion, the sigh of relief was universal. Then the ladder was pulled up, the struggle repeated. And again. And again. Many an hour-long minute passed before the last caver topped out at the Jumping-off Place, shaking with tense fatigue.

Around the campfire, revived cavers excitedly began to plan future assaults on the forbiding cave. Maybe it connected to Seneca Cavern, to Hellhole, to any of a dozen-odd nearby caves. Already the challenge of Schoolhouse Cave had left its mark.

As time passed, elaborately planned return trips pushed onward—a very little. With more efficient lighting, additional pits and passages appeared amid the pinnacles and sharp-carved cliffs. Many were in locations where their discoverers merely sighed and shook their heads. The Big Room, whose floor Don Bloch had touched so fleetingly, proved at least 200 feet long and 30 feet wide. A human fly, strolling along the smooth, flat ceiling, would have thought it a pleasant chamber. No one else did. Mere mortals found their way barred by a three-dimensional maze of hackly pits separated by knife-edged ridges. Jagged pinnacles strained toward the ceiling. Slippery clay floor and enormous piles of loose rocks compounded the problem. Sheer jagged walls rose up and up and up. Often they overhung wickedly. Nonetheless, the walls seemed less fearsome than the tormented floor of this grotesque chamber.

In time, the Speleological Society of the District of Columbia mushroomed into the National Speleological Society. Its increasing member-

Rope work below the Jumping-off Place of Schoolhouse Cave. Prusik knots are being used. *Photo by Lyle Conrad*

ship located more and more caves to be investigated: pleasant, delightful caves. More and more cavers looked over the Jumping-off Place. More and more they decided they weren't quite ready to unveil the farther mysteries of Schoolhouse. Specialized climbing techniques and unusual experience were necessary to advance farther into this fearsome cavern of peculiar hazards and great vertical distances. For fifteen years only a few scattered members of the society were of Schoolhouse caliber.

By intentional default, pioneer caves abdicated the challenge to a reasonable replica of a swarm of human flies: the highly competent rockclimbers of the Potomac Appalachian Trail Club. Using techniques then largely unknown to American cavers, these intrepid climbers eagerly undertook the conquest of Schoolhouse Cave in February 1940. Clanking with pitons and other metallic climbing gear, carrying food and sleeping bags for trips lasting up to twenty hours, these well-organized climbers fully expected to bare the greatest secrets of Schoolhouse Cave within a few weekends.

It didn't quite work out that way. In this vicious cave, every step had to be planned, each handhold tested. Even the rock climbers found special equipment necessary at almost every turn. Yet tall tales of impossible places with unlikely names began to circulate. The rockclimbers' special brand of humor invented the Ribfiddle, Hell's Belfry, Nightmare's Nest, Orpheus' Snootflute, Charley's Groan Box, the Guillotine. Strangely, most of the tales were true.

Skilled rappelling proved the key to the rockclimbers' advance. Promptly standardized was an 80-foot rope descent precisely between Don Bloch's awesome North Well and the 180-foot Cascade Pit, but the rockclimbers' descent was very different from that of Don Bloch.

Rappelling has advanced far beyond those early days. Then, a rappeller faced the anchor point, straddling the rope on which he was about to descend. Snugly curved under one hip, the rope was brought across the body and chest to the opposite shoulder, then backward and downward to the control hand. Friction of the rope around the rappeller's body provided control as he slid in long, delightful bounces against the limestone wall—reasonably safely as long as the rappeller fed the rope rapidly enough to keep his feet low. Later refinements merit two full chapters in this book, but few pre–World War II cavers had heard of even body rappelling.

Below the Jumping-off Place, the rockclimbers swarmed over the jagged walls and sharp-edged pits like oversized ants. Despite inadequate lighting, many an unreachable opening was probed, each bottomless pit plumbed. Tantalizing orifices often proved mere shadows. Others were only shallow grottoes. Several were of greater importance.

Off from the bottom of the Cascade Pit, a squeeze between flowstone curtains led Lowell Bennett to an eight-foot wriggle downward through the Ribfiddle, a knobby cleft just wide enough for a human chest. There a determined little waterfall runs down each explorer's neck and out his trousers' legs. Forty feet below was the hopeful-looking Grind Canyon, 500 feet long. But both ends of this lowest part of the cave ended unremarkably; the sopping exploring team below the Ribfiddle unhappily found itself somewhat in the position of a cork pushed down the neck of a bottle.

Some of these discoveries were interesting. Here and there in the awful black jaggedness were small areas of considerable beauty. Yet it soon became clear that if anything extraordinary were to emerge from this toughest of caves, it must lie to the south, beyond the chaos of its tumultuous "floor."

Led by agile Leo Scott, the rugged rockclimbers already had begun a remarkable trail across this chaotic chamber. First came a tenuous knife edge of rock—the Nick of Time—between the North Well and the Cascade Pit. Beyond, strategic handholds permitted an upward traverse to a two-foot ledge: the Balcony. Although without much visible support, the Balcony provided a much-needed breathing spot and a chance to analyze the chaos beyond.

In the midst of this unearthly chamber, a huge wedged block—the Big Bite—provided another welcome rest. ("It's kinda square except where something took a heckuva big bite out of its east side.") Then down a hair-raising slope to an arch of interlocked stones bridging nothingness . . . a steep rise to the gigantic hatchet head of the well-wedged Guillotine . . . across an incredible wet mess of rock, clay, flowstone, and mud—the Nightmare's Nest.

Beyond the Nightmare's Nest, a 30-foot overhang—the Judgment Seat—topped a mountainous wall of rock and clay. Was this the end? Spotlights showed a dark opening high on the south face beyond the Judgment Seat. A hundred feet above the climbers, however, it seemed amply guarded by vertical walls and overhangs.

An idea came to Paul Bradt. Maybe the Inner Wells alongside the Big Room . . .

Three months after their first overconfident venture, Paul investigated a flank attack. Retracing now-familiar ledges and crevices with increasing ease, he and Don Hubbard scrutinized the gaping, ragged Gargoyle Well. "Here the climber gets a closer look at what lies ahead," they later snickered, "pulls himself together, looks down into the Gargoyle Well, looks carefully at his safety rope and takes that first long step into the great well."

Separating the Gargoyle Well from the first Inner Well is a strange knife blade of pitted limestone almost 200 feet high. Once several feet thick, it has been eaten away by the eternal acid traces in dripping subsurface water. Ragged "windows" now connect the "upper stories" of these great fluted shafts. When no other handholds could be found, the rockclimbers occasionally knocked new holes through the thin pinnacle.

Advancing with precision and rhythm, Paul and Don alternately belayed each other from alcove to alcove. Around one face of the Gargoyle Pit and up its 70-foot rear face they squirreled, thence into the Inner Well and still further up. A balcony-like grotto, delightfully large enough to stand and stretch tension-numbed muscles . . . upward and a little southward . . . and a little more. A scramble through a fissure without visible floor or ceiling, humping along, arms and legs tautly outstretched for suspension above the void . . . into still another well—Charley's Groan Box—to take advantage of momentary holds. Then back into the Inner Wells for another upward pitch, seemingly without even an eyelash's worth of holds but somehow passable. Then, with every overstrained muscle crying for relief, a window ahead. Unbelievably, it led back into the Big Room, with a southward-leading ledge only a step upward.

But what a step! Sitting in the window atop another knife-edge of etched limestone, Don's left leg overhung a drop of 150 feet into the Big Room. His right hung free over an equal drop into the Inner Wells. Dislodged pebbles rattled until the sound dwindled away in great distance.

The ledge which had seemed so inviting was no real ledge at all. It sloped outward at about the steepest angle at which rubble could cling to its wet, slippery face. Even to Paul and Don, the site seemed a trifle airy. The best belay would be of little value here. If Don fell, he would dangle helplessly at the end of the rope, supported only by its friction around Paul's body. Recognizing the insidious dulling of their judgment by exhaustion, the two experts retreated without setting foot on their cherished goal.

Back in the Potomac Basin, word of the new route swept the rockclimbers like wildfire. Again at the expense of enormous exertion, an eager party advanced their freshest man—Charlie Daniels—to the key ledge. The Angel's Roost, someone called it. Step by cautious step, taut muscles trembling just a little, Charlie crab-walked along its sloping, dirt-smeared outer edge. Unhappily out of reach of the reassuring, balancing wall, he inched on: ten feet, fifteen, twenty . . .

Should he slip, the intrepid climber would arc back beneath the window with tremendous impact, swinging from the rope passed through a carabiner. At a momentarily secure point he paused to drill a hole for another expansion bolt. With ebbing strength, he snapped a carabiner to a ring on the bolt, to carry the rope and diminish his free swing should he fall. But his strength was gone. A careful probe of the distant shadows revealed no obviously inviting corridor. Sadly he began the delicate balance step back to the window to join the long retreat.

This venture into the fierce cave exacted a severe toll from even these rugged outdoorsmen. Only after four and a half months did they return. Discouraged by the previous ventures, the next trip included no plans for an assault on the Judgment Seat.

Leo Scott, however, is not easily defeated. With the short-lived energy of desperation, he reconsidered the slippery clay wall beneath the Judg-

Bill Peters belaying Frank Brown past a deep pit in Schoolhouse Cave. *Photo by Lyle Conrad*

ment Seat. While loose rock and clay rained on those below, he hammered a series of 24-inch rods deep into the clay. They gained him precarious balance as he inched his way nearly to the limit of his 120-foot belay rope. As his light reached the level of the overhang, Leo's voice drifted down:

"There's a hole. It goes—up!"

But neither Leo nor anyone else could conjure a way out to that hopeful hole in the overhang, several feet from the top of Leo's Climb. After nine months, the pendulum had swung full cycle. The once-fond hopes were gone. Tom Culverwell prepared a fine sketch of the fantastic cavern. With it he published a detailed analysis of remaining potentials. Nowhere did he mention anything southward beyond the Judgment Seat, where Hellhole and the other caves lay so close.

The repeated repulses, however, rankled Paul Bradt, Don Hubbard, and Leo Scott. Soon they returned to the Angel's Roost with Sam Moore and Ed Siggers. What security Charlie's carabiner provided! Paul could almost relax as he edged across the fateful ledge. Almost dispassionately he surveyed the problem.

Ahead was a curving alcove, with the ledge narrowing to a couple of inches. Some doorknob-shaped stalagmites might be helpful as handholds. Or were they mud-based? Cautiously Paul tested one, then another and another. Each was solidly anchored to bedrock. The way to the Judgment Seat was open. With the ticklish traverse completed, it was child's play to drive a piton and rappel down to the Seat with a whoop of triumph.

Such an achievement should reward the flushed victor with great lengths of magnificent cavern. Paul readily spotted the hole Leo had seen from below. It merely opened through the Judgment Seat. Upslope—southward—was a miserable 50-foot wall of mud and rock, much like Leo's Climb below.

Paul's spirits sank. Deep in shadow at the head of the wall, the only possibility of continuing lay between a huge fallen slab and the ceiling. Having gained only 30 feet, he was belayed back across the tricky ledge to begin the tedious return to daylight.

The next five weeks saw mass onslaughts on Schoolhouse Cave. Steps were cut in the clay wall above the Judgment Seat. Loose rocks were carefully brushed aside, pitons strategically emplaced. Perhaps most important, the climbers familiarized themselves with the entire route. Loose rocks were pushed aside. Steps in the clay slope were stabilized. Now each step need not be planned separately. Precious energy was conserved.

The Loophole was a precious boon. A crude single-rope ladder was fixed in place. As long as its rope was free of deadly rot, the battle with the Inner Wells was won. But heretics refused to use the ladder: usually off-balance, overloaded climbers swore that the twisting, bucking loop ladder was worse than any possible climb. Cocoons of loops and belay ropes ensnarled many a grunting victim spinning almost uncontrollably on this, the "easy" route. Ascent was slow at best—except for the unwary whose wrist caresses from carbide flames spurred them to extraordinary vigor.

Above the Judgment Seat, Fitzhugh Clark attacked the crumbly wall like a man possessed. Hard-dug steps, pitons, and expansion bolts inched him far toward the ceiling. A tight little wallow opened, unimpressive but pleasanter than his airy wall. Surprisingly, it led him to a new room at ceiling level. In typically Schoolhouse fashion, however, it ended in another overhanging barrier topped by a tantalizing orifice.

High on the west wall of this new Hodag Room are two notorious ledges three inches wide. While their fellows guffaw, cave climbers here hop-slide along the lower ledge in a half crouch, finger-scratching for holds, weight on the right knee, left leg hanging down, foot clawing for balance. "Like crawling along a picture molding," one disgruntled caver avers. But it works.

Beyond the Hodag Room, Hellhole is just a few hundred feet away. Eagerly the explorers slithered along the shallow, sandy passage. Just 80 feet onward yawned a huger black void: the Thunderbolt Room.

Compared with the Big Room's racking struggle, the new chamber at first seemed a breeze. Lightheartedly the exhilarated group began exploring every orifice. Cracking a joke, Tom Culverwell picked his way down a steep, shadowed 70-foot slope to a stony platform. Beyond a void, the rear wall of the chamber sloped upward, but the gap was impressive.

Seeking an easier way deeper, Tom started into a jagged hole, then looked past his feet. They were dangling into 70 feet of wide, wicked blackness. That ended that particular trip.

Twelve days later, Fitzhugh Clark, Tom Culverwell, and Dr. H. F. (Stimmie) Stimson led a determined crew to the Thunderbolt Room. Aerial trolleys were constructed to transport the ever-increasing gear high across the entrance room and Big Room. The fastenings of one enormous load disintegrated 75 feet above the Big Bite: first score for the cave. The trolleys were abandoned.

Rappelling from Tom Culverwell's overhanging ledge, three climbers perched on a dry shelf 40 feet down. A rock-fanged pit barred the way, but a hard kick brought rappellers across this Pendulum Pit. Scampering around an outthrust angle of the wall, however, the agile trio slumped dejectedly. Another barrier of clay and rock again halted progress.

Relaxed and warm at home, the rockclimbers again stirred to the superlative challenge of Schoolhouse Cave. In March, Paul Bradt energetically undertook the ascent of the new wall. Tom Culverwell attributes to it "the solidity of a graham cracker." Clay and rock fragments rained down as Paul cut steps and drove pitons higher and higher on the wall.

One exceptional shower of gray clay included Paul himself, fortunately slowed by a slightly belated belay. No real damage, but another score for the cave.

Unintimidated, Paul returned grimly to his personal battle. More clay rained down. Soon he disappeared behind an outthrust rock. The interminable ping-ping-ping of the hammered drill announced a new expansion bolt. Rope continued to disappear upward from the alerted belayer: 50, 60, 70 feet.

A muffled yell of triumph resounded through the echoing blackness. Something garbled, then, "Come on up!"

Above was triumph indeed: the Great Gallery. Thirty feet in diameter, it continued southward 100 feet, 200, 300. In their exultation the climbers airily ignored nasty pits interrupting its floor. Delightfully glistening flowstone shone white in this first illumination. This was the glory the stubborn explorers had sought so long, so hard.

But as the climbers strode ahead, the floor began to change, to slope upward. Increasingly the true floor was buried by dirt and gravel slumped in from a crack in the ceiling. A small domed room terminated their progress.

This was a new and ominous barrier. The climbers checked every possible orifice. Frenziedly they moved rock, dug vainly with their hands. Air currents told them that the cave continued, but they could not.

Perhaps the blockage was only a few feet long. Soon the climbers were back with even larger loads of equipment. An automobile jack proved less useful than a white dishpan which served admirably as a teapot when not carrying loads of mud from their lengthening excavation. Hours of tedious digging brought 30 feet of progress, but what a battle! Vainly trying to keep warm over candles, those waiting in the little domed room suffered almost as much as the cramped digger.

Hopeful of finding a short cut from the surface, the diggers turned to the epic task of mapping Schoolhouse Cave and its surroundings. Tons of rock were moved to permit clear lines of sight. Several locations required belays equally for the mappers and for the instruments. Station 18, in the crawlway between the Hodag and Thunderbolt rooms, climaxed a week of tribulation. The selected point looked ideal to the fatigued survey team, yet had to be reset several times: it just didn't work. Finally onlooker Fitzhugh Clark suggested gently that the surveying telescope would work better if someone removed its lens cover.

But Schoolhouse Cave won the last word. Appropriately completed on April 1, 1942, the survey showed the dugway a hopeless 70 feet below the most promising sinkhole. After one last frantic struggle with the white dishpan, the rockclimbers turned to the fresh, bright sunshine of the Appalachian crags. Though Tom Culverwell soon returned to complete a formal map, to this day no one has gone farther in Schoolhouse Cave.

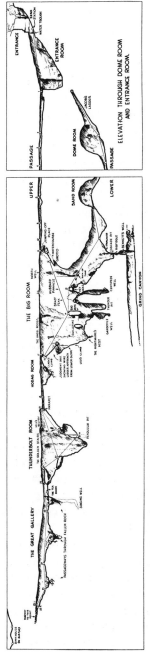

Side view of Schoolhouse Cave. Map by Tom Culverwell, based on a survey by Dr. H. Stimson and Tom Culverwell. Circles are six feet in diameter

Those dreaming of linking underground Germany Valley turned to nearby Hellhole, and more hellish holes nearby. None of the latter have proved worth the effort. Opinions are divided on Hellhole.

While the Speleological Society of the District of Columbia was first daring Schoolhouse Cave, D. K. Harmon of nearby commercial Seneca Cavern was setting up a winch over Hellhole's great black void, about a half mile farther south. Next day, the spelunkers strolled over for a look at his handiwork and its justification: an awesome 187-foot entrance drop into a shadowy natural cathedral 300 feet long and half as wide.

Unlike the jagged depths of Schoolhouse, this cave had been reached by man before: several months earlier, three men had been lowered by just such a winch. Something had gone wrong with the winch, however. One had been badly hurt, and nobody did any exploration.

Undismayed, the Speleological Society of the District of Columbia dropped into Hellhole, Don Bloch again in the lead. Off to the east beckoned a high, spacious tunnel 50 feet wide, then a room smothered in deep musty guano, its bat colony hanging 40 feet overhead. The inviting corridor became ever lower and ended in a narrow crack that went nowhere except straight up. Back to the glowing light of the Entrance Room, and around the margin of its acre of space, seeking less obvious leads . . .

To the north, toward Schoolhouse Cave, a narrower corridor opened. Fifteen paces onward, jagged pits appeared along the east wall. Larger, deeper, and uglier they loomed as Bloch and Stephenson pushed onward along ledges a foot or two in width. A series of short dropoffs, strewn with flood debris, a curious rounded pothole through which they could see the granddaddy of all pits . . .

Then an unsteady call from Don Bloch, who had run out of ledge: "Steve, come help me! I'm stuck! Can't go forward or back!"

The president-to-be of the National Speleological Society rushed forward—too fast. Unexpectedly, a rocky projection nudged him. Teetering on the edge of another unplumbed pit, flailing madly, vainly he sought to regain his balance. In horrified slow motion, he tilted toward the waiting pit. Thirty-five years later he still could not recall the exact motions by which he lunged instinctively into a tiny alcove, helmet flying, light vanishing into the blackness below.

Their fellow cavers soon talked Don Bloch out of difficulty, telling him where and how to move each foot and hand until he could relax in safety. But for Bill Stephenson it was as if every dyne of energy in his trembling body had been exhausted, somehow summoned for the endless split second when he faced black nothingness and won.

Explorations revealed that Hellhole went on and on. Mostly down. By 1946, its depth was listed at 430 feet and cavers had advanced about

500 feet toward Schoolhouse Cave. Although one casual onlooker died here through a sudden decision to try to climb hand over hand down a 200-foot rope, Hellhole turned out to be a forgiving cave, a friendly cave, the source of many a happy tale. "Remember the time Petrie and the two kids got left?" "You think YOU had trouble? We camped at the bottom and got snowed on!"

As the years passed, however, new discoveries in Hellhole slowed to a trickle, then a halt. Until 1970, nobody reported any breakthrough toward Schoolhouse, still about a half mile distant. Then Randy Shipp forced his way through guano-saturated breakdown into a virgin chamber larger than the Entrance Room. Although nobody found the next crucial orifice for another two years, the cave beyond spread out complexly in three dimensions.

Even for skilled cavers familiar with all the new techniques described later in this book, the going was tough. Moreover, the new section swung west before turning north toward Schoolhouse. Only grudgingly did it continue to yield more cave to supercavers like Bruce Smith, Buddy Bundy, and Chuck Byers.

Then, just when excitement was peaking among eastern cavers, the entire history of Hellhole and Germany Valley took a new turn. Compelling scientific studies revealed that Hellhole contained unexpectedly vital biological resources. Careful analysis by John Hall demonstrated that it sheltered the largest known hibernation colony of the endangered Indiana bat (*Myotis sodalis*) east of Kentucky. Once plentiful over a wide northeastern range, this admirable little animal is clearly in danger of following the passenger pigeon into extinction. In other seasons, other parts of the cave held the easternmost known surviving nursery colony of the threatened Virginia big-eared bat (*Plecotus townsendii virginianus*), especially intolerant of human intrusion. Unknowingly, unintentionally, cavers in Hellhole had been causing innumerable deaths of babies and adults of these endangered bats, precariously hanging upside down in a threatening world. Deaths were also occurring among every caver's furry little friends, the little brown bat (*Myotis lucifugus*).

Every serious caver immediately understood the need to protect and cherish these colonies. To their human brothers of the twentieth century, the seemingly odd ways of spelunking bats are familiar, comprehensible. From long experience we relate to every stage of their struggle for survival. On November 3, 1973, the Board of Governors of the National Speleological Society asked all cavers to halt all entry into Hellhole until biologists could determine safe periods when the bats would not be disturbed by caving. Whether or not they belonged to the National Speleological Society American cavers responded overwhelmingly. Until Hall and other biologists can complete the necessary studies, the world of American caving considers Hellhole out of bounds. In the span of a single

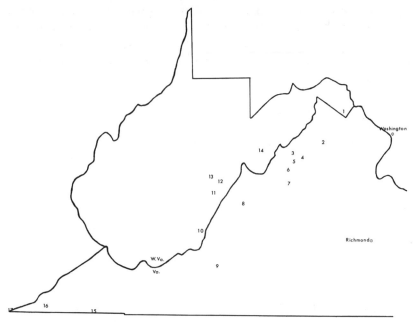

Spelunker's map of the Virginias. (1) George Washington's Cave; (2) Skyline Cavern; (3) Shenandoah Cavern; (4) Luray Cavern; (5) Endless Cavern; (6) Melrose Cavern; (7) Grand Cavern; (8) Butler, Breathing, Clark's, and Blowing Caves; (9) Dixie Cavern; (10) Greenbrier County caves, including Organ Cave system and caves of the Great Savannah; (11) Swago Creek caves; (12) Cass Cave; (13) Sinks of Gandy; (14) Hellhole, Schoolhouse, and Seneca Caverns; (15) Bristol Cavern; (16) Natural Tunnel; (17) Cudjo's Cave

year, the bats began to make a gratifying comeback. Many of us would give much to know whether Hellhole and Schoolhouse Cave connect, as some Germany Valley experts believe. Yet much has been gained and little lost by this unique voluntary moratorium. Elsewhere in the nearby Virginias, innumerable miles of equally inviting cave welcome the responsible explorer.

Today, few gung-ho cavers think often of George Washington's Cave, or Madison's Cave, or even Luray, Endless, or the other great show caves of the Virginias. From his deep-rooted respect for tradition the modern Virginia caver sometimes revisits the caves which began the story of American caving and speleology. But, with three or four candidates for the title of the world's longest cave, and three or four thousand caves beyond the scope of this book, his thoughts inevitably are elsewhere.

Here in the Virginias each new cave, each virgin corridor still brings the most veteran spelunker the eager excitement that George Washington knew before we became a nation.

Plus heartwarming knowledge that cavers' self-discipline is increasingly effective in preserving our hidden treasures for all who follow.

4

The Geologist and the Showman

The Story of Ozark Caves

Long gone was the last shred of the eminent geologist's classroom dignity. Spattered, chilled, and sopping, building a crude dam to hold back endless bucketfuls of water dipped from an adjacent pool, he and his students must have looked like overgrown children playing in the mud.

However ludicrous the spectacle, the participants were in earnest. Some 500 feet inside the spectacular entrance of Missouri's popular Meramec Cavern, the odd team was avidly seeking a breakthrough into the largest part of the cave. Just a few feet away from the well-known main corridor, the hard-sought catacomb beyond had been entered previously only by lengthy circuitous crawling.

The buckets dipped rhythmically, hour after hour. Slowly a low air space appeared above the shrinking spelean pond. When the clearance reached a few inches, Professor J Harlen Bretz called a halt.

"We chopped a rowboat out of the ice along the river's edge," he later wrote, "dragged it into the cave by automobile, and launched it on our lowered pool. Three of us lay down in the boat bottom, and by pushing with our hands against the ceiling we submerged the boat sufficiently to inch our way under and through."

Lester B. Dill, owner of Meramec Cavern, was panting but enthusiastic. "I sure wish we could get some publicity out of this!" he exclaimed.

"Well," drawled Dr. Bretz, "just break the dam after we're through, and you'll get headlines in Chicago: LOCAL PROFESSOR, TWO STUDENTS TRAPPED IN CAVE, SLOWLY FREEZING, STARVING, MAYBE DROWNED!"

Lester Dill manfully resisted the temptation. But it must have wrenched his soul. Mating hard work to inspired publicity, Dill has parlayed what once was just a nice little Missouri cave into one of America's outstanding tourist attractions. Characteristically, within a few days Dr. Bretz's Submarine Garden was drained and blasted out for tourist development. Even earlier, Dill had hopefully blasted an eight-inch hole that echoed when he hollered into it. Beyond lay the undiscovered upper levels of the complex cavern climaxed by the indescribable Wine Table, an intricate assemblage of stalagmites and perched shelfstone. But the overshadowing story of Meramec Cavern is that of Lester Dill's canny publicity.

Many another cave owner probably has the effrontery to call a New York press conference to complain, tongue in cheek, about getting lost in the subway. Only Lester Dill, however, could instinctively create the *mot juste* that front-paged his cave across America. "If it weren't for the ride," he guffawed, "I doubt if anyone would pay their way in to see that subway!"

Dill came by his cave interest early. When nearby Fisher Cave was opened to the public in 1910, the original boy guide was Lester Dill. In 1933 the state of Missouri took over Fisher Cave, now a feature of Meramec State Park. Almost automatically, Dill turned to a nearby

J Harlen Bretz and his spelunking collie Larry silhouetted in the mouth of Smallin Cave (Civil War Cave), Missouri. *Photo by G. Massie—Missouri Commerce*

cavern gaping on the riverbank. In that depression year, he and a friend purchased it for $30,000, mostly on credit. Today Lester Dill avers he wouldn't sell it for $1,000,000—cash.

Missouri speleohistorian Dwight Weaver has come to believe that Meramec Cavern—long known simply as Saltpeter Cave—was the site of pioneer Philippe Renault's saltpeter plant and headquarters of his extensive operations in numerous other caves up and down the river, perhaps as early as the 1720s. The first major cave discoveries in America, he contends, may have been made along the Meramec River. Yet nobody is quite sure of the early history of Meramec Cavern. Dill gleefully admits that there isn't a scrap of documentation for the long rigmarole plastered across four states and recited by the impish guides. Don't get mad at the preposterous signs: WORLD'S GREATEST CAVE—and so on. Everybody knows they're all in fun, and it's really quite a cave.

Long ago Dill concluded that tourists will believe anything about caves except the truth. Early in his chosen career, he worked hard collecting a school of blind fish and keeping them happy in a display pool. But, he chortles, everyone thought that the tiny white fish were fakes—just minnows. "So I tossed a thirty-pound catfish into that pool. As far's I know, that catfish could see like an eagle, and I didn't even say it was nearsighted. But everybody looked at it and said 'see that big blind fish' and went away happy!"

So who wants facts? While excavating the gaping cavern mouth, Dill found prehistoric firepits and weapon points beneath an ancient tree. To those who go behind the scenes, he speaks proudly of such finds. But not to tourists, who hardly suspect how much this unique showman has quietly advanced American speleology. It is the Lester B. Dill Award that grateful midwestern spelunkers annually present to the caver who has made the most outstanding contribution to speleology of the Missouri Valley–Ozark region. More of the serious side of Lester Dill in Chapter 16. At Meramec, it's all in fun.

Local yarns place hideouts of Jesse James's gang in half the caves of Missouri. "How they ever found time for their crimes is a puzzle," Bretz once snickered in print. And so a surprisingly coherent Jesse James tale has evolved at Meramec Cavern, complete with strongbox, padlock, and "26 other items we believe were Jesse James's." As it happens, the Loot Room is the first beyond Bretz's Submarine Garden. The twenty-eight items weren't there when the professor eased the clumsy boat through. But the tourists love it. The imposing entrance in a charming natural park along the Meramec River? It must have been an obvious landmark for whatever *voyageurs* ventured up the placid green waterway—a good way to bring early French and Spanish explorers and miners into the tale. Did they really stop here? Maybe.

Dill it was who jestingly announced that he was preparing to shelter one million Americans from atomic attack. Immediately he was swamped with requests for reservations. With its usual aptitude, the government got into the act. Meramec Cavern, Schoolhouse Cave, and many others now are officially posted as "approved" fallout shelters, often over speleologists' protests.

Behind such buffoonery, however, is a hardheaded businessman who lives his work. For a time, he dressed girl guides in neat uniforms embroidered STALACTITE and STALAGMITE, with appropriate arrows. But it seemed that people preferred park rangers to girl guides. Now he hires malleable youngsters, "half college and half Ozark twang," trains them personally, and dresses them as rangers. It brings the tourists back and back. Once weekly, he distributes searching questionnaires to all comers, and constantly updates his operation from the replies. "Keeps the boys on the ball, and me, too. F'r instance, I found out people wanted more color. So I put in more color. You can't stand still in the cave business." "The P. T. Barnum of the underworld," Dill modestly terms himself. He's right. Purist cavers may bewail the circus atmosphere of Meramec Cavern, but just plain people love it.

Only a few years after his Meramec adventure, J Harlen Bretz revolutionized speleology. Until about 1930, most savants believed that caves were dissolved out of limestone by underground streams. Even then, however, a few notable exceptions had outlined caverns' true genesis. Nearing the end of a long, distinguished career, William M. Davis presented a brilliant deductive study to the Geological Society of America. Davis lucidly demonstrated that caves rarely are formed by ordinary underground streams. Instead, their over-all characteristics indicate that they were largely formed when completely filled with water—below the water table.

Davis urged younger geologists to study the features of caverns inaccessible to his failing frame. This J Harlen Bretz did. In 1942 Bretz showed geologists and cavers how to read the life history of a cave from its sequence of patterns and deposits. In his publications, the University of Chicago geologist often turned to Meramec and other Missouri caverns as ideal illustrations.

After his retirement, Bretz undertook a still more comprehensive study of 133 Missouri caves, which was published in 1956 when the eminent geologist was seventy-five. Usually accompanied only by his collie, Larry, Bretz clearly enjoyed his self-imposed task. Written with dry humor, some sections of his *Caves of Missouri* are fittingly nontechnical. Of Zell Cave he wrote:

Our party, somewhat affected by the heat of a summer's day, stopped at a welcome sign, went into the tavern, and ordered what we thought would be good for us. Said the bar keeper, "Would you like to drink it in the basement?

All but the topmost few inches of the Lily Pad Room of Onondaga Cave will be submerged in muddy water at "flood pool" if the Meramec Park Dam is built. Almost all the remainder of the beautiful sections within 300 feet of this point would also be submerged, including Cathedral Hall, and the Queen's Canopy. *Photo by G. Massie—Missouri Commerce*

It's cooler down there." The "basement" was a cave beneath the building! While we drank and studied the cave walls surrounding our tables, the proprietor joined us.

"A long time ago," he said, "there was a little brewery in Zell, and their product was aged down here. One night a hogshead sprang a leak. It was empty by morning, and all the beer had disappeared down cracks in the floor. About the middle of that forenoon, a German farmer in the valley below came into town wildly excited. 'Mein Gott, Mein Gott,' he cried, 'Mein schpring, she is running beer!' "

The management of incomparable Onondaga Cave delightedly distributes reprints of Bretz's description of that cave, plus a few proud additions. It is their boast that Russell Trall Neville, "The Cave Man" of the 1920s and 1930s, considered their Lily Pad Room "the most extraordinary cave formation" of his experience. Daniel Boone, they say, discovered this "Mammoth Cave of Missouri" but the *cognoscenti* surmised that Lester Dill was a silent partner here long before he took over and (as told in Chapter 16) began to pour his Meramec Cavern bankroll into its preservation from the U.S. Army Corps of Engineers. At Onondaga, even more than at Meramec, the signs long have been all in fun:

"If you want to see the eyes go Ga-ga/ Go straight to Ononda-ga!" No matter. This is the cave that, of all Missouri's 3,100, Bretz chose as the proper site for the 1956 spelean field trip of the august Geological Society of America. Dill and other American speleologists are deadly serious about saving Onondaga Cave, one of America's truly great.

Equally among America's greatest, in a totally different way, is Mark Twain Cave. Every literate American owes much to young Sam Clemens for getting himself lost in what then was just McDowell's Cave, near Hannibal, Missouri. As Mark Twain, he merely drew upon his personal recollections when it came time to set down Tom Sawyer's immortal cave scene. (Should I add that, like Tom Sawyer, Clemens "told the history of the wonderful adventure, putting in many striking additions to adorn it withal . . ."? That whimsical author wouldn't mind. He told all about it in his autobiography: "Our last candle burned down to almost nothing before we glimpsed the search party's lights winding around in the distance . . .")

On the outskirts of his home town, McDowell's Cave—or Tom Sawyer Cave—or Mark Twain Cave—indelibly marked Samuel Clemens. The immortal humorist wove it into *Life on the Mississippi* as well as *Tom Sawyer* and *Huckleberry Finn*. Describing narrow Italian byways in *Innocents Abroad*, he wrote:

> The memory of a cave I used to know at home was always in my mind, with its lofty passages, its silence and solitude, its shrouding gloom, its sepulchral echoes, its flitting lights, and more than all, its sudden revelations of branching crevices where we least expected them.

Much of the power of Clemens' writing derived from superlative ability to re-create such scenes as Tom's. Vividly he portrayed the poverty-ridden, superstitious but intensely human life of a bygone period. His Hannibal prototypes were often interwoven deeply into the spirit of "his" cave. Years later, Twain wrote:

> Injun Joe, the half-breed, got lost in there and would have starved to death if the bats had run short. But there was no chance of that; there were myriads of them. He told me all his story. In the book called Tom Sawyer, I starved him entirely to death in the cave, but that was in the interest of art, it never happened.

Today Mark Twain Cave is a shrine for Americana fan and caver alike: a large, famous cave in a fascinatingly small space. "Three miles of cave under less than 20 acres," announce the bright young guides. "More than 170 passages. Only 23 deadend." J Harlen Bretz termed it "a perfect labyrinth of very similar passages." Mark Twain had to employ very little "art" in *Tom Sawyer*. The passing of six score years here has only mellowed its charm. As my family viewed the cave in 1963, our flashlight-equipped children spontaneously began to play hide-and-seek around its pancaked limestone pillars as did Mark Twain's playmates. The most delightful cave guides I have met bring to life the old, old story so vividly that I had to force myself to remember it wasn't really so:

We call this the Postoffice 'cause Tom and Becky played it here . . . Here's right where they got lost. You can see all five passages look the same. Lot's of others got lost here, too. . . . See that natural cross on the ceiling? Remember Number Two—"under the cross"? [Clemens said, "done with candle smoke," but no matter—W.R.H.] . . . And here's where Tom and Huck dug up the treasure. . . .

Yet it easily might have been otherwise, and the world would never have known the literary genius of Mark Twain. In 1968 three boys disappeared near a similar cave on the other side of Hannibal. Presumably they went spelunking. Despite week-long search efforts of incredible extent, they remain missing to this day.

The hardened veteran traveler, justly wary of flamboyant billboards, may mistakenly lump these great commercial caves of Missouri with equally flamboyant highwayside tourist traps. Such caves are not to be scorned, nor even far more modest commercial developments like Honeybranch Cave (Jesse James apparently really did hide there), Ozark Cavern, and Truitt's Cave. Perhaps lulled by its wildly entertaining advertising, I myself was wholly unprepared for the magnificent immensity of the vast entrance chamber of Marvel (originally Marble) Cave. This awesome hugeness once inspired a canny publicity agent to concoct "the world's highest underground balloon ascent" here. In this vast compartment, every feature shrinks imperceptibly to Tom Thumb proportions. No photograph conveys its spaciousness. Deeper in Marvel Cave lies much more beauty, especially in times of heavy rainfall, but all is anticlimax compared to the awesome entrance chamber.

Marvel Cave has other claims to fame. Its irrepressible guides relate that in 1869 a mining engineer slid down a rope and explored as far as the Corkscrew to the lower levels. They have a nifty explanation, too, for his ability to break the world's record for a rope climb and thus regain the surface. It centers around the fact that a little room at the top of the Corkscrew is perceptibly warmer than the cold-trapping entrance room. The engineer, say the guides, tossed a rock into the 75-foot pit ahead and never heard it hit bottom. Thinking himself nearing the warmish gates of hell, he broke all records getting back to the surface.

Were its ingress 400 feet down on the lowest level instead of at the top of the entrance room, Marvel Cave would well fit the dramatic cavern scene of Harold Bell Wright's *Shepherd of the Hills*. Probably more important, its remoter recesses shelter a valuable remnant of the ungainly, fast-vanishing Ozark blind salamander.

So does the Ozark Underground Laboratory, largely the work of one dedicated caver—Tom Aley—in one southern Missouri cave. About two miles long (which isn't overwhelming by southern Missouri standards), Tumbling Creek Cave had two valuable attributes. It had a wide variety of habitats for more than seventy kinds of cave-dwelling animals, and it was for sale at a price Tom could afford. For a decade Tom has poured virtually all his free time and energy into painstaking effort. Much of his U.S. Forest Service salary is devoted both directly and indirectly to the laboratory. His thoughtful organization was duly rewarded by this, America's first and most noted spelean laboratory.

"The world's highest underground balloon ascent"—an inspired publicity stunt in Marvel Cave's enormous entrance chamber. The actual entrance and stairway tower are also shown. *Photo courtesy Marvel Cave*

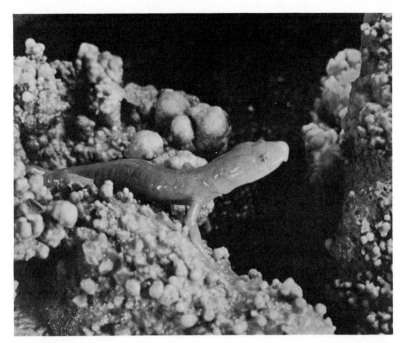

Found sparingly in Marvel and some other Missouri caves, the Ozark blind salamander has neither pigment nor lungs, and its eyelids are fused. *Photo by Charles E. Mohr*

Today it is the site of studies as diverse as the ecological niche of cave snails, the rate of speleothem growth, and the rate of underground transmission of virus and bacterial disease.

Bretz's *Caves of Missouri* far surpassed earlier Missouri cave reports. Yet, even before its 1956 publication, unknown young speleologists were emulating his example. In 3,000 caves necessarily bypassed by the aging geologist and his collie, keen-eyed cavers were similarly building upon Bretz's fundamental studies. All too often under conditions reminiscent of Bretz's breakthrough in Meramec Cavern, their systematic studies frequently seemed to delineate merely the world's stickiest, gooiest mud or its most superfluous water.

On the southeastern fringe of the Ozarks, Bretz noted an impressive karst area in east-central Perry County that seemed surprisingly cave-poor. Typically he identified it as "an unsolved problem in cave histories." Twenty years later, where Bretz found only a single cave to study, the Perryville karst appears to be the most cavernous area of all Missouri. With no end in sight, latter-day Missouri cavers have mapped more than 70 miles here, in some 200 caves. Most of this surprising total is in true giants of the earth: Crevice Cave (currently Missouri's longest, with 20.8 miles mapped), Mystery Cave (more than 13 miles, and not to be confused with the similarly named cave which, at less than 3 miles, is the longest in Minnesota), Rimstone River Cave (more than 14 miles), and Berome Moore Cave (16 miles), with 4-mile Tom Moore Cave only a few hundred feet distant. In parts of the Perryville karst without known caves, enormous numbers of sinkholes hint even greater lengths yet to be discovered.

As I write, speculation rides high about connections which would skyrocket one or two Perry County caves to the list of the world's longest. Midwestern cavers are especially hopeful of soon connecting Berome and Tom Moore caves (shown by stream tracing to be part of the same cave system). Mapping in Mystery and Rimstone River caves has reached points less than 1,000 feet apart. Lost and Found Cave apparently can be added to Mystery Cave by excavation of only 300 feet of clay fill, adding 3 more miles. At the moment, however, the big excitement is about Crevice Cave. Until mid-1974 it was listed at "only" 11 miles, and exploration seemed slowed to a standstill by the endurance barrier. Then the ground gave way on Melvin Rollett's farm. One hundred twenty feet straight down were a room 30 feet in diameter, 4 feet of water, and several miles of passageways, still going like mad. With mixed emotions, however, the Little Egypt Student Grotto of the N.S.S. soon recognized a bag they had discarded some months earlier—and its location. Rollett's pit had opened the entire back end of Crevice Cave. As this book goes to press, its exploration has reached a point within 2,000 feet of Berome Moore Cave, and Missouri cavers are looking remarkably hopeful. "This small part of Perry County is rapidly becoming of national significance," avers Jerry Vineyard.

Yet the finest Ozark caves, in the unhumble opinion of increasing numbers of cavers, are across the state line, in Arkansas.

Until quite recently, much of American speleology thought of Arkansas caves as small caves with big treasure tales—and cave sickness besides. There was some truth in this impression. In the Ozarks and in other states tracked by the conquistadors, tales of bandit caves, buried treasure, and rich silver lodes are widely credited. Few agree on such minor details as who, when, and where. Coronado, DeSoto, Jesse James, Indians, and just ordinary thieves: all are incriminated. Chests of gold coins, bars of gold and silver bullion, golden church statuary—these and more await the lucky finder, the old men say—perhaps in a cave.

For a thousand feet around Texas' Little Blue Spring "every possible cave entrance has been dug into and every foot of loose dirt overturned," reports the Texas Speleological Survey. Missouri's Gourd Creek Cave supposedly contains seven ponyloads of gold coin. At Ramsay Cave, the seven ponyloads are of "dish gold," whatever that is. At Bruce Cave, Missouri, it was "forty pack mule loads" hidden by one Blackbeard, a river pirate. At Texas' delightful Longhorn Cavern it was Sam Bass's uncounted loot. Writing of Money Cave, Missouri, Bretz summed up: "Digging in the floor and hammering on the walls has never yielded anything so far but sad experience and doubtless never will."

But someone always seems willing to try. And somehow the most plausible tales seem to concern particularly objectionable caves like Rocky Comfort Cave.

The people of Foreman, Arkansas, long paid little attention to nearby Rocky Comfort Cave. Its opening in an old chalk mine gradually filled with trash, dirt, and debris. Then a mysterious Oklahoma couple inexplicably turned up with a bulldozer. Grubbing around in the chalk mine, they reopened the cavern entrance. Then they disappeared as mysteriously as they had come. Somebody said they were in the hospital, but somehow everybody expected them back. Suddenly most everyone in Foreman was sure the Oklahomans had a clue to the Spanish treasure! All agreed: "We'd better find it first!"

Three days after the great cavern treasure hunt began, one of the young Arkansans contracted a heavy chest cold. Despite the efforts of his family physician, Dr. E. L. Davis, the youth became sicker and sicker. By evening his fever reached 105 degrees, with a severe headache and stiff neck.

Friends and kinfolk sympathized but continued their dusty digging unabated. Wriggling on their bellies, they had found a long series of small chambers. Largely choked with dirt and dust, any of the little rooms might contain the treasure. It was exciting even to those who didn't take much stock in the treasure story.

The search was thorough and efficient. Clouds of dust restricted flashlight visibility to a few feet. Everyone had a hacking cough from breathing too much dust. But these sturdy Arkansans were unconcerned. Their lives largely revolved around dust, dirt, and soil.

Two days later, three more treasure hunters fell ill: not as sick as their comrade who was fast lapsing into coma, yet seriously enough to start Dr. Davis wondering.

Next day he had an epidemic on his hands. Seven more disappointed cave diggers were sick. And a day later, three more. Within two weeks twenty-one of the twenty-five treasure hunters had fallen victim to a mysterious malady that defied diagnosis. Not every case was severe. Even the sickest gradually improved. But that was the most that could be said. Months passed before all were fit.

Townspeople attributed the bitter little epidemic to some mysterious cavern gas. Dr. Davis doubted this but asked the Arkansas State Board of Health to analyze the cave air. No noxious gases were found.

Other public health authorities called in by Dr. Davis were equally puzzled. Exhaustive studies of the cave and of the patients yielded no definite clue. The disease was centered in the lungs, yet it affected the whole body and was related to the cave. It really didn't make sense.

In 1948 two noted public health physicians joined Dr. Davis in a report in the *American Journal of Public Health*. Entitled "Cave Sickness: A New Disease Entity?" it startled American cavers.

Caves are delightfully healthy places—or so we had long believed. Unless, of course, the water was polluted. Or someone had dumped animal carcasses inside. Or you were bothered for a couple of days with dust pneumonia. Or a rock fell on you. Or you fell or drowned or got bitten by a wildcat or a mosquito or a black widow spider or a rattlesnake or a girl friend, or something else preventable. Despite millions of man-hours underground, America's cavers considered themselves approximately the world's healthiest people. Though one had soon died, some observers had even thought that certain consumptives of an ill-fated subterranean sanitarium in Mammoth Cave had improved temporarily. Cave sickness? It just didn't make sense.

Or did it? A few eastern cavers and physicians vaguely recalled a puzzling epidemic of chest disease among several New Yorkers who had gone to Mexico for cavern-based research a decade earlier. Although one savant had died of the strange malady, no helpful information was gleaned. Too, there were lurid tales of an unexplained epidemic which supposedly decimated the Anglo-Egyptian crew which opened sealed chambers deep in a pyramid. The Curse of the Pharaohs, certain journalists had termed it.

Upon reflection, American cavers began to sweat. In 1952 scant but widely published reports of five cave-related deaths among Mexican guano miners caused considerable alarm. Was there really an unknown menace lurking in our happy underground homes?

The hero of this piece is no caver. The first reports provided a clue for Dr. Michael Furcolow, a competent epidemiologist of the United States Public Health Service. Aware of several similar but less dramatic miniature epidemics unrelated to caves, he soon had compiled data on thirteen such outbreaks. Most had occurred in the Midwest. Bats were not always involved, but all the cases were in small groups exposed to dusts containing bat, bird, or other animal droppings.

From the beginning, Dr. Furcolow suspected the actual cause of "cave sickness" and the dozen other little epidemics. To his experienced eye, they resembled closely the sporadic cases of acute histoplasmosis which had come to his attention.

First recognized in 1906, histoplasmosis often closely resembles tuberculosis. In 1947 this fungus disease was barely known to most American physicians. Following up each of the thirteen epidemics, Dr. Furcolow found the fungus growing in dropping-fertilized dust at almost all the epidemic locations. And the lungs of most of the victims showed healed scars typical of this little-known disease.

Cavers heaved a heartfelt sigh as "cave sickness" was stricken from the books. Now we know that in the areas of greatest risk, most people are already immunized through mild infections in childhood. The Arkansas group was just plain unlucky. Occasionally an American caver— or a whole team—gets histoplasmosis and is sicker than we like for longer than he likes. But now we know that it is only histoplasmosis and not some mysterious cave sickness. None of us has died, and we know not to stir up dust in the histoplasmosis belt.

Gathering material for this book, I found my first visit to Arkansas an entrancing time of constant surprise. Although far less heralded than those of Missouri, its commercial caves are scarcely lacking in fun. Many will recall the entertaining tales of garrulous Ozark guides when nether-

Spelunker's map of Missouri. (1) Mark Twain Cave; (2) Devil's Icebox; (3) Carroll Cave, Jacob's Cave, Bridal Cave, Ozark Cavern; (4) Meramec Cavern; (5) Onondaga Cave; (6) Truitt's and Bluff Dweller's caves; (7) Marvel and Fairy caves; (8) Smallin (Civil War) Cave; (9) Fantastic and Crystal caves; (10) Cave Spring-Devil's Well system; (11) Big Spring; (12) Cherokee Cave

world splendors are long forgotten. And here the U.S. Forest Service and National Park Service are opening to the public two of America's greatest caves: Blanchard Spring Cavern and Beauty Cave, long known to cavers as Half-Mile and Fitton caves respectively.

The story of Beauty Cave did not start with Jim Schermerhorn—it merely seems as if it did. A quiet but avid Tulsa caver, Jim first heard its name in July 1958. A chance underground meeting with Bob Branson and other western Missouri grottoites led to offhand talk of "a whopper of a cave down in Arkansas." In less time than it takes to write it, the combined groups were under way for Beauty Cave. From that moment, the stories of Beauty Cave and of Jim Schermerhorn are inextricably entwined.

Beauty Cave was no fresh discovery. In the pretelevision days of the 1920s, many an Ozark farmer roved the slopes of Cecil Cove for simple pleasure. Zinc, lead, odd rocks, bee trees, ginseng, caves—you could hunt for most anything, and sometimes find it. One bitter day, a septuagenarian by the name of Newberry—his first name seems forgotten— noticed steamy air fuming from a crevice in a little bluff. He enlisted the help of Walter Kingsley. They enlarged the hole and squeezed through. Just inside, great fluted draperies framed a huge black space that seemed a mile long. Their improvised torches outlined gigantic stone columns of weird beauty.

In the deep Ozarks, amusements were rare and homespun. Beauty Cave became a local nine days' wonder. With kerosene lanterns and smoky pine torches, small parties wandered amid the beauties of its main room. Even when paced out at "only" 724 feet, it was mighty impressive.

Then someone found a slippery hole in a huge pile of broken rock. It opened downward to a level of small chambers, low but generously decorated. Beyond was a series of long crawlways on a firm clay floor. The crawling must have daunted some of the early explorers. But a kerosene lantern could be maneuvered through. Overall-clad bodies followed close behind.

At the far end of a relatively comfortable chamber, a narrow meandering stream canyon dropped a dozen feet: the Manhole. Below was another level of small, tubular passages and twisting slots. Carved by long-gone streams, they led on and on into the ridge.

Through the maze crept the ill-prepared spelunkers, then into great masses of dry, dusty broken rock. Without warning, a spacious vaulted tunnel, a dozen paces in diameter, opened before them.

As in a dream, the underground mountaineers plodded on. As they advanced, the tunnel shrank, then expanded into a dim, high-arched chamber with the floor many yards below their vantage point: the T-Junction.

Off to the left, the ripple of a stream was audible. The excited explorers followed the sound to a rushing underground creek. Upstream they hopped back and forth above it through a tight maze of rocky projections. One big room they found, but many crisscross openings barely admitted a man and his lantern. Soon there was no escape from the stream. At least one group found the water up to their armpits, often to their thighs.

After an hour of the raging water, shallower waters roared in dramatic narrow chutes. Those who struggled on another half hour returned to tell their wide-eyed kinfolk and friends of a waterfall "high as a ten-story building." In a mist-shrouded chamber it roared so loudly that "a champeen hog-caller couldn't be heerd a foot away."

Then someone found a climb bypassing the waterfall, which promptly shrank to 47 feet. Seven hundred feet beyond, the stout-hearted explorer emerged blinking from the long-known entrance of Bat Cave. Everyone in the hollows was amazed. To the mountainfolk, bold half-day explorations to view the mighty waterfall were high adventure. Few were concerned with remoter parts of the cave.

In 1958 the Spelunkers Club of Missouri School of Mines found its way into the deeper regions of Beauty Cave. Promptly they went wild with delighted surprise. Climbing up a slick clay slope beyond the T-Junction, they discovered a broad, tunnel-like passage leading straight east for hundreds of yards. Here and there, small pits led downward to a complicated stream area almost 50 feet lower. Occasionally, great funnel-like pits spread across the whole width of the throughway, requiring the spelunkers to balance delicately across steep clay walls. One yawning cavity was bridged by an enormous fallen slab 100 feet long. Sharp-angled and almost knife-edged, it could be traversed only by humping along, legs hanging down each side over 30 feet of blackness.

But it was worth it. Beyond was a moister area where large columns became increasingly profuse. Finally a massed forest of cascading calcite blocked all further passage.

This great East Passage obviously continued beyond the dripstone barrier. Could the blockade be bypassed?

The Missouri cavers turned to squiggly pits and tortuous water-level crawlways below the splendid corridor. Great effort led them farther and farther east. After hundreds of agonizing feet came an upward lead. Eagerly they squirmed upward, not to the corridor but to a black void so vast that their carbide lamps at first failed to illuminate ceiling or far wall.

In so huge a slab-floored chamber, the little crawlway entrance could be lost a few paces away. The cavers smoked a large arrow and OUT before venturing onward. In the way of spelunkers, they called the huge chamber the Out Room.

At one end of the Out Room, the Missourians readily found the far end of the East Passage and tracked it to the maddening blockade. Intensive search revealed no hole, nor any point where one could be made without extensive dynamiting. In the opposite direction from the Out Room they followed an even larger tunnel to a dead end not far away. The end of the cave? Hardly. Hundreds of smaller holes opened invitingly, almost at every step.

Then Jim Schermerhorn came to Beauty Cave. With the Missouri cavers, he toured the waterfall passage and the outer section of the East Passage. It was enough. This was to be his cave of caves. Just seven weeks later, he was back with three fellow Tulsa cavers. With true devotion, they assumed the formidable job of mapping the sprawling cave. First came the rugged route from the main entrance to the T-Junction, waterfall, and Bat Cave entrance—a ferocious 6,812 feet.

But while mapping the Tulsa group remained alert. Near the Manhole, Richard Porterfield became particularly intrigued by an unlovely orifice. "Let's see what's over beyond," he suggested.

The others were more than willing. Their mapping completed, they returned to Porterfield's hole. It opened into mazes of fallen rock and increasingly impressive breakdown chambers. For hundreds of feet painful stoopways cramped their protesting legs. Confusing rock piles everywhere suggested innumerable passages.

The new chambers were the largest underground rooms the Tulsa cavers had ever seen. Momentary exhilaration came with the thought that this was virgin cave. Their dreams were soon dashed. A few footprints had disturbed the dust of ages. But the footprints soon dwindled down to two pairs and ended in a particularly large chamber. The cavers' headlamps revealed a smoked inscription: THE END.

The Oklahomans disagreed. A passage blocked by fallen rock seemed to lie ahead. Like fiends they attacked the rock pile with bare hands. Little resulted.

Jim Schermerhorn squeezed between the cavern wall and the edge of the rock pile at a cost of one shirt and a few insignificant gashes. Attacked from both sides, the rock pile grudgingly yielded. Up they scrambled into a virgin region of surprise after surprise.

Ahead spread a complex of passages amid collapsed rock. Some were low and wide, some merely low and narrow. Broad slabs tilted beneath their light tread, rumbling ominously.

The passage changed again. Ahead was a wide but still lower section, hardly more than a foot high. Mercifully it was floored with dry sand instead of the broken rock through which they had come. Crawling on their backs, their feet driving hard, the explorers came nose to nose with pristine sparkling gypsum crusts.

Gypsum needles bedeck a low crawlway far back in Beauty Cave. *Photo by James Schermerhorn*

Hidden in the sand were less delightful surprises. Occasional rocks jabbed painfully into back or flank at breath-jarring intervals. Many a caver would have quit, and none would have scorned them.

Long minutes onward, the quartet were able to sit almost erect. Pausing to relieve long-cramped muscles, they looked about in awe. From the brown sand sprouted fantastic nests of glistening gypsum needles. Some as large as knitting needles, they projected like glassy jackstraws, angling wildly in all directions. For dozens of feet the thickets continued, then subsided to mere sparkling "sewing needles."

Crawling gingerly around the spectacular clumps, the fast-tiring quartet pushed onward. Anything seemed possible here.

The passage enlarged and sloped upward. Almost at once, the cavers halted in renewed surprise. From the low ceiling, a hairlike braid of white gypsum threads eight inches long waved gently before their panting breaths. Fallen of its own weight was a heaped pile of the "angel hair." Beyond was a larger chamber where still more angel hair and cottonlike masses of gypsum bedecked ceiling and floor.

For many weeks, the explorers planned a lengthy Christmas holiday venture into the cave. Dragging sleeping bags, food, and multitudinous gear in orange crates, they set up Camp I just beyond the Angel Hair Room. Alongside their bedrolls glittered queer complexes of stalactitic gypsum rosettes. Beyond was a corridor they had overlooked. Spacious and deeply pocketed by long-gone rushing waters, it was delightfully easy going, for a while. Then ahead lay additional lengths of tricky rockfall. On and on, aching muscles protested hundreds of feet of squatways. A sudden spacious corridor seemed hardly credible.

But only a few dozen paces farther, a tremendous pitch-black chamber loomed far larger than anything else they had ever experienced. To their amazement, a trail of footprints told of others' explorations here, a long mile from the entrance.

Where under the earth were they? Circumnavigating the huge, slab-strewn room, the cavers followed the spoor into a low corridor atop a fallen iceberg of limestone. Beyond was another blank cavity, almost as large. Here there were footprints everywhere. After the hours of virgin cave, this chamber seemed the haunt of hordes.

Headlamps pinpointed an inscription: OUT. Suddenly the explorers felt very much at home in this well-known Out Room. Delightedly they savored the completion of a loop almost two miles long. Grinning with satisfaction, they smoked a second OUT sign over the egress of their new discovery.

The new OUT sign indeed surprised the Missouri spelunkers when they next returned to Beauty Cave. Eagerly they plunged into the indicated corridor, only to halt at a dead end. Perplexedly, they searched every crack that seemed possible, then turned to the impossible ones. None "went." Disgustedly concluding that they were the victims of an unfunny practical joke, they crossed out Jim's sign and surfaced to berate him. Peace was restored only with detailed written directions.

But, even with the best possible instructions, the Missourians found Jim's loopway only on the fourth attempt, after uncounted hours of search. Despite its beauties, they found it so exhausting that they never returned. To the day of his 1976 cancer death, Jim still counted on his fingers the parties which had struggled through its agonies.

Cavers from all over the central United States joined the exploration of Beauty Cave. Trip after trip probed and mapped its mysteries. Many a complex interconnection yielded its secrets, but additional triumphs were slow to appear.

Then Roy Davis, Jack Herschend, and David Smith appeared. Days later, a thick missive reached Jim Schermerhorn. Pages of description told of a magnificent new passage.

"Right from the first sentence I knew," Jim acknowledged ruefully. "I'd started into it myself. It's a horrible crawl on little jagged rocks, and I thought I could see it shut down. It didn't."

Along low and high corridors, the new arrivals encountered small gypsum needles. Crusts and gypsum flowers sprouted from walls and niches, even from the floor. Through a fine corridor, then over a huge

Large oulopholites in Beauty Cave. *Photo by James Schermerhorn*

pile of breakdown they wandered entranced in virgin cave to the Tennouri Room, as large as the Out Room. In half-hidden grottoes along its lower walls clustered gypsum flowers so huge they lost all resemblance to blossoms. Glowing with transmitted carbide light, their broad, grooved curlicues seemed more the squeezings of some Brobdingnagian toothpaste. Despite the many cavers who have followed in their footsteps, not one of these gypsum masterpieces has been broken.

Much more cave has been explored since the Tennouri discoveries ("Tenn-" for Roy's Tennessee Cumberland Caverns, "-ouri" for Jack's Missouri Marvel Cave). Many orifices once crossed off as "hopeless" have opened with slight effort. Not even smashing rockfall encountered when they were in the cave during the Alaska earthquake daunted these rugged, obsessed cavers. When the National Park Service began acquiring lands for the Buffalo National River (the boundaries of which include Beauty Cave), the master map showed more than ten miles of corridors and memorable chambers. Current plans outline unusual guided tours here, carefully preserving the special wilderness character of this extraordinary cave.

The United States Forest Service is applying a different philosophy at its Ozark underground showpiece, Blanchard Spring Cavern. This equally notable cave came to the attention of American speleology in an unusual way, even for the Ozarks. It seems that for gubernatorial challenger Odell Dorsey, 1964 Arkansas politics presented certain problems. First he had to wrest the Democratic nomination from a certain

Strapping Hugh Shell (in circle) is dwarfed in the Cathedral Room of Blanchard Spring Cavern. *Photo by Hail Bryant and Charles Rogers*

five-term incumbent named Faubus. That rare success would gain Dorsey only the dubious privilege of contesting with Republican nominee Winthrop Rockefeller, hailed by many as the state's economic savior.

Dorsey needed gripping issues—badly. In Half-Mile Cave near Blanchard Spring he perceived a double-edged sword. Rockefeller, it seemed, had met quietly with businessmen interested in concession rights at the cave. And Governor Faubus, Dorsey charged, had been "sitting on the news of the caverns," perhaps so his friends could obtain lucrative concessions.

As the primary election neared, Dorsey's press releases sought increasingly to make Half-Mile Cave a statewide issue. LARGER AND MORE BEAUTIFUL THAN CARLSBAD trumpeted the newspapers: an ideal phrase to make cavers cringe.

For once, skeptical cavers were happily surprised. Even though it has nothing comparable to Carlsbad Cavern's Big Room, it's quite a cave! Faubus turned back Dorsey's challenge with ease, but Dorsey left his mark on American speleology.

Although they were not quite the first spelunkers here, the initial unrolling of this giant cavern was largely the work of Hugh Shell and Hail Bryant. Marine combat veteran Shell was no ordinary caver. It appears that except for Roger Bottoms' initial explorations, which yielded Indian torches and other artifacts, this strapping caver was present on every significant discovery in this whopping cave until the Forest Service took over. Late in 1959 Shell teamed up with Bryant. Their combined equipment and experience made feasible the 70-foot entrance pit of Half-Mile Cave that had tantalized Shell. Even so, the struggle was formidable. Twelve hours' labor at the entrance advanced four spelunkers just four hours into the cave below. In Half-Mile Cave four hours obviously was just a start. Half-hidden by colossal mudbanks, immense corridors called the ecstatic cavers to return.

Two weeks later they did return. Heavy loads of supplies were lowered to them to be cached at a base camp deep in the cave. This time, however, the cave presented them with a new obstacle: the underground river at the base of the entrance pit was at flood stage.

Tippy raftloads of spelunkers and equipment dared swirling waters in hundred-yard shuttles from the base of the entrance pit. One raft capsized, hurling Hail Bryant and a load of equipment beneath a submerged ledge. Hail was salvaged but much gear was lost.

Retreating toward the entrance from a disappointing reconnaissance, Hail spotted a north-lying corridor. A quick look showed it 90 feet high, 60 feet wide, and notable anywhere for richness of speleothems. Beyond a flowstone cascade that nearly choked the passage, a chamber 450 feet long, 125 feet wide, and 90 feet high encompassed the combined width of the new corridor and the parallel Lost River. Sixteen hours after entering the cave, "after progressing about one hour westerly" from the giant hall, they came upon an even more startling chamber. "Several hundred yards long and 150 feet wide," it dazzled with superb speleothems. As they moved onward amid its splendors, a high, sparkling flowstone cascade 200 feet long dwarfed anything else in the cave: perhaps America's most massive single speleothem. When they turned eastward—downstream—the giant chambers and splendors were eclipsed. The awesome Stadium Room proved 600 feet long. Westward on a stupendous higher level were 75-foot stalagmitic pillars which would not disgrace Carlsbad Cavern itself. Here was a variety of more delicate features that would delight any caver. Yet they seemed lost in a corridor averaging more than 100 feet wide. Soon their map reached a length of 10 feet. By mid-1962, round trips to virgin cave required a minimum of twenty-four hours. Operating on four hours' sleep out of twenty-four, parties remained in the cave two and three full days with plastic sheets, sleeping bags, and other gear.

In the next five years, the Shell-Bryant team returned some forty times, exploring, mapping, photographing, studying the cave, and guiding U.S. Forest Service and other officials and influential citizens to its greatest magnificences. In 1963 the Forest Service began the first phase of a long-term development program. Currently about one-third of the planned tours are open for public admiration. Six miles is now on the map of the cave, with the St. Louis University Grotto (inevitably nick-named S.L.U.G.) now 80 percent of the way to nearby Rowland Cave, three miles long and originally considered a mile away. In the process its members encountered additional magnificences.

Crystal Lake, for example. Forcing a tight bend in a high, narrow passage so tight that even Hugh Shell and Mike Hill had turned back, the S.L.U.G. team here faced a glittering 7-foot cascade of immaculately white flowstone. Beyond, beneath a ceiling just 3 feet high, 4-inch whipped cream of milky crystals topped an emerald pool 30 feet wide. "On the frozen lake were white lily pads four to six inches in diameter supported by one-half inch stems. The shore line was composed of a sort of macro-crystal that almost gave the impression of movement . . . the entire impact . . . was awesome," bemused S.L.U.G. cavers reported to the Forest Service, their explorations still incomplete. More and more it appears that Half-Mile Cave was not out of place in that Arkansas gubernatorial campaign. As at Beauty Cave, preservation and public en-joyment of underground magnificence is indeed commendable in this near-wilderness state.

In 1970 and 1971 cave divers from Memphis State University entered a siphon in this cave to determine why dye required more than twenty-four hours to emerge from the little spring cave at Blanchard Spring, just a half mile distant. Soon they connected the two caves, discovering that the slow flow was merely because of the large volumes of water pooled en route.

This cavernous resurgence is a fine example of the karstic springs that provide much of the distinctive character of the Ozarks. Perhaps most famous of these is Missouri's Cave Spring system, where man first made his way into the supply system of a huge limestone spring.

J Harlen Bretz was far from the first to know Cave Spring, just 50 feet from the Current River. Missouri cavers delight in canoeing into its mouth, where 16 to 47 million gallons of crystal water wells daily from emerald depths. Unfortunately the air-filled part of the cavern terminates within a few dozen yards.

In 1956 Oz Hawksley spotted an intriguing newspaper account of a deep pit about a mile north of Cave Spring: Devil's Well. Investigating, he found a deep sinkhole with a 4-foot hole at the bottom. Still farther down was a bell-shaped chamber floored with deep water.

Enthusiastically, Oz and his friends wriggled a canoe through the orifice and lowered it 100 feet to the pond below. Descending in a bosun's chair, they found a unique underground lake 400 feet long and 80 feet wide—and 85 feet deep at a sounding point. From small holes high on the walls, stream-fed waterfalls continuously thundered. Below the surface, the cavern continued to bell out. Above water, however, there was no place to go.

The explorers were deeply disappointed. Somewhere down there was a big cave, but the down-cutting Current River would have to drain it before anyone could do much with it: a matter of a few hundred thousand years.

Then Bob Wallace, owner of the property, spoke up: "I got another one you ought to see, down near Cave Spring. It's got a hole like this, too."

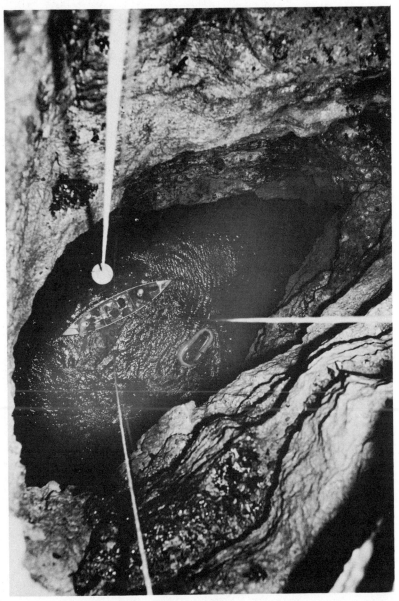

View straight down into Devil's Well. Round object suspended above canoe is a gasoline lantern. *Photo courtesy Salem (Missouri) Free Press*

Water level view of Devil's Well. *Photo by Jerry Vineyard*

"Sure, sure," thought the tired cavers as they fought to drag the unwieldy canoe back through a hole that seemed to have grown suddenly tighter. "Another farmer's wild tale." But next morning found them investigating Wallace's long, tortuous crawlway 100 yards from Cave Spring. Sure enough, 224 strenuous feet inside was a natural manhole. No bottom was visible.

Oz tossed a rock over the edge. Nothing. Then a resounding splash triggered sloshing ripples in a black chamber somewhere deep below.

To plumb this unexpected hole, the cavers tied a rock to their rope and tossed it into the hole. "Who's holding the end?" Jerry queried as the rope uncoiled itself. At that precise moment the last of the rope disappeared into the pit.

The explorers surfaced for another rope and other equipment. A rubber life raft was manhandled to and through the new pit and lowered to the lake. This Wallace Well chamber was found to be much smaller than that of Devil's Well (there's probably a moral there someplace), but soundings eventually proved its water at least as deep.

In routine fashion, Missouri cavers set out deciphering the secrets of this peculiar aqueous trio. Promptly they learned not to underestimate these strange new reservoir chambers. "We put four boats in Devil's Well late one Friday night," recollects Jerry Vineyard. "Next morning I woke up at dawn and stuck my head out of the tent. It was nice—part cloudy. But somewhere way off I heard thunder. By breakfast time we had rain, and it kept coming. Six inches of it in a few days. The water rose so fast that all our boats were trapped under ledges and sank with our ropes, ladders, diving gear, everything. We thought we were doing well to recover two boats."

Soberly reminded of the prodigious forces with which they were toying, the cavers returned more cautiously. Fluorescein dye proved the connection of Wallace Well and Cave Spring, then also that of Devil's Well. Bob Branson, however, may never again go near fluorescein. Somehow a package broke open at the sinkhole entrance while he was directly below. "Had it been St. Patrick's Day, Bob would have been right in style," Jerry Vineyard reported. "He was not only wearing green when he came out of the cave, but he dyed the river green when he washed himself off."

Their exhausts bubbling up in Wallace Well, scuba divers finally found bottom at 150 feet in "85-foot" Cave Spring. Several small passages were found feeding both Wallace Well and Devil's Well above the water level—and huge ones below.

"So we accomplished our first model of a big limestone spring," Jerry Vineyard remarks. "Like Bretz thought, Cave Spring is an orifice of a deep, enlarging system. But it's fed by conduits and reservoir chambers at much greater depth than he thought."

Scuba divers entered Devil's Well to test Jerry's theory. Fluttering up-current in the clear, lightless water, they promptly emerged into another reservoir chamber almost as large as Devil's Well itself—just as predicted. Downstream, toward Wallace Well and Cave Spring the divers found an utterly black, water-filled conduit so huge that they barely entered it for fear of becoming lost. Upstream, the main water-supply passage is even larger. Near the banks of the Current River, a veritable Mammoth Cave may be in the making.

Yet Cave Spring is far from the largest limestone spring in Missouri. An average 276 million gallons "boil" up daily at prosaically named Big Spring—a spectacularly turbulent mass of water. Each day, its enormous volume carries 175 tons of dissolved minerals from subterranean channels beginning at sinks as much as 40 miles away: three or four times as far as at Mammoth Cave.

If all the dissolved limestone emerging from Big Spring came from one lengthening corridor 30 feet wide and 50 feet high, that cave would grow a mile each year. Relatively fast flow through its stygian tributaries suggests that there's a lot of undiscovered human-sized cave along the way.

Ozark caves may be growing faster than the followers of J Harlen Bretz can get them explored.

5

Of Golden Legend

The Story of the Caves of California and Nevada

Twisting helplessly back and forth, the weakening Jack Mitchell revolved slowly in the black void. The rope around his chest cut cruelly into his flesh. Each endless minute increased the torture. Each inexorable turn humiliated his powerful frame with progressive waves of nausea.

Irregular vibrations of the rope told Jack that he was not abandoned. Far overhead, his two companions obviously were working desperately with a jammed pulley. Dangling in the dark, the bold adventurer could only guess at their progress—or lack of it. The narrow slot through which he had been lowered into this unknown desert cavern hopelessly garbled every shout. Could two men rethread the block and tackle without dropping him to whatever bottom lay below in the darkness?

Soon Jack Mitchell no longer cared. As the minutes seemingly stretched into anxious hours, he reached the limit of his endurance. When his frantic friends finally teased the rope into place and hauled him limply to the crisp desert air, he was barely conscious. When Jack returned, it was to pile rocks over the entrance of this Cave of the Winding Stair. The development of his own Mitchell's Caverns presented enough challenges.

Until his death a quarter century later, Jack Mitchell was widely acclaimed one of the great storytellers of the legend-rich Mojave Desert. Undoubtedly he embroidered his harrowing experience—and promptly regretted it. The exciting story spread widely through Southern California. A well-known author retold it in a regional best seller—and re-embroidered it. Mitchell was soon virtually besieged by a horde of would-be explorers of this fabulous cavern. Few were competent spelunkers; even fewer had adequate equipment. Except for a research group which had attempted sonar experiments just inside its twisting entranceway during World War II, Mitchell rightly turned all away.

Year by year, the Cave of the Winding Stair became one of the great legends of the Mojave Desert. Each new version was more fantastic

The Great Pit of the Cave of the Winding Stair. *Photo by William Brown, copyright 1963*

and more fascinating than the last. Soon it was widely accepted that the cave was thousands of feet deep. At the bottom was a river with tidal fluctuations, alternately icy and boiling. Eventually, a new flux of rumors brought an inundation of treasure seekers. Jack's hidden cavern, they somehow had heard, was a secret ingress to the fabulous Cavern of Gold of Kokoweef Peak, a few dozen miles to the north.

When several cavers—including the writer—formed a Southern California Grotto of the National Speleological Society in 1948, Mitchell assisted us in every possible way. Perhaps he was impressed by our prompt successes at Lilburn, Kokoweef, and other California caves. Perhaps we came at just the right moment; Jack clearly was fast tiring of brushing aside insistent treasure hunters. Perhaps he decided it was time that competent spelunkers debunked the uncontrollable rumors. Nineteen forty-seven headlines about Lemurian caverns may have been the last straw. Just a year before our Kokoweef venture, splendidly self-explanatory articles erupted in Southern California newspapers. At first they carried almost scientific credibility:

FINDS MUMMIES OF GIANTS OVER EIGHT FEET TALL

Los Angeles, Aug. 4 (1947). (AP)—A retired Ohio doctor has discovered relics of an ancient civilization, whose men were 8 or 9 feet tall, in the Colorado desert near the Arizona-Nevada-California line, an associate said today.

Howard E. Hill of Los Angeles, speaking before the Transportation Club, disclosed that several well-preserved mummies were taken yesterday from caverns in an area approximately 180 miles square. Hill said the discoverer is Dr. F. Bruce Russell, retired Cincinnati physician.

"These giants are clothed in garments consisting of a medium length jacket and trouser extending slightly below the knees," Hill said. "The texture of the material is said to resemble sheepskin, but obviously was taken from an animal unknown today." The relics have been estimated to be approximately 80,000 years old.

A followup Associated Press account, however, quoted Dr. Russell as describing the supposed discoveries as belonging to the "lost kingdom of Mu." The momentary sensation promptly vanished from most newspapers, but did not wholly die. The *San Fernando Valley Reporter* especially remained loyal:

HILL DEFENDS LOST LEMURIA DISCOVERY:
CLAIMS WORLD-SHAKING FACTS IN CAVES

(San Fernando, Aug. 28) Masonry, at the beginning of the world, virtually formed the keynote of explorer Howard E.

Hill's address on the "Lost Continent of Mu." The prominent Los Angeles speaker appeared before Kiwanis this week, to offer a strong and graphic affirmation of his belief that "the entire history of the Bible" is now justified in the recently found California caverns.

Tells of Find

Hill offering no apologies for the stories printed in the metropolitan press, and offering to "start the expedition right in San Fernando on September 28," told of the fantastic caverns lying southeast of the Panamint Ranges, and told an overflow audience that "Masonic symbols, a Masonic lodge hall and Masonic signets have already been found."

In justification, Dr. F. Bruce Russell, psychiatrist, and Mr. Viola V. Pettit, archeologist, of London University, arose to describe the "Masonic" find, and to tell of "cold fire" which still arose from the subterranean caverns, believed to be anywhere from 50,000 to 152,000 years old.

The People of Mu

Hill made no attempt to hide his own belief in the find, and flatly said that "these are the people of the lost race of Mu." He told of the hieroglyphics, of the clothing, of the size and number of the mummies found; described their color, texture and size, which he set at 8 feet, with footprints 22 inches long and 8 to 10 inches in width. "If the people of this lost race were to wear a modern shoe," said Hill, "it would require a No. 33 Triple E."

The Los Angeles explorer-writer declared that the caverns had been prophesized by Rosicrucian masters, and that two members of a previous expedition fled in abject terror from the caves "when they heard music, laughter and talk and could not be persuaded to ever enter the caves again."

Activity in the Masonic Lodge hall was told, both by Hill and Dr. F. Bruce Russell, who first discovered the caverns 17 years ago.

Expedition Planned

The story, which created a furore internationally two weeks ago, when Hill first told of the find, is based upon an area approximately 80 miles southeast of the Panamints [the approximate location of the Kokoweef caves and the Cave of the Winding Stair—W.R.H.] consisting of 32 caverns, few of which have yet been entered. They are said to be carved in solid rock and that occupants died "possibly from an atomic explosion, as the entire area is covered with a diffused green glaze."

Hill plans to start an expedition to the ancient caverns in San Fernando on September 28, he said, and pleaded for equipment. "We do not need money and we have the scientists," he said, "but desperately need machinery and supplies for breaking into this long lost tombs [sic]."

Commonplace Story

"If you think the story I am telling you is incredible," he said, "wait until we re-enter these caves, for it will be one of the most fantastic, profound and astounding stories that ever will be told, and may change the entire archeological history of the world, clear back to the beginning of time."

Needless to say, the promised press inspection never took place. Yet increasing numbers of eccentrics, occultists, and gullible folk inundated Jack Mitchell. Solemnly they assured him that deep in the cave that only *he* had explored were endless rooms filled with the mummified bodies of pre-Indian pygmies with long red hair all over their bodies. The transmutation of Lemurian giants into dwarfs seems to be part of the normal course of such folklore, but how they ever got into *his* story always remained a particular mystery to Jack.

Still young enough to dream, we assembled a truly formidable expedition at the Cave of the Winding Stair. Laboriously we dragged a 75-pound winch, heavy electrical cable, and 3,000 feet of steel cable up cactus-studded slopes. Once inside, we found them unneeded. Sixty feet down, Mitchell's fissure opened into the large chamber where he had swung so helplessly. Picking our route carefully, however, we climbed down without grave difficulty. Farther back in its magnificent lower levels, a superb pit stopped us for three weeks. Nevertheless, even dynamiting in search of deeper compartments took us barely 300 feet below the surface. We spotted not a single pygmy. And nary a passage ran north toward the lost river of gold of Kokoweef Peak.

As it happened, we had already debunked that fabled connection, from the other end. Our Southern California Grotto's organizational meeting late in 1948 had been as full of the Kokoweef story as of the Cave of the Winding Stair. A lengthy cave 3,000 feet deep, a 500-foot stalactite, and a tidal river with rich placer gold could hardly be ignored. Someone had even looked up the original affidavit in which a wind-tanned prospector named E. P. Dorr swore to all these things and much more.

This was in the grandest California manner. Was there more than legend to the report? Everyone was exceedingly skeptical. Yet we were cautious. Strange things have occurred in the Mojave Desert.

"Let's go talk to Dr. Foster Hewitt," suggested a student at nearby California Institute of Technology. "He's spent all his life out there. I bet he knows about it."

Three of us were given a prompt appointment with the eminent geologist. Needlessly we told him of the tales which had reached us. Hewitt was far ahead of us.

"You cavers should know better," he twitted us. "Dorr might have found more cave than is known today, but certainly nothing like what he claimed. Why don't you go over and see Herman Wallace in Highland Park? He's an officer of the company and can tell you all about the caves!"

Mr. Wallace proved a particular friend. He himself had descended to the bottom of the three caves of Kokoweef Peak without finding the gold. Even more important, he had obtained the incredible story first-hand from Dorr. Wallace's son had prepared a sketch map of the lost river of gold under Dorr's direct supervision.

As Herman Wallace talked, the tale began to make a twisted kind of sense. Clearly, some of these fantastic tales were merely confused with those of the Cave of the Winding Stair. What remained was incredible beyond belief. Yet the story was coherent and filled with plausible details. Dorr had never contradicted himself.

For untold years, it seemed, prowling prospectors of the Mojave Desert had known of a wide, deep cavern on the rocky flank of juniper-clad Kokoweef Peak. In the 1920s weeks often elapsed in this Joshua-tree wilderness without the passing of more than a single prospector and his companionable burro. During those dimming years, hopeful prospectors and other "desert rats" wandered in and out of shack towns at isolated wells along the nearby Los Angeles–Las Vegas road. Even they were few.

At one of these tiny communities, someone announced one evening that he had found another vertical cave on Kokoweef Peak. Maybe it was Dorr. Some say that Dorr had a "treasure map of Spanish or Indian origin," but this seems to have been wishful thinking. In any event, Dorr was fascinated by the new cavern. Soon he was telling of lowering himself on a rope from level to level, exploring uncounted tunnels of great beauty. Beyond one tight hole, he encountered an enormous 3,000-foot chasm. Ledges led onward for 8 miles without a way to the bottom.

Dorr's friends were not particularly impressed. Every desert rat is a practiced spinner of just such yarns.

"Think there might be a river of gold at the bottom, Earl?" someone asked him helpfully.

"Dunno what's down there. But I'm goin' back till I find out," the keen-eyed prospector asserted stoutly.

After his next exploration, it seemed that Dorr had found a way down the formidable underground cliffs. On the banks of the river below were miles of deposits of rich gold-bearing sand.

Dorr's cronies were delighted. His family, however, was more cautious. For years they had laughed at his yarns. He had bragged of hitching up a team of Colorado elk and driving from Cripple Creek to Colorado Springs. No one believed a word of it, of course. Only long afterward did they learn of a Cripple Creek rancher who had trained pet elk to pull a buggy. Other grains of fact had a disconcerting way of turning up in his wildest tales. On the other hand, Dorr told his family of blind fish in the river of gold. Joshingly his brother asked if they were flying fish. Sure enough, after Dorr's next trip they became blind flying fish.

Dorr prepared an affidavit subsequently published in the *California Mining Journal*. No ordinary grubstaker for Dorr—he sought the support of wealthy investors to share his great discovery. A mere 330-foot tunnel might suffice. He was willing to share fifty-fifty with anyone willing to finance his incredible find!

Why was a tunnel necessary? Well . . . for one thing, the river of gold ran beneath his claim, but Crystal Cave wasn't on his land. Besides, he had dynamited shut the secret passage so no one else could get at his gold.

Herman Wallace was one of several Los Angeles investors willing to gamble a little on Dorr's proposition. Most of their investment soon vanished into claim options, tunneling, timbering, and a grubstake for Dorr. Shortly before World War II, however, they struck a rich zinc vein. Dorr begged for more tunnels in new areas, but Wallace's Crystal Cave Mining Company enthusiastically entered the zinc business. Its geologists were as discouraging as Foster Hewitt. As far as the corporation was concerned, the lost river of gold could stay lost. They'd settle for zinc.

"Would you like to have a look and see if you have any ideas?" Mr. Wallace asked in cordial conclusion.

Would we? Ten carloads of cavers and their families swarmed through the Joshua trees the crisp morning of November 13, 1948. In shifts we scurried along the rocky flanks of the barren peak and into the deep little caves.

Seventy feet down in Crystal Cave we found Dorr's name smoked in bold capital letters on the wall of the first chamber. We found it again on the next level, near an area of shattered rock and flowstone. Was this the legendary entrance to the lost river of gold? If so, no one was going through that mess any time soon. In a small alcove nearby we spotted a long, thin trail of ash. It might have been the residual of a dynamite fuse.

We poked into every conceivable orifice, peered into every fissure, and found nothing else. Excavating the shattered area would be a huge undertaking of little promise, and so we told Mr. Wallace. He agreed, reluctantly, plagued by the same nagging doubt. We all know there is no gold beyond. And yet—could we be wrong?

Many a California cavern far from the Mojave Desert boasts equally entrancing legends. To the north in the tangled Siskiyou country, a water-filled pit in little Hall City Cave supposedly contains $40,000 in pioneer gold. Or such was the information extracted from bandits by the posse men who overtook and hanged them nearby. Alas, skin-diving spelunkers have surfaced none the richer. Similarly elaborate tales are told of Del-Loma Cave, deeper in the rugged ranges but discovered in 1849. The forty-niners and their successors long surmised that this rather miserable little cavern was the secret escape route for raiding Indians. Its maddening crawlways must have provided an excellent excuse for the failure of pioneer explorers to find a second entrance "miles away."

Much of the wondrous lore of the delightful Mother Lode country is woven with spelean threads. Cave of the Catacombs, re-identified by the old Stanford Grotto of the National Speleological Society, once was a prime journalistic subject. Here, readers learned, the local Indians incarcerated errant tribesmen or tribeswomen naked, to die miserably of exposure and madness.

In truth, this and several other Mother Lode caves contained human skeletons. The Miwuk and other local tribes attributed them to Chehalumche, a cannibalistic giant who lived underground. Since their terror of Chehalumche kept the Miwuk out of caves, the 1881 legend of Cave of the Catacombs may have been entirely a white man's invention. Perhaps a reporter was bored—or, as occasionally happened, there wasn't enough other news to fill the little newspapers. Celebrated Magnetic and Hayes' Caves of the Sierra Nevada foothills were newsmen's inventions.

Aside from the deep caves of the Marble Mountain Wilderness Area and confusingly interwoven three-dimensional marble labyrinths in the Sequoia area, most of California's caves are rather small. Comparatively large is Samwel Cave, perched high above the sparkling reservoir that is Shasta Lake. That noted cavern contains about nine hundred feet of low passageways and chambers so confusingly superimposed that its moment in history happened almost as an oversight.

To eminent paleontologists gathered at Potter Creek Cave in 1903, the tales of their Wintu Indian workmen seemed romantic but unlikely. Everyone at their camp was pleased with life. The Indians were receiving good wages. Daily they were uncovering a rich trove of archaic animal remains. Yet they continually insisted that this Potter Creek Cave was nothing compared to Samwel Cave, a dozen miles to the north. That cavern, they told paleontologists J. C. Merriam and E. L. Furlong, was the Cave of the Magic Pool, long revered by their tribe. Anyone drinking its magic water was immediately endowed with quintessence of good luck—especially in matters concerning the opposite sex.

So scary was Samwel Cave, however, that few Wintus had ever sought its blessings. Many, many generations earlier, three not-so-young maidens had dared its depths, stimulating much gossip in the tribe. Much more talk ensued when braves continued to spurn the hapless trio.

Finally a toothless sybil spoke up: "There is a second pool in the cave with far more magic than the other. It lies far within, and the danger is much greater. Nevertheless, its waters will make you irresistible!"

To these forlorn women, danger meant little. Patiently and systematically they searched the remote catacombs with smoking, flickering pine knot torches.

Far back in the cave, one of the girls screamed as her footing gave way. The others, terrified, saw her torch vanish into a yawning pit; heard a hollow battering noise, a dreadful silence, then a horrible thud.

Trembling, her companions crept to the pit and peered downward. They could see nothing. Only their heightened heartbeats were audible. Quavering they called. Only mocking echoes replied. The braves they summoned could do no more. Samwel Cave became even more awesome.

Merriam and Furlong were properly skeptical. Potter Creek Cave was proving a treasure trove of the dim past. But they'd better take a look at Samwel Cave, just in case.

In those pre-highway days, travel from Potter Creek Cave to Samwel Cave required lengthy horse-packing. Yet the rewards of test pits in the Pleistocene Room of Samwel Cave brought the eminent scientists again and again. Not so rich a yield as at exciting Potter Creek Cave—that was hardly to be expected—but still exceptionally worthwhile.

In nonworking hours, the paleontologists turned candle-lit spelunkers. Perhaps spurred on by the excitement of their Wintu workmen, they wriggled widely in search of the maidens' legendary pit.

Three expeditions from base camp completed the major excavations. Merriam returned to Potter Creek Cave while Furlong remained to finish the Samwel project. On the day after Merriam's return, an Indian runner burst into camp. Furlong had crawled through a low, overlooked orifice. Beyond lay a low chamber with a pit "90 feet deep." He needed all the ropes and ladders Merriam could locate.

The expedition had only 50 feet of rope ladder. Merriam gathered "all the loose ends of rope to be obtained in camp" and set out at once. Constructing an additional crude, weighty ladder from sturdy branches they hauled it excitedly through Furlong's hole. Clattering, they lowered it into the new pit. Using no belay, illuminated somewhat by a candle held between his teeth, fighting an annoyingly outthrust ledge, Furlong descended the twisting ladder into a jug-shaped room.

Suddenly his shout electrified the tense group above: "There's a mountain lion at the foot of the ladder!"

Vividly imagining a death struggle in the helpless dark, the pit-top group froze, helplessly. Nothing but rocks for weapons . . . Furlong seemed doomed. Merriam later confessed that his thoughts strayed. Could the cougar climb the ladder?

Then came a new shout from the depths. The mountain lion was merely a skeleton—covered with calcite. It had been in place for centuries, millennia. But that was not all.

"Here's the skeleton of the Indian maiden!" Furlong's shout echoed upward triumphantly.

Merriam could stand it no longer. Heedlessly, he too clamped his jaws on a candle and clambered down the long, twisting ladder. There was the calcified mountain lion. A few steps away was a human skeleton, covered by a film of black mold.

But on one side the combined lights of the gawking scientists showed much more: the skull of a large animal with gracefully curving horns, never before seen by man. Nearby was another with widely swept, oxlike horns, equally unknown to these experts from the University of California. Beyond was a veritable Golgotha of animal bones, large and small— treasure indeed for any scientist. Entering or dragged in through a passageway later blocked by nature, they greatly increased man's knowledge of the ancient earth.

The Wintu workmen too found treasure in Samwel Cave. Tenderly they laid to rest the broken body of their tribeswoman in a sunny grave after uncounted years of perpetual night.

For a time, at least. Some assert that the rising waters of Shasta Lake reservoir soon forced a second burial. And that, in the process, the supposed maiden turned out to have been a brave—as happened to "Little Alice," one of the Indian mummies of Mammoth Cave. But that's a different story.

In the 1920s, still more important archeological discoveries came to light beneath the Mojave Desert. This time, the location was a curious little cave in Nevada, not far across the state line from Kokoweef Peak and the Cave of the Winding Stair: Gypsum Cave. That hot, musty cave held heart-warming triumph for M. R. Harrington, cave hunter extraordinary.

M. R.'s cave hunting began far from the Mojave Desert, canoeing along limpid Ozark rivers in the placid years of 1914. Keenly and professionally this young archeologist scanned rocky cliffs, broken bluffs, searching out long-forgotten caves that might have sheltered man hundreds of years earlier. Or tens of hundreds.

Here and there the graceful canoe arched to shore. Patiently, Harrington tested the earth beneath overhanging rock shelters and in the yawning mouths of caves. Almost every trowelful unearthed plenteous traces of aboriginal occupation of the shut-in cliffsides, but none yielded the fibrous relics he sought.

Young Mark Harrington knew full well how slim were his chances. Perishable pre-Columbian artifacts from the eastern two-thirds of the United States were essentially limited to the scant relics of the saltpeter caves of Kentucky and Tennessee. But the gamble of a few days' canoeing was worthwhile. Were he successful, man would gain an insight into the past not possible in any other way. Fifty years later, he pointed up the problem:

Gone, in many areas, are the wood, the bark, the basketry, the vegetal fibers, the gourds, the skin and furs and feathers they used—all long perished through the agencies of decay. The few objects remaining are only such as happened to have been made of durable materials. Who could picture the spectacular glories of the culture developed by the native groups of the Northwest Coast of North America—the totem poles, the huge houses with their elaborate carvings and paintings, the fantastic masks, the great canoes, all of wood, if some cataclysm had wiped out the people before the white man came. Nothing would remain to us but their products in stone, bone, copper and shell, which were few and relatively simple.

Few know better than Mark Harrington how many peoples have vanished in such a cataclysm. In the arid West, perishable materials have been unearthed so widely that we know much about the way of life of the cliff dwellers of Mesa Verde and many another culture. But science had no such knowledge of the dim prehistory of the Ozarks.

Harrington's venture was hardly random. Six years earlier the young archeologist had spotted woven fragments in local collections. Inquiries suggested that such things had been known casually for thirty years.

Unsurprisingly, his 1914 canoe venture failed to achieve its goal. Nevertheless, young Mark was eager to pursue his slim lead. Eight years later, he talked the Museum of the American Indian into a full-scale expedition, its base camp to be at nearby Eureka Springs. Without greater success, the little party spent several days studying caves and rock shelters in the immediate vicinity.

The time was far from wasted. Gradually Harrington broke through the reserve and mistrust of the clannish Ozark people. Overall-clad natives soon led him to remoter caves: bushwhackers' caves, partially walled off by Civil War guerrillas.

Or so the stories went. To Harrington's trained eye, the low stone barricades appeared far older than the Civil War. The narrowed entrance-ways bore the burnish of long, long human passage.

High, dry, and protected from the most severe storm, several sites were particularly promising. The team began the tedious, painstaking details of scientific excavation. Day by day, cave by cave, excitement mounted. Not just occasional Indian relics turned up, but intimate details of the everyday life of an unknown early culture. These long-vanished bluff dwellers had lined dry storage pits with worn-out basketry and matting—and often forgotten them. In a single cache, fifteen separate woven items were unearthed.

As neolithic cultures went, the bluff dwellers seemed backward in many ways. Pottery was just coming into use, and the bow and arrow had not yet rendered the atlatl obsolete (even today this curious spear thrower is used in some isolated parts of the world). Nevertheless, their life was enriched by a wealth of perishables, often finely woven of grass, cane, animal fiber, or willow. Twilled cane basketry was common. A long-abandoned baby back-pack of woven cane was found almost intact. Woven seed bags still contained unsprouted pumpkin or squash seeds. Even some of their food turned up—cornbread burned to a crisp and thus preserved through the millennia. Light, warm blankets and robes were made of cords cleverly studded with turkey down, or of woven strips of rabbit skin. Woven bottles were calked watertight with pitch. Belts were lashed with buckskin ties. The men wore curious grass breechclouts—merely a bunch of long grass knotted at one end, tied to light fiber waist strings behind, then pulled forward between the legs and tucked in behind waist strings. The women may have worn nothing at all during the warm, lazy summers, though one burial showed semblances of a Hawaiian-type grass skirt. There was only one such indication, however, though burials were so plentiful that Harrington could casually report that "a few stray human toe and fingerbones were found here and there throughout the digging."

A single summer was clearly inadequate for the unraveling of this unexpected new culture. When Harrington returned a year later, his heart was warmed by an unexpected dividend. Long winter evenings' gab sessions around the potbellied stoves of little country stores had made up the community mind: the museum man merited their trust.

Not all was rosy. During the winter one promising site had been gutted by pot hunters. But hillmen who once had stared and turned on their heels now volunteered their help. Indeed, it was a previously hostile landowner, Arthur Weimer, who unearthed one of the outstanding ac-quisitions of the expedition. Overturning heavy limestone slabs yards away from the museum workers, he shouted exultantly and rose grasping a rude flint hatchet. Hafted with a heavy oak handle, the head ingeniously held in place with small wooden wedges, it was still usable after two thousand years.

Many another from remote Ozark communities proffered assistance. J. A. (Dad) Truitt, namesake of Truitt's Cave and discoverer and proprietor of several caves in the Arkansas-Missouri-Oklahoma corner, was a particular help. One shallow cave to which he directed the archeologists later proved to have a head-sized hole at the rear of the overhang. A little excavation by the owner opened into a large network cavern with splendid waterfall groovings, now commercialized as Bluff Dwellers Cave. Worthwhile at any time, it is an exceptional spectacle when heavy rains bring the full glory of underground waterfalls.

Many a humbler mountaineer also aided the expedition. The famous Ozark grapevine carried word of the coming of the museum men to hidden moonshine operations—and provided Harrington with warnings for which local revenuers would have given much. Only once did the grapevine go agley. It seemed that one particular still was much closer to a selected site than Harrington had understood. All hands hit the dirt when rifle balls whistled too close for comfort.

Every movement brought determined fire. After long, silent minutes of indecision, a hopeful thought came to Harrington. "If you-all mind your business, we'll mind ourn!" one of his men bellowed loudly.

Nothing happened, but Harrington was hopeful. He wiggled a bush. No shot followed. Rising to his feet a bit shakily, he held his breath, then inhaled with relief. It was the right formula.

At the close of the second monumental summer, Harrington completed the expedition by studying all the local artifact collections he could locate. In one, he observed something unusual—carved animal bones. Puzzling over their ancient markings, he noticed a roughly sketched animal that looked like an elephant. Could such things be?

Generations earlier, speleologist Thomas Jefferson had speculated on "mammoths and megalonyxes" in the unknown American interior. Through the years, however, the pendulum had swung full-length. In the jungle warfare of academic politics, archeology was clawing for its rightful place. Ultrarespectability was the watchword. Few attempted to interpret their findings, for almost anyone's charge of "unwarranted conclusion" could blight the most eminent career. Those who found human artifacts in the Pleistocene bone caves of California at the turn of the century ludicrously sidestepped identifying them as such. Even scholars like Harvard's F. W. Putnam—Harrington's first archeology instructor— were virtually ostracized if they dared suggest that man had come to North America when archaic beasts still roamed its hills and plains.

This ultraconservative viewpoint had some merit. Artifacts of one age can get into layers of other ages in a variety of ways well known to archeologists. No one had ever found a bone of any extinct American animal showing healing around a weapon point—proof that the animal had been hunted by man. A spearpoint near or under a mastodon skeleton or a point embedded in the bone meant little. This might have been a part of magic ceremonies thousands of years later. Besides, some of those who espoused man's association with the early beasts hardly befit the hard-sought new image of responsible American archeology. Consider "Dr." Albrecht Koch of St. Louis. During the 1830s and 1840s he vigorously advocated this later unpopular view in such publications as the *American Journal of Science:*

It is with the greatest pleasure the writer of this article can state, from personal knowledge that one of the largest of these animals had actually been stoned and burned by Indians, as appears from implements found among the ashes, cinders and half-burned wood and bones of the animal.

As it turned out, "Dr." Koch was the operator of a private St. Louis museum and a showman perhaps rivaled only by P. T. Barnum. To latter-day archeologists, his accounts sounded like humbug.

Not all the evidence could be dismissed as the work of charlatans, but most of it was scant and often inconclusive. A Shawnee legend of a great trunked beast was traced to Koch and thus suspect. In 1896 H. C. Mercer of the University of Pennsylvania found burned cane torches with a megalonyx skeleton in Big Bone Cave, Tennessee. Mercer was "practically convinced that man had coexisted with the great sloth in that region." But he and Putnam were exceptions, and their careers suffered for their beliefs. Out of bitter experience, Putnam cautioned Harrington against following his example.

Just prior to the bluff-dweller expeditions, Harrington had found the remains of a ground sloth intermingled with human refuse in Cuba. But Cuba, it seemed to everyone, might well be a special case, where sloths survived late. After all, tapirs and llamas really should have died out with the American camel and the mastodon, their true contemporaries. Not until the 1950s did discoveries in Mexico, Arizona, and Oklahoma finally convince science that early man had hunted the giant proboscideans.

Moreover, archeologists, anthropologists, and other savants are constantly beleaguered by the same problem which annoyed Jack Mitchell. Apparently fathered by John Cleves Symmes some 150 years ago, the cult of believers in a hollowed-out earth recently underwent resurrection by one faction of flying-saucer buffs. Hardly a year passes without fleeting but grand announcements of discovery of amazing, astounding, fantastic cave dwellers. Today these momentary sensations most commonly are alleged to be remnants of a master race, star-tossed far abaft their home galaxy. Not long ago, the "history-shattering discoveries" were of transplanted Aztecs, Incas, or a lost tribe of Israel—or Lemurians. Occultist caverns on Mt. Shasta—by some accounts complete with gleaming steps of pure, untarnishable copper—long were in the forefront of such make-believe. For a generation after newspapers accurately reported the beginnings of modern California speleology, these caves disappeared from the type of story which had previously included them with others in Tibet, Egypt, and vague portions of South America. (Overprecise location of such a South American cave in the best-selling book *The Gold of the Gods* recently subjected author Erich von Däniken to a classically debunking *Playboy* Magazine interview.) Recently, in the new renaissance of the black arts, mystic caverns are again reported on Mt. Shasta—although at the moment they apparently can only be visited through proper projection of one's astral body.

The big feet of the 1947 Lemurians, however, seem to have been permanently lost to California, grafted onto the mythical ape-like sasquatches of recent Pacific Northwest lore. Perhaps the victims of collective wishful thinking, the East Coast editorial staff of *National Wildlife* recently published an article which stopped just short of asserting that these "bigfeet" den up in Washington State's Ape Cave. Actually, that cave is a major tourist attraction visited year-round by thousands (of humans) each year. Needless to say, Northwestern cavers were delighted. Especially the proud owners of certain plaster molds that make magnificent sasquatch tracks.

Gypsum Cave is only sixteen miles from Las Vegas. It seems to have been a favorite picnic spot of the placid 1920s. Local Paiute Indians knew the cave as a sacred spot where offerings were to be left— why, none seemed sure. Local whites knew the cave as a spot to acquire Paiute relics—and for arguments. Peculiar fibrous stuff on the floor of inner chambers led some to tell of horses once stabled inside by marauding Apaches. Others scoffed. The entrance was far too low for horses. Besides, the stuff was dried-up seaweed, not horse droppings. Anybody could see that. O.K., wise guy. How did seaweed get into a cave 250 miles from the ocean? Well . . . maybe there was a pool of water there sometime or other. . . .

In 1924 Harrington undertook the excavation of a large pueblo ruin in the nearby Moapa Valley. Governor Scrugham mentioned Gypsum Cave, but not until John Perkins brought him a segment of "arrow shaft" did Harrington become especially interested. It was an atlatl dart, not an arrow, and any cave deposit of that antiquity merited careful study. In the spring of 1925, he accompanied an officer of the Nevada State Police on the rough journey to Gypsum Cave. Up the rattlesnake-infested hillside they sweated, then down a treacherous rockslide. At its base

Harrington noted that the low openings to the cavern were heavily polished by the passage of animals. But what animals? As he studied the five little chambers of the stifling three-hundred-foot cave, he became increasingly puzzled:

Near the entrance I found a few evidences of relatively modern Indians; in the inner chambers I picked up a few pieces of atlatl darts of the Basket-maker type; these were lying on the surface of a deposit which test holes showed to be dry and fibrous and very much like the layer of dung which accumulates in a neglected stable or barnyard. Later we found unbroken pieces which proved to my satisfaction that the deposit was really dung, but of what animal I could not determine. It seemed too large for horse or burro, and besides I could not see how either animal could have found its way through the low openings into the dark inner chambers.

Harrington could guess no further. In 1929 he seized an opportunity to return to Gypsum Cave. This time he found patches of reddish hair which puzzled him even more.

In the four years between his visits, major changes had come to American archeology. At Folsom, New Mexico, Barnum Brown of the American Museum of Natural History had found unique dart points amid deeply buried bones of a herd of extinct bison. Meticulous studies strongly suggested that early man had dined here on an animal which supposedly had died out with the mammoth. The old concepts were shaken. When Harrington revisited Gypsum Cave in 1929, "light began to dawn. One thing seemed sure: that there was no native American animal in historic times capable of producing such dung; therefore, the probabilities were that it was attributable to some extinct animal."

Not every archeologist agreed that the Folsom discovery required a reversal of their opinions. The stigma of "unwarranted conclusion" was still powerful. When Harrington reported his discovery, he indicated that he had consulted with two noted authorities who agreed that it was sloth dung. Years later he grinned lopsidedly: "It wasn't quite that way. I had no idea what animal it was until I sent some dung and hair to Barnum Brown to analyze. His report came back 'ground sloth.' Suddenly everything made sense."

In April 1929 Harrington's Southwest Museum team staked a claim to Gypsum Cave, and in January 1930 set out in force. In leading roles were Willis and Oliver Evans, Pit River Indians long experienced in archeological techniques—and other matters:

The Evanses arrived on January 13 to put up camp, but on account of a heavy snowstorm the camp was not entirely completed when the rest of the party arrived from Los Angeles on the 20th. On account of the exposed situation . . . and the high winds, it was necessary to arrange the tents in the form of a hollow square and to surround the camp with an Indian fence or windbreak of arrow-brush gathered along the Colorado River.

Spring was hardly pleasanter. Twenty-eight "ordinary" rattlesnakes and eighteen sidewinders were killed in 1930 alone, and summer heat forced abandonment of the excavation in mid-project. The life of an archeologist sometimes leaves something to be desired.

The cave was mapped in detail and survey stakes were set. As trenching began, expedition secretary Bertha Parker Thurston (later Mrs. Iron-eyes Cody, Harrington added when he reviewed this manuscript) donned a helmet and began to explore obscure crannies. On January 30 Bertie stuck her head under one rock and peered backward. To her considerable surprise, she found herself nose to nose with a remarkable skull—not human nor that of any animal familiar to anyone in the diverse party. Ground sloth? No one knew. A volunteer assistant gingerly hand-carried it to the California Institute of Technology.

Meanwhile, the preliminaries continued. Six days later, Mrs. Myrtle Evans (Shoshonean wife of Oliver Evans) and her small nephew Lyman were scratching boredly in the dusty floor. Large bones promptly appeared. Harrington wrote exultantly:

. . . nearby [was] a mass of such bones, as well as a quantity of coarse, tawny hair and a huge claw with its horny sheath still intact—all of which, I was positive, must have belonged to some species of ground sloth. Then came the report on the skull; it really was a ground sloth of the species *Nothrotherium shastense.*

In the wake of the report came two eminent paleontologists, Chester Stock and E. L. Furlong—and a grant from the California Institute of Technology. The first three weeks had hit the jackpot. But it was only the beginning.

Throughout the cave, the layers of remains were remarkably distinct and dry. Not only the hair, but parts of the hide of several sloths were preserved. Like that of humans long buried in certain dry cave soils, some of the hair was bright auburn. Was this the origin of the legend of the redheaded dwarfs of the Cave of the Winding Stair and the Lemurians? I would give much to know.

In the outer rooms the sloth remains were deeply buried beneath later debris of many kinds. In the thin-layered inner rooms, the surface sloth material appeared very recent. Yet a large stalagmite had grown upon some of it, mute evidence of its age: for arid millennia no significant dripstone has formed in bone-dry Gypsum Cave. Before summer heat forced a halt, the story of untold ages was unfolding dramatically in Gypsum Cave.

A generation later, radiocarbon dating revealed that the ground sloths had lived here at least as recently as 8,500 years ago. Today we know that early North American man inhabited several caves in the western United States at that time. In 1930, no one knew, and most doubted it.

As Harrington pieced together the story hidden in the distinctive layers of the hot, dusty cavern, the centuries rolled back. Most recent were a few traces of twentieth-century picnickers. A few recent Paiute relics remained, overlooked.

Below the Paiutes' artifacts were traces of Pueblo peoples related to the Hopi culture of today. Puebloans occupied the nearby Moapa Valley for five or six hundred years. They camped in the entrance of Gypsum Cave for days at a time, making pendants from clear selenite crystals which once adorned the cave. Some were found still strung on fiber neck cords. Unhurriedly, the Puebloans ground corn and seeds, wove baskets, made string, worked flint, hunted, and perhaps gambled and played flageolets (gambling sticks and crude flutes found in the cave have not been dated). Before the Puebloans were the more primitive Basket Makers, who visited the cave while hunting. Sometimes they, too, camped in the entrance room and occasionally made selenite pendants.

Right hind foot and claws of ground sloth uncovered in Gypsum Cave, Nevada, in April 1930. Reddish sloth hair is also visible. *Photo courtesy Southwest Museum*

But, before the Basket Makers, an unknown earlier people seemingly had left fireplaces and abundant artifacts. And the ribs of one sloth bore scratches remarkably like those imprinted on fresh beef ribs by a stone knife from the cave. Not counting obviously jumbled areas like storage pits and pack-rat nests, artifacts were found with or below the sloth and camel remains in twelve separate locations. Before the expedition was completed on January 17, 1931, the mainstream of American archeology had to reckon with the coexistence of man and the last of the great Pleistocene mammals.

As I write, plans are underway for renewed scientific excavation of the cave sections Harrington left intact. As it chanced, radiocarbon dating of Gypsum Cave material has not yet yielded any dates of human occupation even half as old as some from other western caves: Utah's Danger Cave, Fort Rock Cave in the Oregon desert, and Marmes Cave, Washington. Or, for that matter, as old as the sloth dung tested to date. Which was precisely why Harrington—acutely aware that the conclusions of his own particular era would never be the last word—left parts of the cave untouched for techniques of the future.

Unlikely California and Neveda cave legends thus have yielded occasional treasures to others than writers and lecturers. What if these riches differ enormously from those of Dorr's golden tales? To the scientific community and lovers of folklore alike, the real treasures of these glorious caves are vastly more valuable.

Today's California and Nevada cavers happily admit that, aside from the myriad joys of all caving, their cavern treasures are only of this sort. The Lost River of Gold will stay lost, for it cannot exist unless our accumulating knowledge is all wrong. A generation's effort has found only one crevice that "goes" in Kokoweef's Crystal Cave, and it "goes" only into one additional chamber, unhappily as gold free as all the rest.

Yet a nagging thought remains. Before their fateful last exploration, Merriam and Furlong thought of Samwel Cave much as we think of Kokoweef and the Cave of the Winding Stair.

Can we be wrong? Will this chapter someday be rewritten in blazing headlines: SPELUNKERS VERIFY REDISCOVERY OF CAVE OF GOLD?

Probably not. But Gypsum Cave, at least, still holds much of value to mankind.

6

The Greatest Cave

The Story of Carlsbad Cavern

Abruptly, the sparkling cavern floor gaped black before us. The spot beam of my flashlight outlined white, flowstone-draped walls beside the blackness—how far?

Somewhere in these remote corridors of "The Lower Cave," far beyond the familiar chambers toured by every visitor to Carlsbad Cavern, a lost pit reputedly plunged 325 feet. Even a lesser depth could readily yield a new record for caves of the United States. Could this be it?

From the start, my companion—ranger-naturalist Neal Bullington— had been inclined to dismiss the entire tale as a forty-year-old hoax. After plodding hours of systematic search in dark corridors 2,000 miles from home, my own hopes were sinking. I had found myself disagreeing with the speleological tradition that branded Frank Ernest Nicholson a charlatan and his pit a hoax. But, as the relentless hours passed in the timeless dark, I too began to doubt.

Until the sudden blackness yawned before us.

The newspapers' story of Nicholson's Lost Pit began early in January 1930. In the name of the *New York Times* a syndicate of publishers was sponsoring a magnificent 15-man exploring expedition to Carlsbad Cavern, equipped with the very latest equipment, from boats to man-carrying balloons. Fifty-five of America's leading newspapers syndicated daily accounts of Nicholson's exciting discoveries to 20 million Americans.

Perhaps because of the unforeseen debacle which destroyed it, of which more later, the whys and wherefores of the expedition subsequently became as mysterious as the dark, intense Nicholson himself. Apparently it evolved unrecognizably from a plan by young radio enthusiast Eric Palmer to test the penetration of the earth's crust by radio waves. According to the quip of penwoman Jean Cabell O'Neal, Carlsbad Cavern was chosen merely as "the crustiest bit of terra firma available."

For technical advice, Palmer evidently turned to Nicholson, who identified himself as a member of "the cave explorers' club of France," speleologist, and "explorer of caves on five continents." Perhaps he was. For that misunderstood era he was quite a traveler, having worked at least in Alaska with the U.S. Bureau of Fisheries. Later he served as broadcaster on the Hearst Graf Zeppelin Expedition. His were enthralling tales of the glory and excitement of spelunking. More and more Palmer's circle responded to his magnetic personality. More and more the radio test was relegated to the background, overwhelmed by the spellbinding glamor of a magnificent exploring expedition.

Nicholson's basic idea was commendable. Two splendidly illustrated *National Geographic Magazine* articles and other publicity had made Carlsbad Cavern a vaguely familiar name across America. In Congress, careful preparation was afoot to upgrade Carlsbad Cave National Monument into our present national park (which needs even further expansion). Yet the American people were so poorly informed about the cave that their ignorance can hardly be overstated.

New Mexico journalists found "startling" much that appeared in what they termed "the uninformed eastern press." One article referred to passports supposedly needed to travel to the cave. A Chicago newspaper pooh-poohed another's speculations about the possibility of finding dinosaurs in the cave. Not because no westerner had seen a dinosaur for God knows how many million years, of course, but because "dinosaurs were tropical animals, and could not exist in the cold and darkness of the caverns."

"Nearly all of them," one local reporter announced about his eastern colleagues, "are glad that the caverns are to be explored, although some 21 miles of the labyrinths have already been traced out, and 7 miles of it are included in the best lighted, equipped and managed subterranean course in the world." He might have mentioned the new state highway to the cave's entrance, and the convention that had just been held in the Big Room, 750 feet underground, but his home-folk readers already knew all that.

Indeed, ludicrous accounts were all too common outside the immediate Carlsbad area. One wire service spread across America bewitching tales of a bottomless pit "down which weights are said to have been lowered for two and one-half miles, until they broke off of their own lengths." Nicholson himself seems to have done his homework. Yet, even so, he was misled by an account of an explorer traveling "for six days down one tunnel, finding it growing all the while," and forced to return only when his grub ran low.

In actual fact, exploration had been languishing for two years in this greatest of all American caves. From every standpoint, the time was ripe for a breakthrough. Yet, if anyone in the National Park Service still knows why the expedition stirred such enthusiasm in New York and Washington, D. C., he seems unwilling to admit it. As for its embarrassed chief sponsor, the *New York Times* still hasn't gotten around to answering my inquiry of a decade ago.

Even before he arrived, Nicholson changed the course of history of the cavern and the national-park-to-be. Telegrams about the expedition poured from and to New York and Washington. The local staff spent several weeks in somewhat ambivalent preparation. And waited. And

Looking toward the Rock of Ages (the most distant stalagmite in the photo, at right of upper center), even this National Park Service photo shows only a small part of the Big Room.

waited. And waited still longer. The expedition was less than ideally organized, and two minor automobile wrecks en route helped not at all.

Assistant Chief Ranger Cal Miller, Superintendent Thomas Boles, and two other staff members got bored and went out exploring on February 18. Promptly they discovered Lake of the Clouds, not too far beyond the famous Lunch Room via the Left Hand Tunnel, but down two pretty good pitches. Not yet knowing that aneroid barometers need special temperature corrections in caves, they calculated its depth as 1,220 feet below the entrance. Eventually the figure was found to be 1,013 feet (1,024 feet at the bottom of the lake), but even that was a record for North America.

And they did it quite casually.

Nicholson's party arrived two days later and began to grind out daily articles. Almost at once, however, grave problems began to dog the ill-organized expedition. Its famous balloon and much of the rest of its equipment were ludicrously useless. The radio tests were completed on the first day underground. Some two-thirds of the junketeers found themselves unneeded, excluded by order of Superintendent Boles. Equipment manager Edward E. Roberts was caught stealing a stalactite and was duly banished from the cave. Some of the opposition press later lauded him for having relinquished "a career as an actor and shoe salesman to join the expedition" and waxed wroth at such cavalier treatment. He promptly joined the other involuntary exiles at a nearby tourist camp and began to defame Nicholson and his exploring team. Newspapers not served by the syndicate began to hint that Nicholson was exploring only "from his headquarters at the swanky La Caverna Hotel," dictating imaginary accounts to a pretty secretary also housed there.

Although the syndicate's confidential representative, journalist Tom Davin, was among the tourist-camp exiles, Nicholson probably could have survived this venom. The National Park Service rallied to his support. Famed Superintendent Thomas Boles indignantly pointed out how Roberts and his disgruntled allies "spared no effort to discredit Nicholson." As Boles indicated, the newly-streamlined expedition in truth was discovering important new cave; about two miles, he opined.

But an insuperable problem became increasingly evident. With the Guadalupe Room breakthrough a generation in the future, there just wasn't that much additional cave to be found by yesterday's techniques.

The National Park Service had extracted a promise that Nicholson would tell it like it was. Nicholson did his damnedest to keep faith. "These stories have been free from sensationalism," Boles reported to his bosses in Washington on March 10, "and I doubt if we are bothered by such."

The publishing syndicate, however, made it abundantly clear that it

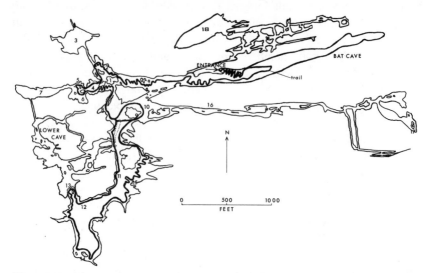

Map of Carlsbad Cavern. 1) Iceberg; (2) Green Lake; (3) New Mexico Room; (4) King's Palace; (5) Queen's Palace; (6) Papoose Room; (7) Mystery Room, with Nicholson's Pit at end of narrow passage leading south to Lower Cave; (8) Boneyard; (9) Elevator shaft; (10) Lunch Room; (11) Giant Dome and Twin Domes; (12) Totem Pole; (13) Jumping-off Place; (14) Bottomless Pit; (15) Rock of Ages; (16) Left Hand Tunnel; (17) Lake of the Clouds; (18) Guadalupe Room. Based on map in N.S.S. Guidebook to Carlsbad Caverns National Park, with additional data courtesy National Park Service

wasn't getting its money's worth of the heroism and stark drama that sells newspapers.

Nicholson had one ace in the hole: a seemingly bottomless pit leading down indefinitely from the Lower Cave—(which he called the 1,025-foot level). Even if it wasn't 2½ miles deep, the syndicate should have rejoiced. A breathtaking descent worthy of record depth led him to a previously unknown level 325 feet farther down. Wired accounts overflowing with drama and heroism described a huge chamber full of eerie forms through which Nicholson wandered entranced, completely alone.

The heat here was stifling and there seemed not enough air to breathe. I glanced back up the shaft . . . equal in height to a 30-story building . . . . I dared no longer stay in this chamber that seemed lacking in air and began climbing the long rope, swaying from side to side against sharp formations on the walls of the pit that brought blood. After what seemed an eternity, but what was in fact a little less than two hours, I pulled myself over the edge of the pit.

A few days later he announced that research had confirmed that his descent had established a new world's record.

It wasn't enough. On the same day that Boles praised Nicholson's accounts to the Washington office, his "business manager" in New York dispatched a fateful telegram:

DELIVER NICHOLSON THESE ARE ORDERS DISOBEYAL MEANS WE LEAVE YOU FLAT IMMEDIATELY APPOINT DAVIN REPLACE SECRETARY APPOINTED WITHOUT MY KNOWLEDGE YOU FACE WRECK OF YOUR LIFE UNLESS OBEY DONT ARGUE ONLY COMPLETE GEOGRAPHICAL SEPARATION PRESENT ENTANGLEMENT ENABLE US SAVE SITUATION CONFIRM OBEYAL SCOUT AND DAVIN MUST GO IN CAVE DO SOMETHING BIG WIRE LENGTH PROPOSED STAY MOVIE FOOTAGE TAKEN NUMBER STILLS HAS BOAT ARRIVED PROSPECTS GREAT OTHERWISE PUTNAM ALSO INSISTS IMMEDIATE CHANGE EFFECT IT OVERNIGHT

The lines were drawn. Boles refused to return Davin and Boy Scout author Oliver to the exploring parties: an act of considerable courage that was vindicated when Oliver's *A Boy Scout in the Grand Cavern* turned out to be 148 pages built around the first three days of Nicholson's expedition—spent mostly on tourist trails. It portrayed Oliver, Roberts, Davin, and other exiles from the expedition as fearless, competent spelunkers. Not once did Nicholson's name appear. While the book did contain good local color, some of its local audience considered it to have added insult to injury. Cal Miller's name appears frequently in the book. Invariably it is given as Carol Miller. That of Superintendent Boles is as missing as that of Nicholson.

If the impasse was to be broken, it could only be done by Nicholson himself.

" 'Getting lost' is always a big stunt for a sensationalist, and we wished to guard against this," Tom Boles wrote later: one reason why Cal Miller or another ranger always accompanied Nicholson. Yet, on March 13, the desperate reporter's dispatch told how he had gotten lost—and rescued himself. The next day the theme was how he had been trapped by fallen rock—and dug himself out. Boles's next report reneged on some of his earlier praise.

It wasn't enough—and, in fact, could not have been. The syndicate cut off the expedition's funds and inadvertently forced the cave's history into a new channel: the legend of Jim White. More anon.

In tracing the history of some of our finest southwestern caves, I repeatedly encountered the spoor of Frank Ernest Nicholson. Intrigued, I visited his home town of Wichita Falls, Texas. There, a few old-timers had boyhood recollections of a brash young journalist who had gotten into some kind of trouble and never came home. One remembered how young Nicholson had overcome a crippling boyhood injury—a serious automobile accident which had crushed his legs—by sheer determination and desire. His last memory was of only a slight telltale limp. Later I found other mention of the limp, but never a hint of self-pity or favor-seeking when the going got tough.

With Tom Boles long retired, I found an understandable skepticism at Carlsbad Caverns National Park. In the pleasant town of Carlsbad, however, I talked with Cal Miller. Cal also was retired, and not nearly as spry as he once was. But he remembered 1930 well, and had been Nicholson's chief guide as well as one of his principal defenders. He hadn't ever gotten around to descending Nicholson's Pit, he acknowledged, but he had seen the controversial explorer a long, long way down. Not as deep as Lake of the Clouds, he guessed, but it was quite a pit. "No, I don't think it's ever been surveyed," he considered, "and I was around during all the surveying, including Lake of the Clouds." No, he didn't think he could find the pit again. It was just too long ago, and he wasn't up to spelunking these days. It was somewhere in the Lower Cave, far beyond the room below the Jumping-off Place at the far end of the Big Room. That was all he could remember. But he wished me luck.

At the El Paso Public Library I studied Nicholson's later booklet, *The Exploration of Carlsbad Caverns,* now so scarce that I am still looking for a personal copy. The sketch map obviously was wildly fanciful, but it showed the pit. If an explorer proceeded from a point directly below the Jumping-off Place, Nicholson's Lost Pit was the second lead to the left. If there was anything to the sketch, of course, which seemed a trifle doubtful.

I discussed the situation with Tom Meador, historian of the Texas Speleological Survey and unofficially chief historian of Carlsbad Cavern. We decided it was worth some effort, and the National Park Service agreed. Unfortunately Tom found himself hospitalized as February 3, 1968, neared, so the search fell to Neal Bullington and me. Neal was pleasant, polite, and helpful, but frankly skeptical. I couldn't really blame him.

From long semicomic experience, Neil and I knew well the discombobulating impact spelunkers have on underground tourists. Finding an unobserved corner of the Lunch Room to light our carbide headlamps we scurried into the hidden stairway to the Lower Cave, thence along the spacious natural tunnel that leads beneath the Big Room. At the entrance of the vast chamber beyond and below the Jumping-off Place, we made extra-sure that the trail lights were off 90 feet above us, then hastened on around the left wall of the great void.

The first alcove was interesting: evidently the bottom of the pit through which exploring National Geographic Society parties descended nearly a half century earlier. Remnants of their ladders still hung amid knob-coated walls. Here too were splendid rimstone pools, the innermost obviously the one in which the famed Jim White momentarily feared he was drowning when lowered blindly on a rope during these pioneer explorations.

Now we needed that all-important second left-hand lead. Passing through the splendid natural arch easily seen from the Jumping-off Place, we found a left-leading orifice almost at once. It led upward about ten feet, then back to a window high on the wall of the spacious room we had just quit.

Unperturbed, we strode onward in a broad-arched corridor dozens of yards wide. Predictably, several small holes evidenced blackly in our headlamps. Systematically we forced our way into each. Each went nowhere. Then came an impressive corridor leading just where we wanted. A splendid pit yawned a few dozen paces onward. Alas! It ended just 20 feet down.

Nicholson's Pit? I suspect that Neal thought so but was much too polite to say it aloud.

Perhaps a little discouraged, I led the way to the next left-hand lead— actually a sinuous, intertwined nest of crawlways. Boot prints and other marks showed previous investigation. Tediously each narrowed to the limit of human penetration.

We pushed on, more slowly now. The Lower Cave abruptly contracted, and for the first time we had to duck as we splashed through a little pool. As the cave reopened, we continued the search. Upward, a jagged orifice angled toward the Mystery Room, far out of sight. We knew that Nicholson had followed that particular route into the Lower Cave from the Queen's Chamber, but that was of no concern to us today. Or was it? Beneath a little natural bridge was a small hole. Not much of a hole, but we checked it anyhow. Beyond the bridge we could stand erect in a little side passage. Immediately ahead, the floor sloped away steeply. Another step and it seemed to vanish until our spot beams picked it out, 25 feet below. Onward, it again was lost in the darkness of greater depths.

I tossed a small rock down the crevice. It hit, bounced, then rolled until the sound was lost.

Our fatigue rolled away as if by magic. The pit was so narrow that I wanted a standing rope to help me back up. We rigged a 120-foot length and I slithered down the first pitch and the steep slope beyond.

About 50 feet onward, the passage was almost blocked by stalactites and flowstone, but a small orifice opened. I slid through without difficulty and found an impressive fissure. It opened upward and downward alike.

The width was just right for chimneying. My lights showed a vertical pitch, then a sloping floor dimly below. I called to Neal, who descended with another rope and a 30-foot cable ladder. We tied the ladder to the end of the first rope. It did not quite reach the bottom but the few remaining feet were easy.

A sloping floor led to a duckunder to the right. I called to Neal, then clambered into an irregular little chamber glistening with heavy calcite masses. At its lowest point was a very tight tube, which echoed resoundingly when I shouted into it. After about ten struggling feet, however, I was glad to have Neal pull me out. Several others have since tried to force their way onward, with as little success.

To our surprise, we found evidence of a 1951 exploration by some well-known Texas cavers, including my friend Carroll Slemaker. What had they known of this spot?

For that matter, where were we? Was this Nicholson's Pit or something he had overlooked? Certainly the little room scarcely resembled the vast chamber he had described, but in view of the pressures which had beset him, that meant little.

We checked the chamber for further clues, found nothing important, and began the long, hot ascent. At the top we found it pleasant to sit and pant until our racing pulses calmed.

As we rested, our headlamps idly turned here and there. Unexpectedly their yellow light revealed faint penciled marks left by others who had rested here. Nor were they recent. An infinitesimal layer of flowstone obscured part of the writing.

As we bent close, a line resolved into an incisive arrow pointing toward the pit we had just quit. With effort the word ROOM was legible. Easily we read 350 FT; with the light just right it became 1350 FT. Eagerly we searched the walls in detail, seeking names that should not have been there, since this was a national park. Yet there they were, in such faint pencil that they were no true defacement: Nicholson, Cal Miller, others in the 1930 party.

We hurried to the surface and I telephoned Carroll Slemaker. Yes, they had explored a pit like that. No, they had known nothing of its history. But near the bottom they had found something puzzling: a crumpled telegram addressed to Frank Nicholson.

Everything fitted together. Nicholson's Lost Pit was found. But how deep was it?

Climbing upward, I had mentally marked off six-foot body lengths and set down my guess as 161½ feet. Obviously a depth of 1,350 feet or even 1,320 feet was impossible, and even the 1,170-odd feet necessary for a new American depth record. This could still be the deepest point in Carlsbad Cavern, however, for I have badly underestimated several other pits (I've overestimated a few, too, but that's part of a different story).

Neal dug through old survey data and found that we had started 790 feet down. We needed a bit more than 200 feet to beat the depth of Lake of the Clouds. So on January 23, 1971, I was back at a survey point Neal had located near the start of our descent, this time with Tom Meador and Claude Smith. At the survey point—#2Y2—we began a precision survey. It turned out to be strictly routine and we surfaced at midnight with the data. Tom calculated the depth matter-of-factly. This time, unfortunately, my guess was so accurate that it was closer to the actual figure than the average cave survey would have been. Barring a still more accurate survey, we now know that the lowest point in Nicholson's Lost Pit and the little room below is 62 feet above the surface of Lake of the Clouds. Which makes the pit almost exactly half as deep as Nicholson proclaimed across America—pretty darn good for those unlamented days.

Its funds cut off, Nicholson's expedition disintegrated. Yet its impact lived on. The Lincoln Library's *Dictionary of American Geography* still gives the depth of Carlsbad Cavern as 1,350 feet—obviously Nicholson's pit. The figure is not universally official. From its first edition, the *Information Please Almanac* has given the figure as 1,320 feet—also Nicholson's pit. The 1974 *Guinness Book of Records* is a bit schizophrenic. On page 139 it uses the 1,320-foot figure. On page 138 it lists Neff Canyon Cave as deepest in the United States, at 1,184 feet—also an obsolete and incorrect figure, as will be seen. The *World Almanac* used the 1,350-foot figure for three years, the 1,320-foot figure in the later 1930s and 1940s, and now cautiously gives no figure whatever. Among major encyclopedias, only the current *Americana* includes the

Jim White as he looked during explorations. *Photo by Willis T. Lee, courtesy Dana W. Lee*

correct figure. The *Britannica* has none at all. Nicholson himself soon switched from 1,350 to 1,325 feet for no apparent reason. I haven't been able to learn what became of the other five feet.

And the national park duly came into being.

Nicholson's greatest monument was much more human than his syndicated articles. To pay his board bill, the suddenly impoverished journalist cranked out the famous booklet *Jim White's Own Story*. Throughout America it spread the legend of a compelling dream. Soon the public forgot the ill-fated man who wrote this most famous of American cave tales and idolized Jim White as the hero of Carlsbad Cavern that he was. Today, all the world seemingly knows the tale of that incredible lone explorer. Many a later account has merely re-described the unlettered cowboy of nineteen whose curiosity was aroused by the cyclone of bats that in those days suddenly spilled out of the mouth of a cavern at dusk, filling the skies. Today's visitors can scarcely imagine the awesomeness of that weird spectacle of yesteryear. Few even suspect the region-wide environmental neglect that already has cost the cavern 99 percent of its once-magnificent bat colony.

Jim White, it seems, was mighty impressed by the bats and the hole

The pattern of the great bat flights of the Southwest often has been likened to that of tornadoes.

they came out of. He had to know more. A knotted lariat made the steep scramble into the cave reasonably safe—by cowhand standards. Down the long, darkening slope he ventured for some 200 feet. There he stopped. At the bottom of a narrowing, funnel-like slide, an overhanging ledge opened into nothingness.

Well, not quite nothingness Jim learned when he returned in full daylight. For a few minutes each afternoon, a dramatic shaft of sunlight finds its directed way into these dim depths. Fifty feet below the ledge, an illuminated patch is surrounded by a gigantic gloom.

Climbing along a shelf above the overhanging precipice, Jim could glimpse stubby white columns far back in a dim tunnel—a tunnel so huge people laughed at him when he tried to describe it. Caves were no novelty thereabouts. There were lots of them much closer to the booming county seat: Endless Cave, Sand Cave, others. Cottonwood Cave probably was not discovered until much later, but local swains had escorted their best girls to McKittrick Cave for an exciting outing at least fifteen years earlier. Anyhow, nobody in his right mind would take seriously tall tales spun by a teen-age ranch hand.

Regardless of his friends' ribbing, Jim White took a few days off to chop juniper and mesquite sticks, which he bound into a crude wire ladder. With little more than a kerosene lantern, he clambered down the overhanging lip into a stupendous tunnel—and an obsessing dream.

This remarkable photo by the National Park Service caught the ray of sunlight illuminating the dark corridor of Carlsbad Cavern. Early explorers were faced with the overhanging dropoff seen at the rear.

Following reflected daylight eastward in the enormous corridor, Jim clambered over huge slopes of fallen, teetery rock. As he advanced, the musty smell of guano became ever greater. His kerosene lantern revealed increasingly deep mounds and drifts of powdery gray-brown granules. In hidden gaps between half-concealed rocks he sank to his knees in the musty stuff, then to his waist. Powdery gray banks shuddered and slid at his passing. Here and there ammonia fumes contributed injury to his insulted nostrils. An incredible ten stories overhead, an occasional distant squeak revealed blankets of bats, each on a square inch of ceiling.

In the innermost, vastest chambers, Jim's lantern showed nothing but blackness ahead or above. As he gaped, he gagged. Underfoot, the very floor seemed to move. It teemed with vermin feeding on dead and dying bats, the guano, and each other. To a modern biologist with a strong stomach, great bat caves are a fascinating ecological community based on a bat's considerable ability to turn insects into guano. Few others share their interest. Jim probably retreated within a few minutes.

Back on the other side of the gargantuan entrance slope, things were pleasanter. Great rock piles were minor obstacles to this hardy young outdoorsman. The incredible natural tunnel continued. A few stony columns broke the monotony. But just where they became more dramatic, the cave seemed to dive precipitously into a stupendous pit.

Bemused by the dim glory shifting from the shadows as he pushed onward, Jim lost track of time—like many another spelunker. Without warning, his kerosene lantern flickered and died, its fuel exhausted.

In the total black, the ice-cold loneliness of the huge cave suddenly shocked Jim White. The darkness was overwhelming. He could taste it, almost touch it on every side. Never had he imagined such a blackness, such an utter void. Only a faraway drop . . . drop . . . drop broke the total silence. Otherwise it was as if he were blind and alone in outer space. Even the echo of his quickened breathing seemed to come from miles away.

Goose pimples rippled along Jim's back. To avoid a general horse laugh at the bunkhouse, he had told no one of his plans. His few precious matches could not be spared for mere illumination. If he could not get the wick lit, or if he fumbled the blind refilling of his lantern, his skeleton someday would be found, but that was scant comfort.

With shaking hands, Jim unscrewed the container cap—and dropped it. Working by the touch of trembling fingers, he spilled most of the precious kerosene. It seemed an eternity before he could produce a welcome slosh-slosh from the lantern.

Now he groped for his matches, one by one. Their momentary glares were welcome, yet blinding. As he had feared, the first flickered out before the wick blazed up. But Jim was lucky. The second was successful.

The warm yellow glow brought him back to life and hope. He slumped, trembling, as the reaction took hold.

But the scant extra kerosene would not last long. Icy fear still half-gripping him, Jim jumped to his feet and charged up the rocky slope, Almost at once he was staggered by a glancing blow from a low-hanging stalactite. Blood began to trickle down his face, but the blow brought him to his senses. More cautiously he clambered steadily onward until a blue glimmer on the inky walls told him that his rude ladder was close ahead. Still trembling, he fumbled his way into the wonderfully warming New Mexico sun of 1901.

Patched up back at the Lucas Brothers' XXX Ranch, Jim got the guffaws he had expected. Like most ranch hands, the whole crew were experienced practical jokers. Jim's excited yarn sounded like the kind of ploy they would have invented to get a laugh at the expense of tender-feet. Underground chambers bigger than the courthouse? Stone icicles bigger than a man? Not even when Jim persisted would anyone return to the cave with him.

No one would go, that is, except a nine-year-old (or fifteen-year-old, take your choice) Mexican boy known to Jim only as Pothead, or the Kid. Still virtually unequipped, the pair re-entered the vast cavern five days after Jim's first descent. Returning nightly to their bedrolls and campfire, they spent three days penetrating ever deeper into the unknown, covering all of today's routes and much more. The odd pair had to clamber over huge piles of rocks for many hours before reaching new passages. Their lanterns dimly revealed chambers in which they were the merest pinprick of light. Stalagmites towered above them to barely visible ceilings. They came to pits so deep that a lantern lowered on a lariat did not reach bottom. And they came upon smaller chambers so gloriously decorated that the early splendors seemed as nothing.

But kerosene caving was never safe for Jim White or anyone else. No matter how deep in the pack, banging and crunching often caused the supply can to leak. Sometimes not only the pack but the cavers' jackets and shirts came to be saturated with kerosene. On what was planned as a three-day expedition, the Kid once came too close to Jim's back and set him afire. Tearing off his own jacket to smother the flames, he quickly extinguished the human torch, but painful burns put Jim out of action for many days.

Jim's burns sobered his scoffing friends, but made them even less enthusiastic about the project. Jim was going loco, spending so much time in the dark with "them bats." And then Jim chanced upon a book containing pictures of Mammoth Cave. His reaction? "His" cave was lots bigger than Mammoth Cave.

That settled it. Jim obviously was the biggest underground liar in

New Mexico. People began to feel uncomfortable around him. If only he'd grin when he told those whoppers! After all, if you wanted to see a big cave, you didn't have to go all the way out to the XXX Ranch. For years Jim urged the surpassing glories of the Big Cave on many who would not listen. Deep in the cave a vision drove him to construct paths and handrails, moving huge rocks for those he knew would come.

These legends of Jim White form a compelling masterpiece. It is almost regrettable that historical research is yielding so different a picture of the discovery period of "his" cave.

As modern speleohistorians see it, homesteaders and ranchers streamed into the rich grasslands of southern New Mexico on the heels of the retreating buffalo. Hardly fifteen years after the Civil War, isolated goat, cattle, and sheep outfits nestled in lonesome limestone draws—by 1885 within two miles of what is now Carlsbad Cavern. Ranch hands probably soon visited the gaping cave entrance, attracted by the nightly three-hour stream of hungry bats spiraling outward at a rate of a hundred each second. Curious spectators close at hand caught the sound of their whirlwind swirling from the black depths of the bone-dry ridge. Bat Cave, or Big Cave, they called it in the matter-of-fact way of the cowhand. By 1892, Bat Cave rated two sentences in the Eddy (now Carlsbad) *Argus*. All in all, it was a spectacular, intriguing cavern, and one soon entered.

We do not know who first ventured over the dismaying overhang at the bottom of the steep entrance slope. Perhaps it was twelve-year-old Rolth Sublett, lowered by rope in 1883 to investigate the portion of the huge entrance tunnel illuminated by the weird shaft of sunlight. In 1900 Sublett showed the entrance of the Big Cave to Carlsbad merchant Abijah Long—or so states an affidavit Sublett prepared years later. Writing shortly before his death in 1934, Long dated his own rather different account as 1903, the first year of contemporary written records on the cave.

In some ways fascinatingly parallel to that of Jim White and thus equally suspect, "Bije" Long began his story with an episode in which he and Sam Evans slid down a rope into a small nearby cave. Long emerged after a struggle. Evans couldn't make it and had to wait eighteen hours in darkness while another friend rode to town for more rope.

Bije Long stood by the pit, comforting his trapped friend as best he could through the long night. At dawn he witnessed a spectacle more awesome than the evening bat flight: the return of millions of bats to the Big Cave. No longer in the familiar stream of their evening egress, the dawn sky was dotted by the fluttering hordes. Spectacularly diving hundreds of feet directly into the great maw of the cave, the weird rushing whirrr of folded wings was clearly audible hundreds of feet away.

Abijah Long did not know that each bat is able to eat its own weight in insects nightly. Nor did he know—or care—that the free-tailed bat catches its prey in its tail membrane. Yet he was well aware that bats are a highly efficient means of converting insects into fertilizer. Guano miners already had staked a claim on nearby McKittrick Cave. To Bije Long, these torrents of bats suggested a fortune in guano. With friends, he soon returned to explore Bat Cave. The awsome, repellent chambers seemed to him filled with musty black gold.

In 1903 Long filed a well-documented claim on this Big Cave, this Bat Cave. The first of several fertilizer companies was incorporated. Jim White forsook cattle chousing to mine guano. The march of history was underway.

Shafts were drilled into the bat chambers. Ladders and a crude elevator were installed—the latter a two-man bucket jerkily raised and lowered by winch. In a quarter century some 100,000 tons of guano were shipped to Hawaii, the California citrus groves, and less exotic points. In places the cavern floor was lowered 50 feet.

None of the guano companies made much money. It appears that practical jokers contributed to several failures. Abijah Long lost an entire crew of Mexican workmen when someone rigged too convincing a ghost in the cave.

In their spare time, curious guano miners probed deeper and deeper along what is now the commercial route of the enthralling cavern. Amid misty uncertainties, two seemingly reliable accounts imply that Jim was a leader in explorations in search of additional guano. Perhaps he had briefly entered the cave as early as 1898—it really makes little difference. "For a long time I thought the King's Palace was as far as the cave went," Jim was recorded as saying in 1925. There, it appears, the enthusiasm of a chance visitor "who knew caves" triggered further explorations. "One day while back in there," Jim added, "we climbed up a steep hill into that big crack and follered it till it quit and there was the Big Room. We spent a night and part of two days wanderin' around over the Big Room, like bein' turned aloose out in a canyon pasture on one long, dark night with only oil torches for lights."

His great new discovery brought Jim increasing delight in the stupendous cavern. Soon he began a rough but readily traversable trail to the greatest splendors. Through the years occasional visitors came away singing the praises of the unexpected wonders of Jim's cave. When the guano business collapsed after World War I, Jim redoubled his efforts.

Slowly the countryside began to wonder if it had overscorned Jim. In 1922 thirteen prominent citizens toured the cave. It must have been an interesting experience. Two by two Jim lowered them in the guano bucket, more than 150 bottomless feet into the black vastness. None had heart failure, and their reports were glowing indeed. Superb photographs by Ray V. Davis created a sensation in Carlsbad. Though some momen-

This view shows the guano bucket at the surface. Shown are Jim White (right) and Russell Trall Neville. *Photo courtesy Mrs. Burton Faust*

tarily claimed the photos fakes, the pendulum swung. At long last the people of Carlsbad overwhelmed Jim. This Bat Cave, this Big Cave suddenly was Carlsbad's cavern, its pride and joy. Jim began to feed his family on guide fees. Enthused visitors began to write faraway Washington that Jim's cavern ought to be a national monument or park. The General Land Office dispatched hardheaded mineral examiner Robert A. Holley, "directing an examination, survey, and report on the feasibility of securing the Carlsbad Cave as a national monument."

For years Jim White gleefully related how he just grinned when the

Because of the overhanging dropoff beneath the main entrance, early visitors had to ride up and down the guano bucket elevator or use rickety ladders 170 feet long. *Photo by Willis T. Lee, courtesy Dana W. Lee*

government man talked of a quick mapping trip. The first eight-hour tour convinced the skeptical Holley. In inspired language which must have startled some of his superiors, he strongly recommended that Jim White's cave be preserved as a national monument.

Others took up the cry, notably El Paso attorney Richard Burges, who knew how to get action in Washington. Perhaps Burges had a part in diverting geologist Willis T. Lee of the United States Geological Survey to the cave from nearby studies. Lee, too, was flabbergasted by Jim White's cave.

Had I been told before entering [the Big Room] that an open space of such great dimensions was to be found underground, I should have doubted my informant's word as frankly as many of my readers probably will doubt mine. Some visitors who claim familiarity with noted caves assert that Carlsbad Cavern surpasses all others in size and in the beauty of its decorations. It seems probable that this claim may be substantiated when an adequate survey and extended examination are made.

Hardly had Lee returned to Washington when President Coolidge proclaimed the establishment of the national monument.

Willis Lee soon was able to make his "adequate survey and extended examination"—a six-month study of the great cavern sponsored by the National Geographic Society. His two superbly illustrated articles in that society's popular magazine brought the glory of the cave into tens of thousands of American parlors. Twenty-three miles of surveys were completed in the tremendous cavern, several miles in the Big Room alone. Exploration struck out into regions where Jim White could not go without a strong team.

The national monument was enlarged repeatedly (at first it excluded the Bat Cave section). Soon Carlsbad Cavern was found to be the deepest cave in the United States, and kept that record for 30 years. Its Big Room was found to have an area of 14 acres, with a ceiling height of 255 feet at one great dome. T-shaped, this single room measures 1,800 feet in one direction and 1,100 in another. Few other cavern chambers anywhere in the world can be compared to it. Eventually, a stairway was constructed down the main entrance. Improved paths, electric lighting, and elevators were installed to accommodate increasing hundreds of thousands of visitors. As every cave-wise American knows, Jim White died several years ago, but his dream marches on.

The Jim White story is vastly more than legend, and the legends began long before the coming of Frank Ernest Nicholson. Except for a few years as chief guide late in the 1920s, Jim always was desperately poor. Throughout his life he seems to have overcompensated for insecurity, and his last years were marred by partially justified bitterness against the government. In the initial burst of nationwide excitement over the matchless cavern, the government and other authorities did not even recognize Jim as an official guide. Obviously he was not of the breed that authority is likely to take unto itself. Without mentioning Jim's name, Willis Lee recorded:

Like other guides before him, he has discovered that tourists appreciate hair-raising yarns. . . . According to his own statement, our guide does not allow dull fact to interfere with a good story. Possibly there will in time appear here a sequel to the volume entitled "Truthful Lies" which will embody the strange characteristics of the Southwest.

And, like other guides, Jim in part simply got stuck with some whoppers when credulous outlanders took everything he said as gospel.

In 1930 park ranger Dixon Freeland attempted to pinpoint the gathering legends. Freeland attributed the "colorful picture of the lone cowboy, the desolate plains and hills, and the cloud of bats" to a 1923 speech of Colonel Étienne de Phillesier Bujac, "an astute Carlsbad citizen and friend of Jim White's." Not until then, Freeland asserted, did anyone claim Jim as discoverer of the cave as well as its first great booster.

Freeland, however, seems to have been overpositive. A 1925 version of the story differs greatly from the Jim White legend we know today. Clearly popular acceptance was slow, for in 1927 a *Nature Magazine* article still urged appropriate recognition of then-overlooked Jim White and his fellow "humble guano miner," the forgotten Dave Mitchell. At least one 1930 document indicates that some then still considered Abijah Long the "real" discoverer of Carlsbad Cavern.

Then came Nicholson.

Yet Nicholson's obvious distortions downgrade the accomplishments of Jim White no more than they downgrade this, our greatest cave itself. Certainly the exploration of Carlsbad Cavern was not the one-man show of legend. Jim White was not the superman-idealist of the hero tales, but it is to his perseverance, determination, and able promotion that we owe our primary thanks. Which is what Nicholson declaimed, even if it was only to pay his board bill.

It is a fitting measure of the magnificence of Carlsbad Cavern, too, that these needless distortions have not tarnished its matchless glory. Here, the visitor's concern is properly the awesome spaciousness, the intricate beauty, the colossal stalagmites: Giant Dome, Rock of Ages, Twin Domes, and much more along the 2¾-mile tourist trail.

Cavers know these wonders, and the different splendors of the underworld wilderness elsewhere in the superlative cavern. Here, without gum wrappers and the pressure of crowds, we glory in the delights of the Christmas Tree Room, the delicate magnificence of the New Mexico Room, the vastness of the newly discovered Guadalupe Room, even if we have set foot there only vicariously. Here man realizes his insignificance and that of every member of the human race.

Several American caves have more miles of passage than Carlsbad Cavern. At least three are deeper. Some may be more beautiful: to each

Thought to be the largest stalagmite in the United States, this speleothem of New Cave dwarfs Superintendent Boles, Jim White, and others near its base. *National Park Service photo*

his own taste. Nearby New Cave—the second of the national park's hundreds of caves to be opened to the public—has at least one column larger than any in Carlsbad Cavern itself. Somewhere there may be a cavern chamber vaster than the Big Room. Nevertheless, no person has lived until he has known its timeless majesty. Few caverns in the world are worthy of comparison with it. No caver knows caves until he has viewed its myriad features. Carlsbad Cavern stands alone.

7

Beneath a Thirsty Land

The Story of Texas Caves

Beyond the hair-raising ledge opened an enchanted wonderland. Deep beneath this thirsty land, a passing touch of lime-charged waters had wrought a crystal magic.

Once this had been just one more west Texas cave, not unusual in its thickets of graceful stalagmites. Then, in the eternal night, the silent curtain of the waters baptized them with a mystic spell. When it withdrew, sparkling forests of miniature crystalloid Christmas trees replaced the stalagmites. Here and there in the silent blackness, low-dipped stalactites sprouted newly intricate tassels. The awe-struck explorers' carbide lights revealed the very walls vanished beneath glittering banks of coral-like calcite projections.

As the dazzled spelunkers pushed deeper beneath the mesquite-studded hillside, their bulging eyes encountered still greater splendors. Glassy helictites glistened, like crazily contorted crystal worms suddenly frozen in place, their hair-thin nutrient canals visible throughout their sinuous courses—just a few, then writhed masses covering entire chambers so thickly that the limestone bedrock could not be seen.

Now and again, crystalline wings hung folded from the bellies of weird helices. Glass-clear tomahawks, giant fishhooks—inconceivable splendor piled glory on glory as the young adventurers tiptoed forward in the silent, brittle corridors. Here was a 6-inch butterfly with folded wings of palest yellow. And here another stone lepidopter, crystal-white and widespread as if sunning itself in this sunless realm. And, at the farthest point of the main corridor, a low-level chamber embraced tall,

Heligmites of Cavern of Sonora. *Photo by James Papadakis*

writhing heligmites, sprouting wildly from the floor like stone snakes halted in mid-dance.

This incredible cave was no new discovery. More than fifty years earlier, a sheepherder spotted a manhole-like orifice on the Mayfield Ranch. Occasional parties of curious folks probed its depths. They didn't find much—one spacious chamber—if you could locate it again—two major corridors, reuniting at a vicious-looking pit. A few small stalactites and stalagmites and some popcorn-like deposits, but that was about all. Much finer caves could be enjoyed nearby beneath this thirsty land. Mayfield's Cave didn't amount to much.

In the patient, undramatic way of cave surveys, Bob and Bart Crisman came to Mayfield's Cave in mid-1955. Like dozens before them they halted at the jagged pit a quarter-mile inside.

Bob and Bart, however, were a different breed. From their perch they could see orifices 50 feet away on the hackly far wall. A mere descent into the pit was bad enough. The far wall looked impossible. Any cave beyond would be virgin.

The Crismans were engaged in a routine, lightly equipped scouting venture. The intriguing, overhanging 60-foot rear wall would not yield easily. Perhaps a narrow ledge, high on the right-hand wall . . .

But time was running out. Half a day somehow had passed, and Mayfield's Cave was not even the main target that weekend. Reluctantly they left. Today's tourist crowds, safely conducted on a sturdy bridge, exclaim at the vicious spectacle which confronted Bob and Bart.

The Crismans reported their hopes to the Dallas Grotto of the National Speleological Society. While Bob and Bart pursued investigations not many miles away, the Labor Day weekend of 1955 found the grottoites present in force. For half a day they struggled, unsuccessfully attempting to scale the tantalizing wall. Finally they sought a new viewpoint: the alternate corridor which reaches the pit at roof level.

From this aerie, the passages beyond appeared even more inviting, 50 empty feet away. But what a void below! The pit looked even deeper, blacker, more menacing. Well they knew what rocks lay concealed in the dark shadow, 60 deadly feet below.

But perhaps . . . A ledge ran halfway along the east wall, high up near the ceiling. Dirt-covered, it sloped alarmingly pitward. No wider than a foot or two, it disappeared around a beaklike projection halfway across the pit. Perhaps it ended there. No one could tell. But the intriguing openings lay not far past.

His emotions deeply stirred, Jack Prince rose to the challenge. Carefully balancing along the rim of the shadowy void, he inched onward to the overhanging prow. With relief he called back that the touchy ledge continued around the bend. Tautly he inched onward.

The ledge narrowed and slanted upward, but an opening was just ahead. Cautiously advancing one foot, then the other, he inched into reach of a sturdy stalagmite. Jack tested it. It seemed firm. He swung up —into paradise.

Today, a decade later, flood lamps reveal much of the glistening beauty Jack's carbide lamp first illuminated that unforgettable September day. Cavers and tourists alike come away shaking their heads at the matchless splendor of this Cavern of Sonora.

In the Cavern of Sonora the work of underground water gave Texas perhaps the world's greatest display of underground beauty. The very hollowing of the rock, the slow drip that formed its ancient stalagmites, the coralloidal glories, the calcite film that forced the incomparable helictites into being, the helictites themselves, and now a little dripstone and flowstone: all exist because of subterranean water.

Yet the glory of the Cavern of Sonora is perhaps the least part of the story of water beneath this thirsty land.

Sometime late in the nineteenth century, ranch hands spoke furtively of a volcano erupting nightly from the bone-dry plain. On May 21, 1876, Ammon Billings found its source while tracking marauding Indians: "a

The Ledge in Cavern of Sonora. *Photo by James Papadakis*

helluva hole in the ground" with a million bats tucked away deep below. The profanity was perhaps justified: a careless hand could have ridden his horse into it before he saw it. Right plunk out in the God-forsaken mesquite flats, it opened freely downward from a deeply undercut hole. Sixty feet across, the shaft spread tremendously, then shot straight down. At midday a beam of sunlight illuminated an earthy patch far down in the purple shadows. How far? Hundreds of feet maybe. To mightily impressed Anglos it was exciting enough. As recently as 1960 an Associated Press dispatch termed it "a 180-foot deep crater left years ago by a falling meteor," which it isn't. Among the *paisanos* arose tales of ghosts and devils and hell-fire and the mother den of all rattlesnakes. It indeed was "a helluva hole."

But Lucinda Billings, Ammon's bride of a year, was firm. "The Devil's Sinkhole" was the most vigorous name she would allow.

Someone saw the huge shaft as a natural freeway to the elusive water which underlies these wind-swept limestone plateaus. Where there was water, ranches prospered. Whole cities were springing up along the Balcones Fault, which brings this water to the surface: Austin, New Braunfels, San Antonio. Their lifeblood was the liquid gold disgorged from the limestone which swallowed whole rivers farther upstate. But here even goats died. The water lay too deep for contemporary well-drilling tools.

One wall of the great pit was not quite vertical. A now-forgotten rancher blasted a hole through the overhang of its southern lip. A short ladder brought him to the top of a slope passable with the aid of a lariat. Kicking loose rocks booming into the shadowy depths, he half slid from ledge to ledge. Zigzagging back and forth, he progressed deeper and

The Butterfly, Cavern of Sonora. *Photo by James Papadakis*

deeper into a shadowed realm, 50 feet, 100 feet in the wide, rounded shaft. It seemed much more.

As the bluish daylight dimmed, the brave venturer began to see that the airy shaft was merely the smaller part of a greater void below. Hidden from vantage points on the rim, a colossal chamber arched away from the incredible pit. The flat bottom he sought was in truth merely the top of a gargantuan underground mountain centered in a cavern so huge that the whole Alamo could have been tucked into just one corner. His loyal friends on the rim high above suddenly seemed antlike and terribly far away. Abruptly the sinkhole seemed sweltering, stifling, stinking of guano, and plagued with previously unnoticed bugs.

That first explorer almost certainly did not reach his goal on the initial attempt. Looking much longer than its actual 40 feet, a terrifyingly overhanging final dropoff was no simple obstacle. Yet the persevering rancher triumphed. At the foot of the underground mountain, 303 feet below the sunbaked plain, an emerald pool gleamed far back in an obscure alcove. In some herculean manner, the rancher extended a water pipe into these black depths. The windmill which drew it to his thirsty stock was mere routine. Spelunkers now rappelling into the suddenly lonely vastness of the Devil's Sinkhole marvel at his endeavor. But water means much in such a land.

This remarkable pioneer conquest of the Devil's Sinkhole was not man's first utilitarian venture into a Texas cave. Spanish Well, a narrow 120-foot pit cave north of Austin bears an eighteenth-century date, a Maltese cross, and an ancient arrow: common markings of the conquistadors.

In later times enormous cattle drives were watered at "09 Water Well," strategically located on a bone-dry divide between the Concho and Devil's rivers. The Case Cattle Company is said to have reaped a small fortune from its cavernous depths. Not even the greatest drives of Kansas City–bound cattle could drink dry its daily flow, pumped from 127 feet underground. For many years, a stage line stopped here daily, its dusty passengers hardly dreaming that a later generation would explore more than a mile of stream passages deep below.

Attempting to reshape his shattered life, Frank Ernest Nicholson turned to the cave regions of his native Texas. Clearly he had learned his lesson well—too well, in fact, in the minds of many later speleologists. Perhaps because his later Texas-size cave yarns got out of control, speleohistorians have not yet traced his entire spelean trail. His spoor appears at the Devil's Sinkhole, whence he reported blind fish and running streams which have eluded all later cavers. In time he came upon Hester's Cave, near Boerne in the hill country.

Nicholson was far from the first to scan the walls of Hester's Cave.

For generations local sweethearts had journeyed underground with lanterns to view a stygian pool 500 feet from the entrance. Or so recounts an old brochure in the San Antonio Public Library.

In 1932, Nicholson concluded that Hester's Cave ought not to end at that pool. To him, the corridor merely appeared blocked by low-hanging drapes which touched its surface. Perhaps his thoughts flashed back to

The awe-inspiring surfaceward view from the top of the great rock mountain of the Devil's Sinkhole. *Photo by Bill Helmer, courtesy of Mills Tandy*

the "cave explorers' club of France." Perhaps he consciously hoped to emulate the great Norbert Casteret, who dove blindly beneath a seeming barrier to spelean glory.

"My assistant and I decided to find out," Nicholson wrote. "We stripped off our outer clothing and waded into the icy, black waters. The floor was irregular and we gyrated considerably as each step led us into deeper water. Yet we traveled cautiously and sure-footedly as there was the possibility of a jump-off."

For 50 feet Nicholson ducked amid stubby, coarse stalactites, floundering in the watery stoopway. Flashlights sealed in fruit jars provided weird, cumbersome illumination. Nicholson opined the distance to open cave as 200 feet. It probably seemed even farther as the unearthly minutes wore on. The water lapped their chins, their noses. Then, all at once, pitch-black space opened ahead.

The little chamber beyond this watery wallow, however, hardly seemed worth while. A few delicate terraces, some minor flowstone—that seemed all. But close ahead was a thicket of delicate white soda-straw stalactites, reflecting these first lights like diamonds. A high-vaulted passage with huge ceiling potholes was punctuated by a splendid group of stalactites. Just around a bend to the left was a sonorous cathedral-like chamber more than 200 feet long. Below its altar rock a pit led the explorers to a fine lower level with other large chambers.

Nicholson was beside himself. He had done it—the American Casteret! He promptly bought the property, drained the lake, and soon opened his discovery to the public with great fanfare under the name Cascade Cavern. This time he conformed, exaggerating dramatically in contemporary journalese. Again newspapers across America published a widely syndicated Nicholson article. This one included "blind albino frogs with translucent bodies, pale grey bats and milk-white crickets . . . queer blind fish . . . Spanish oak and hackberry trees defying the laws of nature, more than a mile from daylight . . ."

It was too much. Despite an honored place in the creation and popularizing of Longhorn Cavern State Park five years later, Nicholson vanished into obscurity, tacitly banished from the company of the founders of modern American speleology. A generation later, some of us wonder if they were too harsh. Despite vandalism and a lack of speleothemic glory, Cascade Cavern subsequently became a popular attraction. Though never yet found in caves by anyone else, two Texas blind fish are now known from artesian wells near Cascade Cavern, and in 1946 a salamander new to science was identified in that cave. Some of his tales were not entirely Texas whoppers.

To those intrigued by the weird life of extensive cave areas, the crowning jewel of the cave waters of Texas lies eighty feet beneath the sun-parched town of San Marcos. Here alert speleobiologists pinpointed

The incredible Texas cave salamander, one of the threatened life forms of Ezell's Cave. *Photo by Robert Mitchell*

an urgent need. Paying off the resulting deficit in its funds as contributions accumulate, the Nature Conservancy in 1967 headed a coalition that purchased, rehabilitated, and today zealously guards the biological treasure house that is Ezell's Cave. Crystal-clear waters of this remarkable little cave have inspired perhaps the most compelling appeal of all time:

. . . suddenly, floating somewhere through this eerie film [of water], a transparent shrimp, seemingly in thin air, passes the flashlight beam. Then another, and another, drift effortlessly behind the first, their inch-long antennae waving ahead of their inch-long bodies. . . . Master of the cave is the famous Texas Blind Salamander, iridescently white, delicate, toothpick-legged, red-gilled, eye-remnanted, about four inches long. This creature is one of but two such highly specialized salamander species in North America, a living monument to the unusual modifications required of vertebrates that live entirely in caves. Its life, its very existence, depends on the unique invertebrates and peculiar environments of Ezell's Cave . . . a unique biological phenomenon, worthy of stringent protection. . . . In the small world of Ezell's Cave, thirty-six or more species are known. Of these, ten are aquatic, including six which are known to occur only here. . . . Only a unique series of geological, zoological, and human events has produced and spared this rarest of rare ecosystems, Ezell's Cave, an emerald cameo of life. You can help preserve this fauna.

Texas caves are not unanimously popular with the citizenry. By some colossal miscalculation, Medina Lake Reservoir was designed precisely across one of the most cavernous belts of Texas. Local cavers aver that in its eighty years of quasi-existence, the reservoir has filled twice—briefly. Bedding planes and joints downstream from the dam spurt water as new caves grow around the barrier. Great springs two to three feet in diameter announce caves washed open by the diverted waters. A half-mile difference in the dam site would have avoided the entire problem, but engineers are only beginning to listen to speleologists. "That reservoir usually doesn't slow the river up at all," Texas cavers say, wondering if the same fate will befall the controversial Amistad Dam Reservoir on the Rio Grande River, which destroyed several fine caves.

As thoughtful conservationists, however, Texas cavers do not oppose all dams or damn all engineers. The dam that floods Indian Creek Cave meets their total approbation.

Except during flash floods, Indian Creek is a dry, cactus-studded arroyo. Once or twice a year after heavy west Texas storms, tremendous gully-whoppers thunder along its deep-gouged banks. Local ranchers long knew that part of these floodwaters eventually sank to the water table, to emerge at Uvalde, or San Antonio, or elsewhere on the Balcones Fault. But much of the water ran on into the Gulf of Mexico, or evaporated.

On Fred Mason's ranch in the hills above Uvalde, there was a small crack in the bed of Indian Creek. Hunters had noted that it steamed in winter. Fred wasn't really a spelunker, but he was kind of curious about caves and often had ranch hands at work, excavating some blocked entrance.

"Three months two Mexicans dug at that crack in 1955," a wondering neighbor recounted years later. "Got down quite a piece, too, with crowbars and sledgehammers. Found one little room. Then they wouldn't dig no more. Seemed like there was ghosts down below. When they hammered at the rocks, somebody way down below hammered, too!"

To Fred Mason, the "ghosts" sounded like rocks falling into a deep hole. Anchored by safety ropes, his Mexican crew returned to work. Not long thereafter, the entire floor roared 70 feet straight down, leaving the diggers spinning in mid-air. Their *compadres* hauled them up, built ladders, and found an extensive cave 120 feet below Indian Creek. "Looked like a lot more cave was blocked by clay fill, too," the neighbors insist.

In 1956 Uvalde County was searching for methods of increasing its water supply—ways to divert water underground where it wouldn't be lost. Indian Creek Cave seemed ideally located. The county spent $50,000 constructing an 8-foot dam across the 225-foot dry gulch.

In May 1957 an evening flash flood filled the gully behind the dam. All night the limestone vibrated. The roar of plunging water was audible half a mile away. By dawn 250 acre-feet of liquid gold had vanished underground, and the stream bed was dry.

As an extra dividend, that and later floods washed away much clay in the cave, opening vast sections to exploration. Bill Russell, Mike Pfeiffer, and Bob Benfer first came upon it in February 1960. Within two years 17,623 feet of passages had been mapped, and Indian Creek Cave became Texas' longest. For affable Fred Mason the cavers conducted a special survey. "Drill here," they told him. They missed the main cave stream by two feet, but a single charge of dynamite brought plenty of water for thirsty stock.

These water caves of Texas are not always friendly. In H. T. Meiers' Cave, fate once hovered dangerously close to some excellent cavers.

University of Texas biologist Floyd Potter unsuccessfully sought blind salamanders here, then filed his data with the Texas Speleological Survey. With Tom Evans and Roger Sorrels, Bill Russell promptly headed for the new cave. "It turned out to be more difficult than we had anticipated," the sheepish trio later confessed. Not until now has the real story appeared in print.

The usual pre-exploration chat with Mr. Meiers confirmed the cavers' impression: just a little cave, mostly a single medium-sized chamber with lots of breakdown. Parking their car along a road one hundred yards from that cave, they vanished into the netherworld about 8 A.M. one fine morning.

Texas cavers pride themselves upon their friendly relations with ranchers. The pleasant Mr. Meiers was no exception—he was looking forward to the "University boys' " report on his cave. What in the little cave could be keeping them so long? Their car was still there at noon. And at 6 P.M.

By nightfall Mr. Meiers had begun to worry about his boys. By midnight, every nearby rancher was wondering what to do. Finally someone summoned the Del Rio Fire Department from thirty miles away.

Around 4 A.M. the young explorers were bone-tired but wonderfully elated. Deep amidst the breakdown they had found a hole 15 feet deep. Twenty feet onward was another ladder pit, twice as deep. Just 15 feet farther was a third pit, 26 feet deep. That seemed to end their discoveries. Obviously, floodwater coursing into the entrance poured still deeper, but the fissures it followed below the third pit would not admit a caver.

But on one wall, a jagged little hole was just barely caver-sized. Curving upward into a 15-foot chimney, it wasn't particularly promising. Without much enthusiasm, Bill Russell wriggled to its summit, then stared and shouted for his companions. Breathing hard, from the ascent, they too stared—down into a pit-like chamber 70 feet deep. Below, they came upon additional pits and a thousand feet of walkable passage.

The delightful cave went on and on and on. The tired trio were jubilant as they neared the surface after twenty hours. But what was that strange groaning r-r-r? Almost like an electric generator . . . Silly thought; let's get the heck out of here. We must be tireder than we figured . . . twenty hours . . . Is that dawn already? Sure looks funny . . . MY GAWD!

Blinking dazedly, the trio emerged into the glare of floodlights. A major rescue operation was fast shaping up.

Red-faced and stammering apologies, the young Texans braced themselves for a verbal storm that never came. The rescuers were too delighted to be angry—delighted to see the remorse-stricken explorers safe and well, delighted that they would not have to venture underground after the "lost" spelunkers.

Overwhelmed, the trio overdid their apologies. Emphatically they insisted that no matter how long they were in his cave in the future, Mr. Meiers was not to worry. In such a cave, expert cavers could care for themselves better than untrained rescuers.

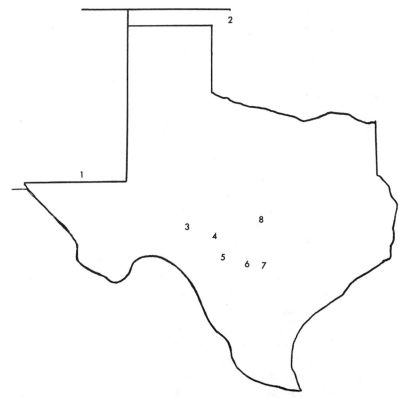

Spelunker's map of Texas and vicinity. (1) Carlsbad Caverns; (2) Alabaster Cavern; (3) Cavern of Sonora; (4) Devil's Sinkhole; (5) Indian Creek Cave; (6) Cascade and Century caverns; (7) Ezell's Cave, Bracken Bat Cave, Wonder Cave, and Natural Bridge Cavern; (8) Longhorn and Cobb caverns

They convinced him a little too well.

Expedition after expedition returned to H. T. Meiers' Cave. Lengthy, tiring trips revealed more and more to be reported to the genial rancher. But as cavers became more familiar with the remote recesses of this surprising cavern, they became more cautious. Flood waters had wedged sticks and branches against the ceiling of the highest parts of the cave, 100 feet above the floor. "There would be little refuge if you were caught in the cave during a six- to eight-inch rain," Tom Evans concluded in the *Texas Caver*.

In west Texas, six- to eight-inch thunderstorms blow up out of nowhere within a few hours. The weather was delightful when Jim Reddell, Terry Raines, David McKenzie, and Sharon Wiggins entered H. T. Meiers' Cave. Many hours later as Jim re-ascended the 70-foot ladder, he heard an ominous roar of a rushing torrent.

Aghast, he rushed to the top of the discovery chimney. Fifteen feet below, a ferocious boil of water rose perceptibly as he stared—already almost a siphon. When it was completely siphoned, no one could force his way through such a current.

This was the real thing. The cave would soon fill completely. Charging back to the ladder, Jim shouted the ominous news to his companions. "Get up as quick as you can! No belays! No time! QUICK!"

Those ladder climbs may have been the world's fastest. Abandoning their equipment, the quartet plunged into the funneling water. The torrent dragged mercilessly at their straining bodies, but each burst through into the wider cave beyond. Perhaps their haste was needless. Jim later concluded that there had been a minute or so to spare.

"After that other time, Mr. Meiers wouldn't have come to help for a week." Jim grins lopsidedly. But Texas cavers now are extra-careful to leave detailed instructions with some reliable person topside.

Nor is H. T. Meiers' Cave the only Texas water cave which has troubled spelunkers. The senior generation of modern Texas cavers fondly recalls the 1956 skin-diving trip to Devil's Sinkhole, but at the time some wondered if it was worth it. Billed as the biggest cave exploration project in the history of southwestern speleology, it is still recalled by Bill Helmer as also the most fouled-up venture in said history.

"We were going to confirm or disprove, once and for all," he recently avowed in the *Texas Caver,* "the rumors that another part of the cave lay beyond the lake room, and we were going to do this by sending down scuba divers. . . . The cave would be lighted by a large generator, step-up and step-down pole-type transformers, high-voltage lines running into the cave, 110-volt lines running down to the lake room, a wooden diving platform built out over the lake; a suspended cable system for winching heavy air tanks down to the lake room entrance, field telephones and provisions for about 20 people for about four days."

The Inevitable Law of Inverse Perversity inexorably went into action. With the 1 A.M. temperature at 20 degrees and the wind velocity 40 m.p.h., the big expedition tent blew down as soon as it was erected. Everyone finally took refuge beneath the creaking, overloaded expedition truck. Matters were no better in daylight. All humans were safely lowered by a pulley system using a heavy manila rope tied to an axle of Lynn Allman's car. An icebox didn't fare quite so well, annihilating itself and several bags of groceries after 150 feet of free fall. A hundred-pound transformer was next to go ahead of schedule, with a hundred feet of slack rope still between it and the car. Helmer happened to see it all from the base of the huge underground mountain:

"When people on the mountain top heard all the yelling and looked up, and saw that big, black object getting too big and too close too fast, they looked like the high divers at Acapulco. . . . When the rope came up taut it stretched like it was made of nylon, then plucked that trans-

Great quantities of gear were lowered—or dropped—into the Devil's Sinkhole in an attempt to learn what might lie beyond the lake. *Photo by Bill Helmer*

former back into the air like a yo-yo. Allman's car was jerked backward about ten feet but the rope didn't break and nobody got killed. Awhile later a 10-pound rock got knocked loose at the entrance and shattered in the middle of the people on the mountain top, but by this time it was just another in-coming mortar shell that barely interrupted the conversation."

In due or undue course, the divers proceeded into the black, chill water. Apparently entering into the spirit of the junket, the prized kilowatt generator broke down. All underwater exploration had to be done with jerry-rigged lights. These worked well enough for the divers to work their way through unstable boulders and piles of breakdown to a depth of 48 feet. As for additional caves beyond, however, only one diver groped his way upward into a great black air-filled void. "He looked like Christopher Columbus about to plant a flag until he saw me standing there," wrote Helmer, who happened to be in a remote part of the lake room engaged in a vital function which should have been restricted to sanitary facilities. After this great 1956 expedition, Nicholson's conquest of the Devil's Sinkhole without modern techniques and equipment somehow looms more impressive.

No matter how important are the cave waters of Texas to its thirsty economy, American speleology tends to think equally of its great bat caves. Here it was that Ed Bryan was buried alive in guano. It seems that Don Widener had a bit of difficulty, teetering precariously on a toehold of lumpy guano. On the ledge to which he was climbing were unmistakably fresh traces of a large bobcat. Well up a touchy slope in a funnel-shaped room of Snelling's Cave, Don's tenuous perch suddenly seemed remarkably undesirable. Texas bobcats are notably irritable when cornered in caves.

Don shifted his balance to retreat. It was enough—too much. His toehold collapsed. With a ludicrous howl, he and a roomful of guano began to gain momentum.

It didn't seem funny to Wally Martin and Al Malone, farther down the slope. They dove aside as the part-human slide whizzed past into the chute of the natural funnel. Gray clouds of horrid dust mushroomed upward from the corridor below.

Hearts in mouth, Al and Wally sped through the choking mist. Their carbide lights revealed Don's posterior, so violently agitated that their fears lessened. A hearty pull on his belt brought Don unsteadily to his feet, glasses broken, a sizable lump on his cranium but otherwise none the worse.

But where was Ed Bryan? While his friends explored above, he had stationed himself at the bottom of the chute.

The cavers stared at each other, momentarily paralyzed with new

The short-lived diving project gets underway. See the text for their finding. *Photo by Bill Helmer*

concern. A plaintively muffled cry and a waving arm emerged from one edge of the gray, musty pile. Moments later Ed too was free, spitting mightily but unharmed. Even Dante had not considered such a fate.

Today guano is primarily regarded as a noisome nuisance—one more obstacle for cavers to overcome. Speleobiologists demur, considering it a base of cave ecology, but they are a tiny minority. Several promotional campaigns have failed to create the desired image of guano as a super-fertilizer. Perhaps they were merely overdone. Bat guano is a good fertilizer, yet commercial guano operations in the United States are marginal at best.

It was not always so. The pattern of settlement of Texas required little homemade gunpowder, but the niter content of its great bat caves was known at least as early as 1856. Someone built a fire in Blowout Cave to smoke a bear from its guano-rich depths. The guano exploded and burned for two years. The fate of the bear and the hunter were not recorded. In 1863, Texas' Confederate ports were effectively blockaded by the Union fleet. Great quantities of guano were hauled from its bat caves to produce saltpeter for Confederate armies of the West. Frio Cave alone yielded one thousand tons. Oxcarts were used wherever possible, burro pack in more difficult locations. Working in fetid, ammoniacal caverns by day, sometimes standing tense watch for nightly marauding Apaches, many a guano miner must have longed for the pleasant battlefront.

After the Civil War, lesser quantities of Texas guano were bagged for fertilizer—perhaps not collected even as fast as it was deposited. During World War I, when chemical fertilizers were suddenly unavailable, the demand for guano skyrocketed. A Texas Bat Cave Owners' Association was created and about two million pounds were shipped out of the state. Several enterprising Texans constructed comfortable cupola-like bat roosts—initially to control Texas' voracious malarial mosquitoes—and started growing their own guano.

The colossal guano deposit within Carlsbad Cavern is dwarfed by several in Texas. Fifty-foot depths of guano are mute witness of incredible numbers of bats over untold millennia. In some regions, moister postglacial days supported more insects than today's relative aridity. Those insects in turn supported a greater bat population. Bats convert bugs to fertilizer very efficiently, yet each contributes less than an ounce of guano weekly.

Nevertheless, even with today's populations of "only" a few million bats, the great bat caves of Texas are among the world's most awesome spectacles. In 1957 my family and I stood amid the massive bat flight of Frio Cave, whose gigantic entrance room would engulf a football field. As we left the evening sunlight, the musty smell of guano welled toward us. Somewhere near at hand whispered the flutter of bat wings. Already a few dozen advance guards were circling the lower part of the entrance room.

Most plunged back into the pitch blackness of an inner chamber, but a few broke off from the main pattern in swirling vortices.

We crossed to the mouths of the lower chambers, marveling at the growing circle of whirling bats. A strangely distant, hushed roar began to build and echo.

Looking back through the gaping entrance, we saw sunset colors fading from the clouds. As we watched, a small stream of bats left the revolving circle, flowed along the right-hand wall, cut back to the entrance, and swarmed out into the dusk. But that initial burst was infinitesimal. An incredible cloud of bats was erupting out of the blackness below.

We were out of the main pattern of flight. Nevertheless, bats were everywhere around us. Some were circling back into the roost chambers. Others emerged in constantly renewed swarms. Circles formed within circles, then broke off into constantly changing patterns. The roar deepened, like rapids in a narrow gorge. Other sounds broke through the overall roar. The bats were chattering, squeaking. Often they flapped lightly against the walls or against one another as their individual sonar was lost amid that of thousands of others. No longer were they silent animals of mystery.

The entire entrance chamber had come alive with a flickering, streaming torrent of circling bats. Increasingly they dared the outside world. Small bursts, then waves whipped from the cavern. Against the sunset

The bat colony of Frio Cave, composed of Mexican free-tailed (guano) bats. *Photo courtesy United States Public Health Service*

sky they formed a tornado-like stream five to ten feet in diameter. As we watched, their hordes held closely to a single, wavy, undulating course: a gigantic whipping rope. We could see the black column far past the point where any individual flutter could be detected. Perhaps it was a matter of miles before it dwindled wholly from our vision.

Still the roar welled up from the innards of the cave. The stream of bats seemed endless. For half an hour the main flight poured forth into the increasing dusk, its volume undiminished. I have no idea how many bats we saw. Experts have estimated the population of this single cave at many million. We could not disagree.

Gradually we could see that the stream circling the vast chamber was lessening. Then, abruptly, the room seemed almost empty. It still held hundreds of fluttering bats, but the great swirls were gone. In the dusk the massive undulating rope was replaced by individual wings.

Why were we so interested in bats? Much of mankind dreads the little creatures—but cavers are not ordinary people. Perhaps knowing them better than anyone else, we are the first thin line of defense against the ignorant who periodically urge their extermination. We know, as did Mark Twain, that they are not the malign creatures of *Tom Sawyer* and many a wilder novel. Occasionally we find ourselves inundated with light-dazzled or confused swarms. In such emergencies we respect the bats' problems and find that they respect ours.

Fear of bats is cultural, not innate. Many Chinese consider them a symbol of good luck. Nor, despite the batwings of the devil in many a lurid old engraving, is bat panic inherent in our Judeo-Christian culture. The ancient laws that became Deuteronomy and Leviticus merely classed them as unclean: unsuitable for food. Before they were five years old, our children distinguished *Myotes* from *Plecotuses*. With glee, they upset the fixed ideas of several elementary schoolteachers as Mark Twain did his mother:

[Aunt Patsy] was always cold toward bats, too, and could not bear them and yet I think a bat is as friendly a bird [!] as there is. My mother was Aunt Patsy's sister and had the same wild superstitions. A bat is beautifully soft and silky; I do not know any creature that is pleasanter to the touch or is more grateful for caressings, if offered in the right spirit. I know all about these Coleoptera [!!] because our great cave, three miles below Hannibal, was multitudinously stocked with them, and often I brought them home to amuse my mother with. It was easy to manage it if it was a school day, because then I had ostensibly been to school, and hadn't any bats. She was not a suspicious person, but full of trust and confidence; and when I said, "There's something in my coat pocket for you," she would put her hand in. But she always took it out, again, herself; I didn't have to tell her. It was remarkable, the way she couldn't learn to like private bats. The more experience she had, the more she could not change her views.

Terror of bats probably stems from two sources: the alienness of something flying that is not a bird, and the vampire legend. The latter is probably more important. Today's version of the tale relates that the vampire can change at will from a bat to a man, both living on human blood. Centuries, perhaps thousands of years old, the basic lore is of central European origin. Not until the discovery of the American hemisphere, however, did bats enter the legend. The conquistadors found blood-lapping bats occupying a huge Spanish American range extending as far north as north-central Mexico. Wild animals and cattle are their usual target. Nevertheless, these small bats have been known to alight upon a sleeping man, painlessly gash him with razor-sharp teeth and lap the blood that wells from the cuts. Some natives of eastern Mexico insist that the telltale cuts are the bites of *brujas* (witches), but the Spaniards recognized their true origin. They took the tale back to Europe, and most of our civilization has feared the furry little beasts ever since.

It was inevitable that people should wonder how bats managed to steer through the total blackness of caves with such ease and accuracy. Early experimenters blinded bats and found that it did not hamper their navigation. Today humane hoods are used in similar experiments.

The next step was to plug the ears of bats. In a lighted room the bats were still able to fly with fair accuracy, but not so well as before. When both sight and hearing were blocked, however, the bats fluttered to the floor, helpless.

It was obvious that bats' ears served as substitute eyes. Modern electronic equipment revealed how. Like "silent" dog whistles, flying bats emit noises at a pitch so high that they cannot be heard by human ears. From their echoes, bats are able to judge the distance and shape of nearby objects, and perhaps their texture. This is not a phenomenon limited to bats. The cave-dwelling guacharo bird of South America has a similar system, and some blind men are able to develop it to an amazing degree through tapping of their canes. Nevertheless, it is an outstanding example of the adaptability of nature.

Some years before we visited Frio Cave, rabies—hydrophobia—was found in American bats. Not the vampire bats of the tropics, in which it had been known for years, but in our ordinary winged little fellow cavers. Though not nearly so common as rabid skunks or foxes, rabid bats promptly were found in nearly every state. In many states no one had investigated before, but the problem still seemed new and ominous. Cavers suddenly hesitated to approach our little friends.

Usually incurable once its terrible symptoms have begun, rabies is one of the few truly dread diseases. No American caver has died of the bite of a rabid bat—or from other rabid cave animals. Yet several noncavers have died after bat bites.

As we hiked from Frio Cave to our car beneath the fading Texas sunset, we little guessed that it was soon to be closed to the public. Eighteen months earlier, a Texas State Health Department entomologist had died of rabies after merely studying the bats of Frio Cave. No animal had even nipped him. A year after our visit, a mining engineer visited Frio and nearby caves of potential fertilizer value. He also contracted rabies without any bite.

There was little doubt of the circumstances. Both men were reliable observers. While still alert and rational, both denied being bitten. The entomologist "had probably rubbed a chronic skin eruption on his neck with contaminated gloves while working with bats in the cave," reported the Texas Speleological Survey. All that the engineer could recall, however, was being "nicked in the face" by a flying bat.

Cavers were hardly consoled with the thought that this tragic phenomenon was new to science. The United States Public Health Service immediately began to investigate. For once, the government did the right thing: Denny Constantine, D.V.M., was named chief of the Southwest Rabies Investigation Center. Denny is a graduate of the Southern California Grotto of the National Speleological Society as well as of leading universities.

Denny pondered the problem. The rabies virus is present in saliva—everyone knew that. But it was also present in the urine of sick bats. Could it be floating lethally in the urine-clouded atmosphere deep inside Frio Cave? Or could the deadly saliva from rabid bats "foaming at the mouth" also float free in the cavern atmosphere?

Denny placed special animal pens in one of the roost chambers of the vast cavern. After a week the carnivores were removed and kept in isolation. Of thirteen animals protected from bat bite in the pens, one fox, two coyotes and one ring-tailed "cat" promptly developed rabies.

Perhaps the rabies was spread to the test animals by cave insects or bat vermin. More animals were caged in Frio Cave, this time under a variety of conditions. Some were penned without protection from bats. Others were protected from bats but not from insects. Several were protected also from large insects. And two sets were protected from everything except the cavern atmosphere—theoretically, at least.

These animals were left in the cave for almost a month. This time every fox and coyote in each group developed rabies. Similar "control" animals penned under observation in Denny's laboratory continued healthy. So did those caged in other caves, and in shelters two miles from Frio Cave.

The very air of the great bat caves thus seemed able to transmit rabies to foxes and coyotes, and perhaps to man. Even though no guano miner is known to have developed rabies, cavers and biologists working in bat caves began to seek immunization.

The theory outlined by Denny Constantine's alarming facts is far from fully acepted. Few unimmunized American cavers hesitate to venture into the great bat caves—occasionally. But most of us do so a little uncomfortably. If we return often to the fetid depths, somehow we soon find ourselves seeking immunization, and urge it on our guests, even though we do have one special cause for relief: Denny's own good health. Even before his own 1955 rabies immunization, that amazing young scientist breathed more bat cave atmosphere than any dozen other humans. Some may contend that Denny has a specially charmed life (*Time* has gleefully related how Denny once took a hibernating bear's rectal temperature deep in an Alaska cave while pacifying the torpid monster with sugar lumps). Nevertheless, if anyone were to contract rabies from inhaling the atmosphere of bat caves, it should have been he.

When we now see a sick or erratic bat, we avoid it as if it had the plague or worse, for some of them do. Unless we are engaged in a specific study we regretfully avoid even healthy-looking bats. No more pleasant little pets, purring happily in our palms or hiding in our draperies; their tiny bodies can harbor rabies smoldering unsuspected for weeks or

A bat coasts above the pens of Denny Constantine's Frio Cave studies. *Photo courtesy United States Public Health Service*

months. Even one chance in a thousand of contracting rabies is far too many.

This, of course, is vastly different from tourists visiting Carlsbad Cavern and other commercial caves which shelter bats. Unless a bat—or a skunk—or a coyote—bites you, there or elsewhere, danger is far greater on the highway, merely getting there from your home.

If the story of the great caves of Texas begins with Cavern of Sonora —in my opinion America's most beautiful—it properly should end with Natural Bridge Cavern, Texas' most magnificent. Perhaps the latter's brand-new history portends the shape of things to come beneath the Lone Star State.

Like the Cavern of Sonora, Natural Bridge Cavern is no new discovery. Artifacts of considerable age and bones of animals no longer inhabiting central Texas indicate ancient users of the entrance area. At least thirty years ago a few brave souls ventured underground here, but soon re-emerged, unimpressed. Texas University grottoites were hardly more successful in the early 1950s. Two interconnected levels were recorded, with about 2,000 feet of small passageways. Some of it was quite attractive. More was horrible.

In 1960 the St. Mary's University Speleological Society interested itself in this Natural Bridge Cave. The entire entrance sink was the remnant of a huge collapsed room. Surely there ought to be a big cave somewhere inside!

Three expeditions in January and February 1960 were pleasant but only moderately productive. On March 27 Orion Knox returned with three companions. Their plan was to map the tortuous South Fault. Instead they became interested in a tight, rubble-choked crawlway. After a grubby 60-foot struggle, their head-lamp beams were lost in a black void.

Crawling thankfully out of their miserable tunnel, the cavers found the room smaller than on first glance. Still, it was larger than anything else known in the cave. And below a steep slope was a huge, canyon-like corridor prefaced by a dramatic thicket of thin totem pole stalagmites as much as 20 feet high.

Here tourists today walk dry-shod high above the floor; the valiant 1960 spelunkers wallowed through soupy mud. Wearily they ascended a nasty slope, and gasped at colossal splendor beyond. In an airy auditorium, rimstone slopes led their delighted eyes to magnificent columns which would not be scorned in Carlsbad Cavern. Past other dismaying struggles was an even larger chamber splendidly bedecked with stone hangings as symmetrical as the famed veils of Luray. Beyond was much more.

Each logged in detail, 33 expeditions entered Natural Bridge Cavern in the two years that followed. All agreed that so great a cave must be opened to the public. Work was begun in 1963. With exploration still incomplete in June 1964, convention-going members of the National Speleological Society were treated to the first commercial tour of magnificent Natural Bridge Cavern.

The opening of the commercial development does not quite end the story of Natural Bridge Cavern. Except for dramatic seasonal floods on Purgatory Creek, the tourist sees little water on the tour. Mud-covered Texas cavers, however, can assure him that there's plenty farther back.

Like most Texas ranchers, Mr. and Mrs. Harry Heidemann, the pleasant owners of the cavern, needed water. Could the cavers help once more?

Texas troglodytes considered this new project. Using electromagnetic sensors on the surface and a loop antenna transmitter in the Lake Passage, they homed in on a chosen pond with hardly a foot of error. That was plenty close enough to obtain 40 gallons per minute for this thirsty land.

The story of Texas caves and their lightless aqueducts is still being written. Casual discoveries are still routine. In 1962 a single weekend's work by 45 co-ordinated cavers mapped 3.6 miles and shot Powell's Cave into the position of Texas' longest.

One condition of the 1962 study was an agreement not to request permission to return. Mr. Powell just doesn't like being bothered, which is his prerogative. But 15 minutes from the end of the 48-hour project, a soggy team crawled out with the announcement of a large virgin waterway—the first in the cave.

Mr. Powell too needed water. In 1964 more than 100 Texas cavers descended on Powell's Cave for a 72-hour assault.

". . . 72 cavers were in the cave at one time, and one team hardly ever saw another. There were eight mapping teams going at one time, and four or more exploring teams. . . . Only one survey was ever completed," announced the proud but melancholy *Texas Caver*. Though 4 miles was mapped, so much more remained to be done.

So, with special permission from Mr. Powell, 36 cavers returned in May 1967. Project 36 brought the total up to 8½ miles. One year later, 15 extra-valiant spelunkers braved unseasonable storms in what is unenthusiastically recalled as Project Washout. Much of their effort was in streamways where a few inches' rise might have meant disaster. But with the exception of one torpid rattlesnake, whose chosen hideout required excavation of an entire bypass route, the problems were all on the surface. Project Washout brought Powell's Cave past the 10-mile mark and strongly suggested that the water of its streams may sink as much as 200 miles away.

Yet at this point the current generation of Texas cavers simply got tired of the muggy waterways of Powell's Cave. "The cave still goes and goes," admits Tom Meador, smiling, but no major effort has been made since 1968.

Beneath this thirsty land, water has wrought more than any man today can even guess.

8

Tall Tales and Icy Water

The Story of Northeastern Caves

The little cavern hid its secret well. The settlers' grimly tightening circle had failed to entrap the vicious she-wolf which was terrorizing northeastern Connecticut. Perhaps the wily quarry crouched at bay in the dim recesses of the fissure-like cave where the hunters converged.

No one was eager to investigate the menacing black crack. Eyes turned to Israel Putnam, whose leadership already commanded wide respect in these pre-Revolutionary days.

Putnam was no reckless young hunter. Indeed, it appears that he was no more enthusiastic about crawling into a wolf den than any of his fellow townsmen and farmers. But no one else was willing. Head-first he inched into the crevice. Birchbark torch in hand, he hunched forward, one painful body length after another.

In the darkness there seemed a formless movement. Then the flickering torchlight fixed upon a pair of blazing animal eyes.

Putnam yanked hard on the rope tied around his waist—the prearranged signal. His worried friends outside responded with vigor. Their hero came shooting backward out of the wolf's lair with the loss of considerable skin and more clothing.

Manfully straightening his clothing, Putnam devoted several moments to detailed profanity. Calming, he loaded his gun with "nine buckshot," borrowed another torch, and crawled back into the cave. Gingerly he approached the snarling wolf, crouched at bay against the far wall of the tight little cavern.

The gun's reverberation was stunning, the smoke choking. Putnam found himself again jerked feet first from the cave, so painfully and unceremoniously that even he was momentarily speechless.

A few swigs of good hard cider remedied that unprecedented situation. When the powder smoke had thinned, he again crawled into the little cavern. When his friends felt a new tug, they pulled more discreetly. Putnam's feet appeared, then his body—then the wolf's carcass, dragged out by the ears.

Colonial and even London newspapers recounted the exploit in terms suggesting that newspapers have not changed much in two centuries. "The story grew more sensational each time it was published," later noted Colonel David Humphrey, sober aide-de-camp to the dashing Revolutionary hero-to-be. Such embellishment was needless. To this day, Israel Putnam's feat remains one of the great cave stories of America.

Much of the story of the caves of the Northeast consists of such tall tales of small caves. Some are true, some should be. A few northeastern caves are sizable, but their importance is hardly related to mere size. Here a fantastic medley of races and old-world cultures for three centuries interwove ancient traditions of half-whispered underground mysteries. Superimposed were new tales of frontiersmen, superstitious Indians, and braggart pirates.

Even small northeastern caves possess unusual beauty and interest. This banded marble is in Eldon's Cave, Massachusetts. A caver's bare head, near the center of the photo, reveals the scale. *Photo copyright 1946 by Clay Perry, courtesy Paul Perry*

This is truly the country of literary giants in the earth. Here, Rip Van Winkle fell asleep in a "hollowed out rock." Here are Nathaniel Hawthorne's Cave and John Burroughs' Cave. Here Dorothy Canfield Fisher owned a glaciere—Skinner's Hollow Cave. Here Oliver Wendell Holmes could weave a metaphysical romance of a snake-woman cave girl—Elsie Venner—from misty legend. With author's license, James Fenimore Cooper enlarged the narrow cleft of Natty Bumppo's Cave into a dramatic locale. And here, in 1936, Madison Avenue did likewise. Many an amused caver still recalls the flashlight battery advertisement centered on Sam's Point Caves—a New York State fissure cave where daylight is rarely far from sight. "Mile after mile we had wormed and twisted and crawled our way into the blackness of these caverns," the dramatic ad began. "Fished up through eight feet of water, these fresh strong Brand X Batteries maintained the brilliant beam that led us over the long, slow route to daylight. . . ." Somehow northeastern cavern fiction has deteriorated in the twentieth century.

Perhaps half the stories of northeastern caves began in such a way. Perhaps the Anti-Renters of the Patroon Days of New Holland took refuge in Haile's Cave and other caves nearby. We have no proof, only tall tales. "King Philip" of the Narragansetts should have watched his Indians sacking the fertile Farmington Valley from King Philip's Cave high above. Probably he did not, though northeastern Indians did seek shelter in many a cave and rock shelter as late as 1833.

Through the patient research of Henry W. Shoemaker, several early cave-centered traditions of central Pennsylvania have been preserved for posterity. One of the most noted concerns Stover Cave in Centre County.

Stover Cave, it seems, was well known to the young people of Penn's Valley in pre-Revolutionary times. In that strongly religious milieu, dancing was forbidden by the elders as a pastime of the devil. Rebellious youths promptly constructed a dance floor in the spacious main chamber of Stover Cave, accessible only by ladder. Meeting secretly on Saturday nights, "the young folks danced far into Sunday morning to the strains of the dulcimore, the dudelsok and the geik."

One complication developed. Thoroughly familiar with Stover Cave, the local Indians started dropping in on the dances. Race relations were no better in the eighteenth century than today.

One young Indian, Abendunkel, became enamored of a dark Hugue- not belle, Casella Dolet. When his whispered proposition was indignantly rejected, Abendunkel burst from the cave, shouting threats of exposure. The dancers fled in confusion. Casella was not seen in Penn's Valley for decades. Waylaid and further blackmailed by Abendunkel—some say via a connecting passage from a nearby cave—Casella chose to go north with her lover rather than face her irate Huguenot parents. Only at

Abendunkel's death a half-century later did she return home—about the time a new generation revived the ill-fated dancing club!

The Seneca Indian legend of Penn's Cave places the Huguenot shoe on the other foot. More than two hundred years ago, according to legend, young Malachi Brown was executed here by an Indian band. His crime? He ran away with Nitanee, namesake of the Nittany Valley and beloved daughter of Chief Okocho. Nitanee's seven brothers promptly overtook the fugitives. They thrust Brown into Penn's Cave naked and spent an enjoyable week hurling tomahawks at him each time he approached either entrance. Tradition says the husky youth survived five days seeking an unguarded exit, groping in dark passages and splashing in the chill black stream. But the vengeful brothers of Nitanee kept up their watch for a full week.

Acording to another Indian tradition, Pennsylvania's Woodward's Cave contains the lime-encrusted bier of their fearless warrior Red Panther. In later years this spacious cave and Historic Indian Cave both claimed to be the rendezvous of Davey Lewis, the "Robin Hood of Pennsyivania." This dashing outlaw supposedly met beauteous Daltera Sanry and other belles underground. This particular tale, however, seems badly strained. Woodward's and Historic Indian caves might have served Lewis' band for temporary shelter, but hardly amorous couples. Bats breed happily in chill wet Pennsylvania caves but humans find almost any other locale more congenial.

Since fires can be built therein more safely than in true caves, rock shelters saw considerable use by northeasterners. Some of the first settlers of Philadelphia sought temporary refuge in shallow Delaware River shelter caves: old Philadelphia families sometimes are still termed "cave dwellers." At Connecticut's Judges' Cave, history emerges from cloudy tradition. Three of the grimly hunted judges who signed the death warrant of Charles I found refuge here. Only scholarly eighteenth-century research by an early Yale president gives authenticity to its scarcely credible three-hundred-year-old tale of transatlantic alarms and hairs-breadth escapes.

During the Revolutionary War, Tories and patriots alike sought shelter in many an uncomfortable northeastern cavern and rock shelter. At least one Bucks County (Pennsylvania) cave was a depot for Tory-printed counterfeit Continental currency—and perhaps its print shop. Moody's Rock in New Jersey is named for its Tory troglodyte, James Moody, notorious spy and leader of a Tory band which terrorized North Jersey. On one occasion vengeful patriots thought that they had cornered Moody in Devil's Den, a two-hundred-foot limestone cave about a mile from his rock lair. Confidently they set about starving him out, only to find him raiding as usual. Legends promptly sprang up of a secret cavern connecting Devil's Den with the rock shelter of Moody's Rock. Eventually added were such elegant details as hidden doors opening to opulently furnished chambers and a superb wine cellar—all from a 200-foot cave and a 50-foot rock shelter!

These cave shelters were not forgotten at the close of the Revolutionary War. Daniel Shays of Shays' Rebellion is said to have stabled his horses in a large "hanging rock" cavern near Amherst. Upon the collapse of that revolt, one of his officers named Peter Wilson sought asylum in Peter's Cave in Lee, Massachusetts, today "much frequented by students and lovers." Slave Cave, supposedly later the last stop on the Underground Railroad before Canada, traditionally hid William Johnson during the forgotten "Patriot's War" of 1838.

In more peaceful years, many a fugitive from the law sought shelter underground. Counterfeiters particularly seem to have been drawn to northeastern caves. One of the earliest and perhaps most famous was Gil Belcher. At gunpoint, this well-researched rogue was caught red-handed in a small cave on Bung Hill on October 30, 1772. "To counterfeit is death," declared the New York provincial shinplaster currency Belcher printed. All appeals failed, and he was hanged. In New England he is remembered as the man who printed his own death warrants. At least one nineteenth-century desperado installed a stolen stove in his hideout cave. Few, however, seem to have been enthusiastic troglodytes. When state troopers arrested a cave-dwelling outlaw in 1960, he offered no resistance. Jail, he had decided, was much preferable to raw squirrel and marrow-freezing bedrock bunks.

Hardy northeastern hoboes seem to make better cave dwellers than do bandits. Some who plan well have found an underground life fairly pleasant. Celebrated cave hermit Charles Hill occupied a rent-free Niagara Falls cavern for several years.

Strangest and most famous northeastern troglodyte was the Leatherman. The odd name of that mute itinerant derived from his bulky suit of creaking, wide-pocketed squares of thick leather, roughly joined by thongs. The Leatherman continuously traveled a mysterious 365-mile circuit every 34 days. Never did he alter his curious pilgrimage. Housewives at remote farms could almost set their clocks by his calendar-marked arrival. Only once was he late, and that when the blizzard of 1888 piled twenty-foot drifts across his inexplicable Connecticut—New York route. Never did he converse, or halt for any kind of job. Always gravely polite, this shaggy, sloppy, and superficially repulsive itinerant soon became the object of widespread pity. That pity turned to a queer affection. Proud housewives rose early to fix sumptuous meals on Leatherman Day. Husbands complained mildly that the Leatherman got better fare than they did. Wives retorted that if they failed the Leatherman, he'd go somewhere else—and their spouses had no such choice. Thousands of people came to coordinate their activities to the Leatherman's relentless schedule. Even today, it is with pride that aging northeasterners say, "My grandmother fed the Leatherman."

The mysteries of the Leatherman have never been explained. Never

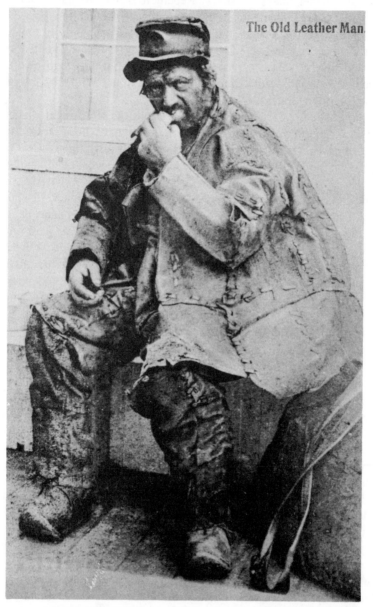

The Old Leather Man.

Old commercial photo of the Leather Man. *Courtesy Leroy Foote*

did he speak. Supposedly one Jules Bourglay of Lyon, he carried an old French prayer book. According to one story, he once was surprised perusing the *Ladies' Home Journal*. Most peculiar of all was his troglodytic impulse. Even in the worst New England blizzards he politely declined invitations to sleep cozily in barns or sheds. Instead, he chose to pass the nights in carefully selected caves and rock shelters. Partially sheltered by pole leantos thatched with leaves, the Leatherman brushed aside the coals of his nightly fire and slept warmly on their bed. Despite temperatures as low as −20 degrees, he prospered for thirty-one plodding years. Finally, on March 25, 1889, the Leatherman did not appear for his breakfast in Pleasantville, New York. The alarmed townspeople sought out his rude shelter on a nearby farm. There they found their faithful mystery man, his inexplicable secrets carried to the grave. Thousands of warm memories and a bronze plaque in the Sparta Cemetery in Ossining, New York, equally commemorate this oddest of American troglodytes. At least one other Leatherman attempted to step into his bootprints, following an erratic wider circuit at least until 1893, perhaps as late as 1911.

However, the story of northeastern caves is much more than this glamorous legendry. Little came of isolated seventeenth- and eighteenth-century northeastern spelunking. But if the mainstream of American spelunking flowed westward from Virginia, American speleology leapfrogged out of the nineteenth-century Northeast. New York, Vermont, and Massachusetts saw the first state-wide geological surveys. Though speleology was still undreamed of, such methodical geologists as W. W. Mather gave due consideration to caves by 1825—the same early period that saw the first crossing of Mammoth Cave's Bottomless Pit and the walling-up of Wyandotte Cave to keep cows out.

Others expanded Mather's painstaking spelean tradition, sometimes at considerable risk. In 1853 Professor McFail of Carlisle Seminary fell and died in the New York State cave which today bears his name. Even this tragedy failed to daunt the conscientious geologists. Systematic turn-of-the-century studies of John H. Cook and Amadeus Grabau amply laid the groundwork for modern New York speleology.

Beginning in 1871, several vertical caves along the southern boundary of this northeastern cave province yielded very different data. From a series of ancient natural traps came much knowledge of the succession of animal life of past millennia and hence our changing climates. Recent studies have concentrated attention on Maryland's Cumberland Bone Cave and the New Paris sinks of Pennsylvania. First discovered and perhaps most important, however, was Port Kennedy Cave, just below Valley Forge on the Schuylkill River. Before flooding and contamination

Peg Palmer and Jack Freeman stride along the canyon passage of McFail's Cave.
Photo by Arthur N. Palmer

halted operations, remains of fifty-four mammals—forty-one extinct—were identified. Included were an extinct bear larger than a grizzly, two species of sabre-tooth "tiger," four species of ground sloth, a mastodon, and much more. In truth, the work of dedicated early northeastern speleologists largely set the standard for all America.

Especially in New York, the intensity of the nineteenth-century intellectual curiosity stimulated a surprising wave of cave exploration. Schoharie County was a particular center of activity. Though it was necessary to lower a boat forty feet through its narrow entrance shaft, Ball's Cave (later Gebhart's and Knoepfel's and now Gage's Cave) was explored by 1831. True to New York State tradition, it was described in the *American Journal of Science* within four years.

With a handsome, spacious main corridor, which Professor John H. Cook soon measured as 4,411 feet long, nearby Howe Cavern soon was commercialized. By 1882, when spelunking pastor Horace C. Hovey published his famous *Celebrated American Caverns,* it probably was the most noted of the lesser caverns of America as well as the largest in the Northeast. Widely renowned for its pleasant underground boat ride, its clean, inviting walkways, and its restrained advertising, it remains equally celebrated today. Few visitors catch even guarded hints of bitter northeastern legendry which here has spawned a century of controversy.

Lester Howe would scarcely recognize his cave today. Visitors now enter through a 156-foot elevator descent at what was once the far end of Howe Cavern. A quarry has eaten into Howe's entry section, sacrifice to the Northeast's insatiable need for cement. What remains, however, is small only by cyclopean standards, excellently prepared for visitors who stroll high-heeled on brick paths, ride flat-bottomed boats on an underground lake.

Either his celebrated development was premature or Lester Howe proved a poor manager. Broke, he sold out to the Albany and Susquehanna Railroad, which enjoyed a greater success. In later years from his Garden of Eden Farm across Cobleskill Valley, Howe could watch tourist crowds at "his" cave. The sight increasingly embittered him. To some trusted confidants and supposedly again on his deathbed, Lester Howe uttered an ambiguous statement which has haunted northeastern caving for a century. The exact words seem lost, but most agree on its context: "I have discovered a cave more valuable than my first. I call it my Garden of Eden and have disclosed its location to no one."

Hopefuls who searched Howe's Garden of Eden Farm found no Garden of Eden Cave. A decade later, Professor Solomon Sias of the now-defunct Schoharie Institute described a nearby cave that he thought might be the missing cavern. But there's a bit of a problem. Laselle Hellhole is at the location listed by Sias. But it doesn't fit his description. More of the Garden of Eden Cave in this chapter.

In line with the broad Southeast Passage of McFail's Cave is the equally large main passage of world-famous Howe Cavern, where visitors dress somewhat differently. Unfortunately connection of the two caves seems unlikely. *Photo courtesy Howe Cavern*

After three generations, the first great wave of Empire State exploration inevitably ebbed. Many caves were "lost." Others acquired confusing new names. About the time of the Floyd Collins debacle, Arthur Van Voris extended the pioneer studies of the Schoharie County underworld. The new flowering of New York speleology, however, awaited the coming of the National Speleological Society and its organized units. The first grotto of that society appropriately was officially organized in Massachusetts' Pettibone Falls Cave, technically before the society itself was chartered. In the first postwar mushrooming of American speleology, more than a dozen northeastern grottoes sprang up from Pittsburgh to Boston. Spelunkers began to savor and enjoy the intimacy of northeastern caves, equally delightful for banded marble, "impossible" squeezeways, superb flowstone, and mud baths. A wealth of information began to flow into the society's files for the use of fellow cavers. Lights returned to caves "lost" for generations.

Investigating recollections by caver Phil Johnson's grandfather, Tri-County grottoites found a bush-hidden entrance on the property of a Mrs. Dewey. Inside were 800 feet of walkable passage, then a half-mile of crawling, climbing, and dunking in chill, stygian ponds. One water-floored crawlway was dubbed the Bawl Crawl—"as you soak up the water, you bellow and bawl!" In this pockety fifteen-inch space, bulky Howard Sloane once found himself atop a large, indignantly agitated catfish for longer than either desired. Despite six-hour round trips in such surroundings, cavers felt well repaid by this Onesquethaw Cave. Deep inside, black-swirled pink marble contrasts superbly with cascading flowstone.

Sometimes the northeastern cavers overdid it, however. On July 3, 1960, the legendary fate of cavers overtook a well-known Harvard caver in a tight New York cave. About 200 feet from its entrance he became thoroughly stuck in a keyhole-shaped orifice. The threat of thunderstorms made it undesirable to merely leave him until he lost some weight. Five companions tugged mightily at his heels, without success. They then tried pushing. Like a cork, he popped through into a small chamber beyond the keyhole.

This was hardly the ideal situation. With a friend pushing from behind and three others pulling, the keyhole seemed even tighter in reverse. Successively, each layer of clothing was shredded off—without freeing the hapless caver.

Rescuers accumulated as time passed. Someone produced a bucket of axle grease. It did the job. Six and one-half hours after getting stuck, the luckless caver emerged, thickly clad in axle grease and nothing much else. In the best Harvard tradition he donned a proffered coat, surveyed the milling scene, politely thanked the grinning crowd and apologized for his attire.

Elsewhere in northeastern caves, problems have been somewhat different. Howard Sloane never did live down the fiasco of New Jersey's Fasolo's Cave. In the November 1952 National Speleological Society *News* he wrote of a planned trip to a New Jersey cave "several thousand feet in length. A four-hour exploring trip failed to reveal the ends of the main passage."

Unfortunately, Howard was relying on information that proved less than precise. "Fifty-four cavers showed up Sunday morning, all eager to see the new cave," Ross Eckler recorded. The crawlway length of Fasolo's Cave proved only 44 cavers long. Though persistent search and excavation have brought New Jersey a cave over 1,000 feet long, Pennsylvania cavers still snicker that New Jersey caves are measured in "YOU's": "you, you, and YOU can all get into them at once!"

Still lacking a cave with much more than a mile of passages, Pennsylvania cavers' own delights are wacky enough. The numerous caves of the Keystone State are the home of perhaps the world's tamest, most inquisitive pack rats. Consider just one of the myriad accounts that delight northeastern cavers. Peculiar crunching noises puzzled intent photographers awaiting return of a furry family to its snug nest:

> Turning around they beheld Mr. Rat contentedly gnawing a neat hole in Jim Walczak's camera case. No sooner had the rodent been shooed away than he returned and diligently tried to drag his new-found food away from the human disturbance. Meanwhile Bob Higgs received an odd sensation as Mrs. Rat bridged a chasm by nonchalantly walking across his head en route to her spouse.

Consider, too, a girl caver's introduction to the pleasures of Kooken Cave:

> Mud, mud, nothing but mud—crawling in it, walking in it, sliding down it, sitting in it. We crossed traverses with handholds and footholds literally carved in the everloving mud and always a pit for you to roll into if you lost your balance. . . . We had mapped about 12 stations when I jumped down six feet into what seemed like a shallow pool of water, only to sink up to my knees in mud and eight inches higher in water before sitting down rather suddenly in the whole mess. With much squishing and squashing and tugging and pulling, I grabbed each leg with both hands, finally winning out over the suction power of that censored mud. This brought to an end the mapping, and we headed back for the entrance with me imitating a slimy eel the whole way.

After Kooken Cave, many a more famous Pennsylvania cave somehow seems a trifle dull.

But northeastern caving is much more than fun and games. The chill waters that appear so friendly in Howe Cavern and many others are deceptively deadly. Beautiful little Schroeder's Pants Cave, near Dolgeville, New York, was widely appreciated as an exceptionally pleasant cave despite the way it got its name: high school principal Herb Schroeder, president of the old Adirondack Grotto of the N.S.S., twice lost the seat of his pants in its tightest crawl. Then on February 13, 1965, James Mitchell—vice-chairman of the Boston Grotto—lost his body heat to its wintry waterfall, and with it his life. Ascending competently, using Prusik knots, and with the additional precaution of a skilled belayer, he died of

Nibbling a tasty morsel, this *Neotoma* failed to tuck its tail into its snug Pennsylvania cave home. *Photo by Charles Mohr*

hypothermia after a few minutes' exposure, just a dozen feet from the top of the waterfall and safety.

Although sometimes better known for setting cavers at each others' throats, the modern search for Lester Howe's Garden of Eden Cave is looming ever-larger in the annals of American speleology. Some surmise that the mysterious cavern saw light a generation ago. The ever-questing speleoauthor Clay Perry learned of plans for a major 1950 expedition "centering upon Garden of Eden Cave." Cliff Forman has recorded the disappearance from a 1949 Tri-County Grotto outing of Charles Hanor and Chet Laselle, and their ragged, mud-smeared reappearance fourteen hours later, close-mouthed and smug. The 1950 expedition was soon canceled as quietly and mysteriously as it was planned. Like Howe and Solomon Sias, Hanor and Laselle are dead, their secret untold.

Although Hanor spoke guardedly to Clay Perry of traveling two miles underground "somewhere in Schoharie County," my own guess is that the 1949 mystery marathon was just across the county line, in Albany County. There, today's cavers characteristically emerge from even longer marathons in Skull Cave looking quite as bedraggled as did Hanor and Laselle a generation ago.

Skull Cave is no Garden of Eden. For over a century it was known vaguely as an unpleasant streamway crawl longer than most people wanted. Except for cavers less than five feet tall, only one tight crack permits explorers to stand erect in its first 2,200 feet. All but 10 feet of this is about 2 feet high and not much wider. Some points are half that height, with ponds as much as a foot deep. Local storms sweep flood-killed animal life throughout the entrance section. A self-described *aficionado* of the choicest miseries of this unfriendly cave, speleomasochist Art Palmer gleefully nauseates unwary friends with some of his memories. Especially vivid are his accounts of a certain 300-foot length of deep-ponded water with the appearance of noodle soup but the smell of extremely dead worms. "At times it has been truly awful," he acknowledges.

Over the years, the prominent entrance and occult legendry drew some very odd visitors. Theirs were tales of a mysterious entrance room full of human skulls and those of cattle unlike any on today's earth. Strange runic letters carved deep into the wall, too, were part of the mystic lore.

Records of those furtive early visits are few. Palmer has been unable to document any acknowledged exploration before 1936. Curious neighbors who left a date of September 15, 1948, and the bizarre inscription JESUS SAVES in bold white paint 2,800 feet inside may have been the second group to conquer all of the menacing entranceway.

Officially, organized speleology arrived a little more than five years later, when a mapping crew wrung nearly a mile of ugly, dangerous passage from its innards. Years later it became evident that the mappers had been so benumbed that their compass bearings became wildly inaccurate. Deep inside were large chambers and intricate networks, but even among extra-hardy northeastern cavers, few have ever waxed enthusiastic about Skull Cave. Even its spine-tingling lore vanished with its first scientific exploration. The skulls, alas, were merely those of ordinary cows; the runes, mere natural solution along the numerous cracks in the limestone walls.

Doggedly, Art Palmer, Hugh Blanchard, Marlin Kreider, and other ultra-dedicated northeastern cavers like Ernst Kastning, Wayne Foote, and Frank and Larry Mullett, kept up the effort and the cave is still grudgingly yielding virgin passage. With more than 4½ miles already mapped, it is probably the longest in the Northeast. Yet new discoveries here are met with less than normal enthusiasm. Some have required carefully prepared underground stays of 24 hours or longer, amid some of America's nastiest subterranean surroundings. Still unmapped are 1964 discoveries of a system of fissures so muddy that chimneying is like climbing on grease. At their bottom, as might be expected, are waterfilled pits. Evidence of 50-foot annual floods has deterred further exploration here. Even in routine sections of the cave, exhaustion and hypothermia have caused several near-tragedies.

Faced with numerous rescues of ill-prepared explorers, the local fire department is reportedly so irate that many wonder if Skull Cave will remain open long enough to be completely explored and studied—potentially a surprisingly serious loss to those who doggedly continue to face its miseries year after year. Inevitably they remember the day in 1971 a newcomer first tackled Skull Cave, without closing his mind to the barely tolerable outer part of the cave. Graduate student Mike Queen came, crawled, looked behind a rock everyone else had seen zillions of times, squeezed into an inconspicuous crawlway behind it, and immediately found himself in a virgin section of high fissures totaling more than a mile in length.

As for Hanor and Laselle, it is quite likely that they talked with neighbors who accomplished the 1948 penetration. Once past the stinking, backbreaking 2,200-foot entranceway, it would have been no great feat to extend their investigations to something resembling two miles. Chuck Porter and Art Palmer mapped that much in just two 1962 trips. For the few who think it's worth it, there's a lot of cave in Skull Cave.

If Skull Cave was not the site of the mysterious 1949 venture, Hanor and Laselle may have stumbled into the confusing saga of Professor McFail and what is today known as McFail's Cave. And perhaps also into that of Howe's Garden of Eden.

The history of McFail's Cave is almost as confused as that of Howe's missing cave. Hanor was not the first to speak of having traveled two miles or more beneath Schoharie County. As already mentioned, McFail fell and died here in the pit-type entrance of a cavern three or four miles northwest of Howe Cavern. Twenty years later, published accounts mentioned his explorations "in a beautiful cave some three miles long." That alone meant little. Speleologists snicker that a mile is best defined as any underground distance more than 100 feet long. Here matters were different.

Through the inexorable years, history blurred. Inhabitants of Schoharie and Albany counties forgot all but the general area of McFail's fatal fall: somewhere in deep woods thick with pits. Late nineteenth-century accounts specified that he had fallen and died in Selleck's or Sellick's Cave, known since at least 1841. In one pit cave which some knew by that name, he had placed his initials and the date: 1844. Others claimed a different pit as Selleck or Sellick Cave. Both were considered pretty but very small. Both had lakes just off the bottoms of their pit entrances and not much more. Nobody really cared which was which. Even those seeking the Garden of Eden Cave seem largely to have bypassed them as unimportant.

Professor Cook came to Schoharie County in 1906. At the bottom of the pit cave he was told was Sellick's Cave—the one with McFail's initials—he found running water but no penetrable passage. More importantly, he talked with one R. J. Roscoe, one of McFail's old companions. From Dr. Roscoe he learned that from the bottom of the other McFail's or Sellick's Cave passages extended northeast and southwest, and a stream led in the latter direction.

But the passage of fifty years had blocked this shaft. Cook's time and resources for cave digging were limited, and another pit nearby appeared more promising. "Cave Disappointment" he dubbed his vain effort. Evidently he never returned to what he called McFail's Cave: a sad decision. In the 1920s, a party led by Arthur Van Voris removed a few stones and fragments of wood, easily dislodging the natural plug. Below they found "waterway passages . . . extending . . . for a considerable distance." Van Voris himself was less than vigorous in exploring these watery stoopways, and for decades the reopening of the cave remained virtually unknown. Yet in later years, one of his teammates tantalized latter-day cavers as much as had Lester Howe—and much more cheerfully. Gleefully Edward A. Rew announced that a letter to a local editor was "actually to needle them by giving them just enough information . . . to make them really get down to business . . . [The Garden of Eden Cave] is richer and more spectacular than any hitherto unexplored cavern in Schoharie County, or, for that matter, any other in the North. I personally and alone, on a secret expedition, discovered and explored it one night in 1931 to the extent of about two miles. . . . In wishing them good spelunking and success, my feelings are mixed, for until they do make an entrance, it's mine, all mine!"

At various times, Colonel Rew happily planted intentionally puzzling hints. Some may have been red herrings. Often he referred to the geological structure of Terrace Mountain, southeast across the valley from Howe Cavern. On occasion he mentioned the relative ages of nearby valleys. At least once he hinted broadly that he had forced a seasonal siphon in little Van Vliet's Cave, on the southeast side of that hill. If so, that siphon still seems to be his, all his.

Somewhere along the line, an especially haunting phrase further complicated matters. "The Finger of Geology points to the Garden of Eden Cave," he is said to have pontificated. "Read the geology and keep at it and you'll find it just as I did in 1931!"

The Finger of Geology . . . leading northeastern cavers puzzled the phrase for years. A fingerlike limestone outcrop? One on Terrace Mountain points straight at Van Vleit's Cave, VeenFleit Cave, or whatever spelling one chooses. But no matter how the name was spelled, no one today seems able to make it yield the Garden of Eden. Innumerable alternative solutions were suggested, investigated, rejected. Many came to believe the entire tale a hoax. Perhaps it is.

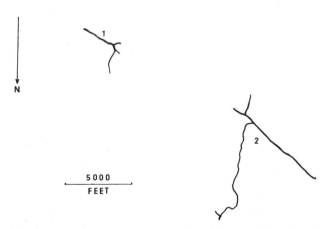

The main passage of Howe Cavern, N.Y., as the mysterious "Finger of Geology." Based on sketch maps by Steve Egemeier. (1) Howe Cavern; (2) McFail's Cave; (3) VeenFleit's Cave (name also spelled in several other ways)

Brand-new members of the mushrooming National Speleological Society first came to the pit-pocked woods around Howe Cavern earlier in 1949. Investigating the area to the northwest, evidently they knew nothing of McFail, nor of anything called anything like Selleck Cave. North of Carlisle Center, Charles Hanor spotted a small cave. Since it seems never to have had a name, it remains Hanor's Cave to this day. Working northward through the brush, they soon encountered two blind pits. "Ack's Shack," they ebulliently named the northernmost, for their own enthusiastic Ernest Ackerly. It was no compliment. Like that of "Featherstonehaugh's Flop," 100 feet farther southwest, the name was inflicted as a result of the long, strenuous ascent.

In 1960 the area around Howe Cavern came under detailed speleological study. Norman Olson and others of the Howe Cavern Project had learned of Professor Cook's investigations and started looking for the pit Cook had called McFail's Cave. When they located it, it turned out to be just 200 feet northeast of Ack's Shack. That aligned it neatly with Featherstonehaugh's Flop, and almost exactly with Hanor's Cave, too. Again it was blocked by nature, but again the floor roared out from under Peter Van Note—properly on belay—as he probed the most promising corner. Below, they found the streamway much as Van Voris had described it. With one exception. At each end of Van Voris' east-west passage the cave turned a tight corner and continued.

Off to the northeast, the avid spelunkers scrunched under a low spot and kept on until the icy black water reached their chests. About one half mile of cave eventually turned up here. Off to the southwest, things were almost as bad as in Skull Cave for 400 feet. Beyond was a slightly more spacious criss-cross with several hundred feet of streamway. Promptly published in the N.S.S. *News*, their preliminary map of more than 1,000 feet of the cave showed this end blocked by a siphon.

With a length of almost half a mile, "McFail's Hole" was one of the largest and hence one of the most important caves in all the northeast. Quite a few extra-hardy cavers more or less enjoyed its moist pleasures. In 1961, however, Fred Stone of Cornell University spotted a bypass around the "terminal" siphon. Excitedly he led onward for several hundred feet. Alas! It seemed nothing more than the typical New York cave: wet, barren, and small.

Eventually the team encountered a dismayingly deep pool 60 feet long. The low ceiling dipped even lower, to a bare two inches of air space. Nose to ceiling Fred grimly wallowed on—into the main part of McFail's Cave.

Thousands of feet of high, sinuous, waterfall-fringed canyon drew the cavers back and back through 1961, trip after misery trip. The new discoveries led them southwestward, just as Professor Cook had been told a half century earlier. In time the cave angled sharply to the right, along a prominent cross-joint. Tighter and tighter it squeezed the exploring team. These were ominous signs familiar to all northeastern cavers. Obviously the cave was going to end, at long last.

But this was McFail's Cave, obstinately different from its fellows.

Eighty-foot rappel in "Ack's Shack." Shaft continues down as tight crack behind Matt Humphreys. *Photo by Arthur N. Palmer*

Instead of ending, it continued, expanded, amazed. The next few thousand feet of cave was surprising enough for its very existence, but beyond was a series of large rooms. In the best New York tradition, huge blocks of limestone hung menacingly from their ceilings. Vast accumulations of new-shattered rock amply confirmed that the ceilings were as loose as they looked. Softly the explorers trod.

Beyond these tiptoe chambers lay still more cave, as low and wet as in the entrance section. Surely this would end the cave. . . . Yet still the cave "went." Still farther inward, the cavers halted, truly awestruck, at the key junction of the entire cavern. Eight thousand feet from the once-terminal siphon a spacious, tubular cross-passage engulfed their miserable little waterway. In their yellowish carbide beams it looked amazingly like the splendid main corridor of Howe Cavern, not far distant.

This great Northwest Passage, however, led the awed explorers precisely away from Howe Cavern. Splendidly spacious, it extended straight northwest for more than a mile before ending in an impasse of breakdown and fill. In the other direction, straight toward Howe Cavern, it momentarily took on Kentucky size—as much as 60 feet wide and 25 feet high. But that wouldn't do beneath New York. The lowering ceiling led to a siphon after "less than a half mile"—perhaps the only time that phrase has seemed suitable in a New York State cave! Five miles of the finest cave this side of the Virginias, the exhilarated explorers told their friends. And, nineteenth-century tales or no, it appeared wholly virgin.

Yet these responsible cavers were ultra-cautious in their reports. McFail's Cave—or Hole—or Selleck Cave or whatever it was, was a beautifully pristine cave in an area where unprotected caves routinely were robbed of their natural splendor as soon as their existence was known. Furthermore, biologist-cavers like Frank Howarth vigorously acclaimed the near-virgin cave as an ideal site for study. Perhaps the deepest major New York cave, it theoretically offered a refuge to cave life exterminated in shallower caves by the glaciers which deeply overrode this verdant region not many millennia ago.

Safety, too, was a major consideration. Nobody wanted another McFail death. Alarmingly loose rock along their chutes was characteristic of the pit entrances of these central New York caves. The Van Voris party had been incredibly lucky when the floor vanished almost underfoot. In 1949, Duane Featherstonehaugh correctly read the signs a trifle belatedly in Hanor's Cave. Dictatorially he ordered an entire party out of the cave moments before loose rocks injured two of its members and almost wiped out the entire team. By 1962, some began to worry about the McFail shaft. Preparing to descend the final 32-foot vertical pitch, Art Palmer was actually in mid-sentence: "No, this cave isn't considered to be at all unsafe . . ." Before he could finish, the boulders on which

Ernst Kastning and Jack Freeman consider precariously wedged breakdown in first chamber of main passage of McFail's Cave. *Photo by Arthur N. Palmer*

he stood roared from beneath his feet, leaving him swinging one-handed on the rappel rope. With cavers' luck, no one was below. Later that year, Fred Stone was ascending on a standing rope at almost the same spot. A faint grating noise reached him from somewhere overhead. With time for nothing but an involuntary grunt he kicked violently against the wall. As he swung, a quarter ton of loosened limestone vroomed past him, straight at Frank Howarth. Properly alert, Frank also lunged desperately. As he glanced off the far wall of the pit, the great rock shattered ear-splittingly precisely where he had stood.

Definitely impressed, Frank set the N.S.S.'s Cornell Grotto to excavating a bomb-proof alternate entrance: a gravel-filled crawlway at the bottom of Ack's Shack. At this point, names became somewhat less confused. The new entrance led into McFail's Cave even if some thought the original entrance was Selleck Cave or Sellick's Cave or something like that. Or McFail's Hole, for that matter. After this 110-foot crawlway was cleared, nature cooperated. By 1964 the old entrance had closed itself completely, and it was all McFail's Cave.

Outside northeast caving circles, very little was said about this cave for several years. In 1964 Fred Stone sparked a quiet drive which led to its purchase and donation to the National Speleological Society. The N.S.S. prepared rules for visits, then installed barbed wire, and NO TRESPASSING signs to discourage those without permission. Then it notified the state police and the local justice of the peace, who began levying fines of $25 to $50.

Yet all of these precautions were inadequate. On March 16, 1968, one of four trespassing local youths died, stuck on a rope under the waterfall in Ack's Shack almost exactly as happened in Schroeder's Pants Cave three years earlier. When they left the hospital, his three chilled companions were routinely greeted with trespass fines. Either because of the tragic publicity, a locked gate the cavers soon installed, or increasing knowledge of the enthusiasm of local law enforcement, few further trespasses have been recorded. And no additional deaths.

In the proud tradition of New York caving, scientific study proceeded quickly in McFail's Cave. Cave diving through the new "final" siphon yielded 800 feet of stoopway, then another siphon. This one is still "final," but this particular word has become a trifle ambivalent in McFail's Cave. Albeit sometimes at the extraordinarily slow average rate of 178 feet per hour, mapping proceeded apace. By the end of 1968 2.42 miles was on paper, and in 1970, 4.92 miles. Each spring, northeastern speleologists re-excavate the Ack's Shack crawlway to map a bit more, study a bit more on half a dozen ventures each year.

And, as speleology came to the spacious cross passage, a curious finger of geology became apparent. Steve Egemeier studied the underground drainage of this entire section of New York State. Soon he suggested a reason why the "immense" cross passage so closely resembled the main corridor of nearby Howe Cavern. Less than two miles apart, both had formed along what he called the VeenFleit Fault. Originally a single continuous unit, only ordinary cave fill—two impenetrable miles of it, it must be admitted—seems to separate these two ends of what once was a throughway corridor that would have done justice to Mammoth Cave itself.

On Egemeier's map, the Howe Cavern end of that corridor points along the fault, like an index finger. Other passages of Howe Cavern form the outline of the remainder of a right hand.

Although the average northeastern caver doesn't seem to have even considered the possibility, I'm inclined to believe that this is the Finger of Geology that proves that McFail's Cave is Lester Howe's long-lost Garden of Eden Cave.

True, this particular finger of geology may be fickle, the fault faulty. Some recent studies even question whether the faults seen in the two caves are the same fault. Although the fault passage in Howe Cavern certainly lines up with the fault-determined Northwest Passage of McFail's like a finger, it may not be the original Finger of Geology. Ideally the "hand" of the Howe Cave map should be reversed so that the finger points northwest instead of southeast, which happens to be toward Veen-Fleit's Cave. But perhaps that would have made the riddle too easy. . . .

We avidly await the next chapter in this drama. In this fabulous land of spelean history, legend, excitement, beauty, extra-hardy cavers and even a gaggle of notably large caves, your guess about the next exciting development is as good as ours. Pirate treasure, age-old bones, the Underground Railroad, the Garden of Eden: here in the unique netherworld of the northeastern United States, the unlikely is mere routine.

9

Of Men and Caves

The Story of Southeastern Caves

November 4, 1967, was a busy night for Terry Tarkington. As nearly as I can reconstruct it, his part in the action began about 7:30 P.M. The place: Huntsville, Alabama, later to become official headquarters of the National Speleological Society. The event: a telephone call from Troy Watts near Anvil Cave. For several hours a search had been underway for five boys lost in that maze of all mazes: almost 13 miles of cave beneath some 18 acres of sink-pitted pastureland!

Knowing the cave well, Terry responded immediately. Nearing the cave area at 8:20, he encountered the tail end of a stupendous traffic jam. Later he recorded selecting "a virgin, four-wheel drive route around [it]." In plain English, he drove clear around the mess. The fun was about to begin.

8:25. Terry locates and announces to the National Guard, the Flint Rescue Squad and assorted bystanders that they can now relax. The Cave Rescue Unit (himself) has arrived and will have the boys out within a few minutes. Blank looks, no relaxing—but no interference: the power of positive thinking.

8:26. Terry looks around for competent help and recognizes Jimmy Moore, a local high school footballer who has been in the cave a few times. Terry hands him a hard hat and light and the duly constituted search party is off.

8:30. At Entrance #1 of Anvil Cave, Terry and Jimmy encounter twenty or thirty weeping parents and other kin of the lost boys. Patiently they explain that there are no pits or other particular hazards in the cave and that they will be back in minutes. Credibility is greater than at rescue headquarters.

8:35. Terry and Jimmy enter the cave. Immediately they encounter "a mile of string [in] the first couple of hundred feet inside the

Central portion of Anvil Cave, Alabama. Map of Huntsville Grotto of the National Speleological Society. Map by Huntsville

cave. We had to fight our way through it, crossed and criss-crossed in every direction as though we were in a gigantic spider web."

8:40. Terry and Jimmy encounter the human spiders—several would-be rescuers with a single flashlight and a large ball of twine.

8:45. The "spiders" are persuaded to return to the entrance. Terry and Jimmy break out of the net of string and set off at a fast walk, shouting at each junction as they "sweep" the cave systematically.

8:52. The lost are found.

9:00. The lost are restored to a hysterically happy crowd. Reporters interview Terry and tremendously overdramatize the event.

9:05. Somebody mentions that there are still two search parties in the cave.

9:10. Terry and Jimmy go hunting again.

9:20. One missing party is found, its members denying vigorously that they are lost.

9:25. The search party is escorted to the surface through a convenient sinkhole entrance.

9:30. Terry and Jimmy return to headquarters. The other group is still missing, but someone suggests that everybody wait a while. Which is O.K. with Terry and Jimmy, who have covered a lot of cave in a hurry.

11:00. The missing are still missing, so Terry and Jimmy go after them.

11:20. A long sweep of the cave locates them far to the southeast.

11:30. Everyone out of the cave.

11:45. Coffee with the Moores; sure tasted good.

Midnight. Homeward bound from a fun evening.

Later. Reynolds Duncan muses in print: "Being lost is not unusual and might even be a normal state of affairs for the average caver in Anvil."

The cave lands of the southeastern United States are indeed a fantastic region. As in the Mammoth Cave country, some sinkholes here are so large that whole farms are located inside. One—Grassy Cove—is three miles long and a thousand feet deep: perhaps the largest in America.

Most American cavers adjust with some difficulty to caving in southern Appalachia. Elsewhere, caves "just naturally" belong out in the hills. Not so here. In these southeastern mountains, caves are a casual part of the life of city dweller and farmer alike. Here are caves by the hundred. In one 1958 week, H. H. Douglas located 112 previously unrecorded caves in the far western tip of Virginia, where the natives speak of the Shenandoah Valley as "back east." Three years later, John Holzinger returned for another week's outing. In the same small area, he found another 127 caves—44 of them during one 10-hour period. Whole encyclopedias might be written about the caves of Tennessee and Kentucky alone. Kentucky's Carter Caves State Park contains some fifty caverns. Short Cave and a few others permit a limited amount of caving by automobile.

In general, southeasterners seem to enjoy living with their caves. True, the authorities of Huntsville, Alabama, had to cement part of Big Springs Cave. They claimed the First National Bank was sagging somewhat, but perhaps they merely looked askance at cavers crawling under their vault. "Once there was talk of commercializing it," Jim Johnston drawls. "Too much sewage dripping in there now, though. We mapped it eighteen months ago through a manhole. The old map said it didn't go under the courthouse. We said it did. The city drilled a hole and found we were right. One chimney goes within a couple of feet of the courthouse basement. I guess we're spoiled. Our caves are big and beautiful—and they're right here!"

Ruby Falls Cave is hardly ten minutes' drive from downtown Chattanooga. In 1928 spelunker-promoter Leo Lambert began a 420-foot elevator shaft to long-lost Lookout Mountain Cave. Originally opening near river level, that historic cavern was sealed by nearby railroad construction in 1905. After 99 days' blasting through crystalline limestone, Lambert reached the cave he sought. At the 260-foot level, however, the shaft had intersected a tubular cavern then about 18 inches in diameter.

Spelunker Lambert needed no urging. With friends he forced his way into the rock-ribbed crawlway. As they grunted along, it broadened slightly into a triangular tube about twice as wide as Leo. On and on it led the explorers. Gradually it expanded into a clay-narrowed squeezeway, then much more.

At length the faraway swishing of an underground waterfall met Leo's ears. Eagerly the little party strode into a broadening vault fluted like a rich Egyptian tomb. In their lights gleamed a sparkling waterfall plummeting from impenetrable darkness.

Ruby Lambert was as impressed as her husband. "Leo," she announced, "you simply must open this cave for people to see. It's beautiful and clean, and this, this . . ." Words failed Ruby Lambert. Ruby's Falls they called it: a glorious 75-foot waterfall which roars in winter, whispers temptingly in summer.

Leo demurred. The lower cave was much larger, although a trifle sooty from train smoke. It was historic, with legends of the lost Cherokee national treasure and a signature of Andrew Jackson dated 1814. (Unfortunately, Andrew Jackson's letters indicate that he wasn't anywhere near Chattanooga in 1814. Maybe it's a leftover election sign.)

The Lamberts compromised. During part of the 1930s, both caves were open to tourists. The visitors vindicated Ruby's judgment. They much preferred Ruby's Falls Cave. Today so many visit it that miniature red and green underground pedestrian lights are needed.

In such a region many superb caverns have been opened to the tourist almost automatically. In some, modern spelunkers are still advancing man's frontiers. Tuckaleechee Cave (once Great Smoky Mountain

Ruby's Falls is extremely difficult to photograph. This picture is more than slightly retouched. *Photo courtesy Ruby Falls Cave and Chattanooga Convention and Visitors Bureau*

Cave) was reopened to the public in 1953. Two months later Tom Barr and Roy Davis arrived. In routine operations only seventy-five feet from the commercial route they squirmed into a high hole. Beyond was a superlatively ornamented chamber where man had never been. "We didn't measure it. We were almost afraid to," Roy says, grinning. Their discovery is now a feature of a splendid tour.

Thirty-five circuitous miles from Huntsville is Cathedral Cave. Always a trifle wary of *Reader's Digest* humanizations and familiar with local cavers' early reports, I came to Cathedral Cave somewhat skeptically. "The mud reminded me of what a June bug would look like wading in a chocolate pie," remarked Kenneth Bunting in 1953. A decade later I was flabbergasted by the magnificence that opens inward from its gigantic, fog-spanned mouth. Thick level beds of limestone, regularly spaced wall cracks, vast flat ceilings, an over-all clean smoothness and unending indirectly lit spaciousness all contribute a magnificent Gothic illusion. Deeper inside are superb embellishments climaxed by forests of dripstone columns—and Goliath, one of America's greatest stalagmites.

Still a trifle dazed next day, I welcomed Roy Davis' commiseration: "Only southeastern cavers will ever know what miracles Jay Gurley has worked in Cathedral. It used to be the muddiest in Alabama. Now it's clean, neat, orderly—and magnificent."

The human interest emphasized by the *Reader's Digest* article also seems unexaggerated. Jay Gurley indeed fell victim to the spell of Cathedral Cave—then just large, muddy Bat Cave. Today Jay Gurley does not speak of the terrible months of penniless struggle against overwhelming odds. Nor does he speak of despair when a sudden flood wiped out his hard-built trail. He merely rebuilt it in a difficult spot, high above flood level. Nor does he mention the illness and injuries that came from his "impossible" struggle. Jay Gurley never lost faith in himself, his cave, or his fellow man. Today his Cathedral Cave is a nationwide attraction; he grins about the present and dwells on a bright future. "I heard a couple of ladies talking this morning," he recounted gleefully in 1963. "One said 'When the Indians built something, they really built it right.' The other came back with 'Yes, but they should have built it closer to the highway.' I'll go along with that last one!"

Jay Gurley splendidly typifies the brave new breed of cave operators in love with their caves. Few have an easy life. As Gurley remarked: "An Argentine couple said that down there their government would run anything like this. I told 'em that our government's smarter than theirs. It lets us think we own things. *We* get the ulcers, the frustrations, the problems, the short lives, and the government gets the profit!"

Cavers commonly deprecate the promotional techniques of commercial caves—often justly so. Perhaps we should devote more sympathy to their side of the problem. How would *you* answer the questions people put to the guides in all sincerity: "Are the ladders natural?" "How long did it take for the lights to form?" "Are we breathing air down here?" "Do you wear long underwear?" "Is this cave all underground?" "What makes stalactites attract stalagmites?" "Is it like this in the daytime?" or "What happens when the sun goes down?" "Is this water natural?"

Goliath in Cathedral Cavern. *Photo Courtesy Jay Gurley, Cathedral Cavern*

"Did you model the cave after Disneyland?" Tourists wearing sunglasses often ask to have the lights turned higher. Every day guides are asked "How much of the cave is unexplored?" Variant queries include "How much of the cave is unexcavated?" That one is easy: "Just the solid part, ma'am. But not many people go there anyway." Such men as Jay Gurley deserve real praise—and a greater profit.

Spelean history of this cave-rich region partially antedates that of the more famous eastern seaboard states. Southeastern Kentucky's Great Saltpeter Cave was perhaps the site of America's first detailed speleological study. On February 7, 1806, Dr. Samuel Brown described this cave and its then current saltpeter technology before the American Philosophical Society. Sixty-six years earlier, Sir Alexander Cuming visited a "petrifying cave" (probably Sinking Creek Cave) near Tennessee's Tellico Plains.

The great Cumberland Gap gateway to the bloody ground of Kentucky was already known as Cave Gap when the first official exploration penetrated it in 1750. In 1935 the cave of Cave Gap was commercialized as Cudjo's Cave. This famous cave in the very tip of western Virginia has had a curious succession of names. By 1835 a grist mill, sawmill, and iron forge were recorded in operation at "Cumberland Gap Cave."

In the 1840s this cave was called the John A. Murrell Cave after a notorious outlaw, just as many a cave which never saw Jesse James bears his name. During the Civil War the upper level of the cave became known as Soldier's Cave. In 1892 balls were held here in "King Solomon's Cave."

In 1935 developers chose the present name—Cudjo's Cave—from a famous Civil War novel in which Cudjo, an escaping slave, sought shelter in a cave. The narrative fits Cudjo's Cave surprisingly well, considering that author J. T. Trowbridge never saw the Cumberland Gap region.

A young mountaineer by the name of Daniel Boone may have been the first non-Indian to set foot in nearby Natural Tunnel. This curious tubular cavern averages more than 100 feet in width throughout its 900-foot length. In 1882 its existence simplified trackage for the South Atlantic and Ohio Railroad. The company simply ran its rails through the inviting passage. The Southern Railway still uses the cavern, leaving plenty of room for visitors.

Kentucky has several Boone caves. If Kentucky geologist-historian Willard Rouse Jillson is correct, Daniel Boone spent the winter of 1769–1770 in a small cave near Harrodsburg. Another Boone Cave near Camp Nelson in Garrard County has one opening atop the bluffs of the Kentucky River and another at a lower level. Boone is supposed to have used the cave as an escape hatch when hotly pursued by Indians. Still another cave is his supposed refuge after his famous grapevine swing across a ravine. We will never know if these delightful tales are true. They might have happened. His escape cave is particularly close to Boonesboro.

Innumerable caves here were mined for saltpeter, and visited by Civil War soldiers from both sides. Perhaps the most famous and most hapless Civil War spelunker was Major General William Starke Rosecrans, commander of the Union's Army of the Cumberland in 1863. His nasty assignment was sweeping a large, well-entrenched Confederate force off Chattanooga's Lookout Mountain, thus clearing the way into north Georgia. Late in August, 1863, he visited famous Nickajack Cave, without recorded incident. Early in September, he tried to outflank the Rebels. On September 3, his forces captured Long Island Saltpeter Cave. Two days later, the general camped at nearby Cave Spring. With the road ahead blocked by wagons, he and many of his men took the day off to tour much of that cave's intricate maze. Some 800 feet inside, they left many names commemorating their visit, thus qualifying as DamYankee vandals.

The inscriptions, however, also commemorate a well-documented wedging of luckless General Rosecrans. One Civil War historian noted that, with the portly general thoroughly stuck, "for a few minutes it was a question whether the campaign might not have to be continued under the next senior general." Which might have been just as well. Wily Confederate General Bragg was maneuvering for time to obtain reinforcements. Spelunker Rosecrans played into his hands and saw his army crushed a few days later at the battle of Chickamauga. The fall of Atlanta and the Confederacy were delayed many tragic months.

Some southeastern caves still serve an archaic role. Local cavers become extra-polite when informed firmly: "I don't care what nobody says. They ain't no cave up theah. Understand?" Well they know that during their Tennessee Cave Survey, Tom Barr and Bert Denton encountered eight underground stills producing corn squeezin's—and more than two dozen others smashed by revenuers. In the live-and-let-live country, cavers don't argue with moonshiners.

Moonshining, of course, plays as insignificant a role in the story of southeastern caves as it does in the everyday life of this relaxed region. Yet such periodic encounters confirm the closeness of its caves to its traditional way of life. So, for that matter, do certain legal distilleries. Perhaps you have noted the advertisements of the oldest registered distillery in the United States:

DEEP IN THE LIMESTONE HEART of a Tennessee hill flows the kind of water that Whiskey makers dream about. It's water from a "freestone spring." . . . absolutely free of any trace of iron . . . streaming pure and constant at 56

General Rosecrans' near-fatal signature in Long Island Saltpeter Cave. *Photo by Bill Torode*

degrees. It's water from Jack Daniel's Cave. This rare possession and our slow, old-fashioned methods largely account for the sippin' smoothness of Jack Daniel's Whiskey. . . .

Jack Daniel's Old Time No. 7 Brand Quality Tennessee Sour Mash Whiskey thus has become the caver's own bourbon—though some find it odd that any noncaver would enjoy a product of a southeastern cave. When cavers tipple, instead of a Tom or John Collins they loyally tipple a Floyd Collins: a jigger of Jack Daniel's, a dash of cave water. Only once were spelunker-distiller relations strained here: a hapless spelunker fell squarely from a crumbling ledge into the "streaming pure" water. Aghast, he found great clouds of mud swirling brownly around him. According to the storyteller, the cavers beat the mud to the entrance but the plant had to shut down for three days.

Cavers here have learned to be extra-polite, too, when solemnly

Jack Daniel's statue and cave, located behind the Jack Daniel distillery at Lynchburg, Tennessee.

informed about "ha'nts" and "ha'nted caves." The story of the Bell Witch Cave of Robertson County, Tennessee, bothers us all.

For more than a century, the Bells lived in reasonable amity with Kate, their private witch, whose manifestations were first recorded in 1817. "She could be hateful one minute and pleasant the next," an old account quaintly related. "She knew the age of the earth, and of future happenings. Many believed and witnessed her existence."

Kate, it seems, appeared as a dog, a large bird, or a lovely young girl. For those who missed church services, she thoughtfully recited the entire service, word for endless word, out of thin air. Disobedient slaves were pelted with flying kindling. When annoyed, she commonly ripped the covers unceremoniously from sleeping Bells. Even skeptical President Andrew Jackson is said to have visited the Bells, observed the apparitions, and gone away convinced.

In 1956 Nashville cavers mapped the cave, which—according to some versions of the tale—was the home of the famous witch. Another fine old legend debunked. No witch anywhere.

Except that there was a bit of a problem. One of the group emerged missing his trusty canteen and belt.

Another found his newly filled carbide can empty and outside Kate's cave, his pack still in place.

And in Tumbling Rock Cave, the name Ghost Crawl commemorates the occasion when Bill Varnedoe almost gave up caving. In that low, sandy crawlway, Bill inexplicably began to howl for help. Some animal—

Bill thought it was Phil Zeittler-Seidel's spelunking dog—was pawing and lunging at his back. The mystified cavers behind him in the crawlway suspected that Varnedoe had lost his mind. The dog was outside, and nothing whatever could be seen touching Bill's back.

In such a region only an extraordinary cave and an extraordinary caver could stand head and shoulders above all the rest. Yet such are Cumberland Cavern and Roy Davis, who accidentally made it great. Or perhaps it was Tank Gorin who made it great, even more accidentally.

A seasoned practical joker himself at nineteen, Roy should have been on his guard. But the veteran Tank—officially Standiford Gorin—was his master. Tank insisted that a magnificent new extension opened deep in Roy's favorite Higginbotham Cave. Four miles of tremendous corridors with huge columns that made the much-admired Monument Pillar look like nothing. He had spent fourteen hours underground with a bunch of Boy Scouts, who had discovered the New Extension, and still hadn't seen it all. And, since they had sworn him to secrecy about its location, he couldn't break his word, even to Roy—his best friend.

Every word of it was pure invention. "He fed us a real line of foolishness," Roy recalls happily. "And we really fell for it. We kept at him so he'd 'accidentally' drop hints. We got a pretty good idea that it was off to the left at the far end of the Hall of the Mountain King—the Mountain Room, Tom Barr calls it—so far in that nobody'd really searched. So off we went on our wild goose chase, just as he'd planned. Knowing us, he figured we'd come back with some story even wilder than his when we couldn't find anything there."

So Tank guffawed appreciatively when Roy returned with Kenneth Bunting, David Westmoreland, and Albert Wyatt. He was too canny for the boys' ecstatic tales of fantastic gypsum flowers, of endless miles of virgin corridors, of crystal chambers with millions of transparent gypsum "knitting needles" studding the floor.

Soon, however, Tank was a trifle worried. Other cavers returned, raving equally about the "imaginary" crystal-filled caverns.

"Finally, after about six weeks of this, Tank decided he'd better go have a look," Roy relates with a wicked grin. "All he'd say on that first trip was 'Good heavens. Good heavens!' "

Thanks to Tank's tall tale, the brash young Nashville quartet had realized the lifelong dream of every caver. Searching beyond the rock-strewn nether slopes of the well-known Hall of the Mountain King they stumbled upon a not-so-well-known passage. Soon it narrowed down into a rather ordinary crawlway, apparently ending in a massive rockfall. Only because of Tank's tale did they begin a rock-by-rock search of the unappealing area.

Three hours later Kenneth Bunting slithered upward into a small opening. Roy and the others eagerly followed, popping into a broad new

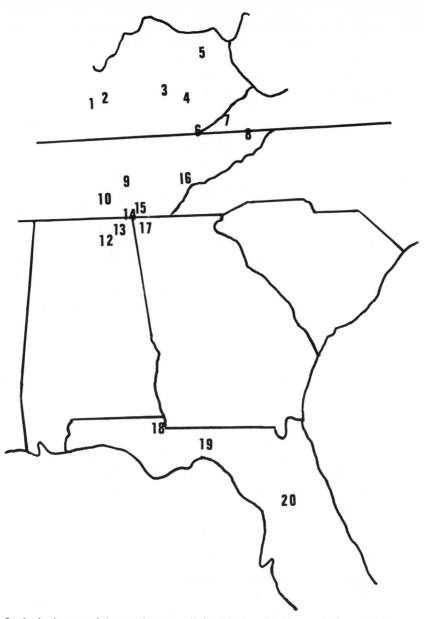

Spelunker's map of the southeastern United States. (1) Mammoth Cave; (2) Horse Cave; (3) Sloan's Valley Cave; (4) Great Saltpeter Cave; (5) Carter Caves State Park; (6) Cudjo's Cave; (7) Natural Tunnel; (8) Bristol Cavern; (9) Cumberland Cavern; (10) Jack Daniel's Cave; (11) Anvil and Fern caves; (12) Cathedral Cave; (13) Russell Cave; (14) Nickajack and Long Island Saltpeter caves; (15) Ruby Falls Cave; (16) Tuckaleechee and Bull caves; (17) Ellison's Cave; (18) Florida Caverns; (19) Wakulla Spring; (20) Silver Spring

chamber 200 feet long. Beyond, thousands of feet of airy, level corridor intoxicated the youths. Losing all sense of time and distance they charged along, shouting ecstatically, jostling each other in their eagerness, stopping only to admire some new grandeur.

Here and there in this glorious Great Extension, waxy crystals of gypsum and coral-like concretions added sparkling beauty to flowstone-decked walls. Only after many hours did oft-ignored fatigue force the avid cavers to turn back. Nevertheless, accumulated exhilaration still compelled the four youths into one last sloping side passage. Picking their way down a steep, muddy slide, the explorers stopped abruptly at the bottom, their eyes incredulous at the frozen loveliness ahead. In a little chamber, a pristine glory of crystalline curls and flowerets of glistening gypsum sparkled in the headlamps. Broad, waxy petals and feathers swirled incredibly, inches long. Only with difficulty was each individual wonder defined from its encroaching neighbor. For a full thirty minutes, the little band paused, awestruck, almost motionless, their eyes traveling from one marvelous creation to another still more sublime.

"We were sure nothing could be finer," reminisces Roy, a faraway look on his young face. "And it was just the start."

Almost delirious with excitement and fatigue, the young explorers cautiously picked their way forward amid weird miniature forests of brittle, needlelike crystals of gypsum. In little grottolike pockets, thousands of the clear needles crisscrossed like jackstraws. Many were a foot long; a few three times that. Usually angled upward or flat on the brown sandy floor, a few jutted outward from the wall and low ceiling like porcupine quills.

On one wall occasional masses of gypsum fuzz closely resembled masses of absorbent cotten, but neither cotton plant nor planter had passed this way. A few feet away clumps of similar threads hung down like long white hair, while gardens of thousands of breathtaking crystalline flowers provided the crowning glory.

Here perfect gypsum lips eight inches wide curved amiably. There a perfect white orchid two inches wide boasted a long irregular pistil curving out and upward for several inches. In this small chamber, a few feet wide and less than a hundred feet long, all the underground glory of creation seemed concentrated. Such things occur once in the lifetime of only the most fortunate caver. The four youths emerged numb, hoarse, and trembling, quite unaware that they had been underground twenty-eight hours.

The story of Higginbotham Cave began six generations before that epochal 1953 day. Roy Davis obtained a shorthand account from old Tom Barnes shortly before the latter's death—folklore, of course, and thus suspect. In the mouth-to-mouth tradition of the hill country of

Tennessee, details undergo rapid change. The essential facts of their ancestral dramas, however, are not readily distorted. In 1819 old Tom's great-great-grandfather died in an explosion in nearby Powder Mill Cave. Source of the fatal saltpeter was Henshaw Cave, less than a mile from Higginbotham Cave's miserable little entrance. Old Tom himself grew up in the tradition of the cave. At the turn of the century he served as guide whenever an off-track tourist came its way. A man of surprising culture, in late life he was acknowledged unofficially as county historian. Minor details varied with each telling, yet his tale seems near enough to the truth:

It was in 1810 or thereabouts. Aaron Higginbotham was a surveyor, coming up the old Chickamauga Trail from Georgia. They camped on the limestone flat where all the arrowheads are, just above the cave. I dunno what he was doing down the slope in the poison ivy, but he felt the cold air about 50 feet off the trail and found the cave. Couldn't get anybody to go in with him, and I s'pose they had to go on anyhow. But he took a mule and came back. Tied the mule to a tree and went in and didn't come out and they had to go rescue him. That's how he discovered Higginbotham Cave. His family settled down in the gap. Lots of his kin still around.

We cannot know what compelling emotions led Aaron Higginbotham to return to his newly found hole. Perhaps his first hurried glance had showed him intriguing spaciousness beyond the crawl-hole entrance. In any event, he lit his pine torch and boldly slithered through. Ahead lay a broad, airy corridor, only occasionally interrupted with intricate natural colonnades—or by rock piles. Only his own moccasin prints marred the dust.

Many wondering minutes inside, a blank wall challenged Higginbotham. Was this the end? He searched the shadowed alcoves, his pine torch tilted to dispel the glooming shadows—vainly.

What about upward? The smoke from his torch rose freely into a narrow, chimneylike cleft barely out of reach. He stooped and shoved a square limestone stepping block beneath the hole. Should he slip, the fall would be negligible.

Stretching his full length upward from the slab, Higginbotham grasped a convenient projection with his free hand. His shoulders squirming upward into the hole, he extended his torch stiffly upward out of harm's way. Feet scraping madly, he sought desperately for a tiny boost. Now he could glimpse a broadening opening above. But there was no foothold.

Giving up the struggle, Higginbotham dropped lightly to the floor. His ankle turned slightly on a loose stone, and he tumbled, his torch striking a rock. Only a few fleeting embers broke the utter darkness.

Sudden icy fear taught Aaron Higginbotham what foolishness cave fever can produce. Here he could only wait and hope—and pray.

Legend has it that Aaron Higginbotham's hair turned white in the three days before friends came hunting him. But Higginbotham was of stern stuff, befitting the future patriarch of the coves of Cardwell Mountain. His family settled, he dragged a notched pole to the site of his fall. With friends he climbed onward into "his" cave, time after time. Up a second slope beyond the crack lay a higher level populated by strange pale crickets, their marvelous antennae waving gently at these outlandish intruders. The corridor beyond was narrower and low, then re-expanded comfortably. A broader chamber loomed ahead in the torchlights. A few bats squeaked protestingly. Side corridors beckoned. Some led nowhere. Others opened to a maze of small passages, and some not so small. Aaron Higginbotham had found himself quite a cave.

His cave fever unabated through the years, Higginbotham led party after party farther and farther into the maze: up and down treacherous cliffs, through sparkling rock piles flecked with tiny gypsum crystals. Alike they penetrated tight crawlways and spectacular underground canyons. The hardiest spent many hours in the dark cavern. Returning from "miles back in there," they spoke of distant rooms so huge their crude torches and hard lamps could not illuminate them.

Most people attributed such tales to more cave fever. Besides, every normal person "knew" by that time that Henshaw Cave was the really important one, "even if it wasn't nearly as big as Higginbotham's Cave."

Not many months after Higginbotham's great discovery, the nationwide 1812 search for saltpeter had touched Cardwell Mountain. A pleasant little cave near the Henshaw Farm contained valuable peter dirt.

Ebullient Roy Davis researched the long-lost trade:

Miners were set to work with wooden shovels and gluts separating the dry earth from the limestone rocks that impregnated the surface of the cave floor. Four wooden hoppers were built in the center of the cave, and these were lined with cedar shingles. The bottoms of the containers were of halved and hollowed logs, interlocking and forming a series of troughs which were to catch the liquid as it was processed. Water pipes were made from lengths of hickory and yellow poplar logs—by tapering one end, bevelling the other, and drilling a hole through the pith—probably with a red-hot auger. These were joined together and conveyed water from a nearby waterfall to the hoppers. Water was percolated through the earth (which had been dumped into the hoppers) and a thick mud was formed. The liquid eventually drained out of the bottom of the hopper, carrying with it in solution the nitrate material desired. This liquid was poured over wood ash and boiled in huge iron cauldrons—evaporating the water and leaving behind the potassium nitrate, or saltpeter. This was carried out in burlap bags, and hauled by horseback across Dark Hollow to tiny Powder Mill Cave where it was mixed with sulphur and charcoal and ground into fine powder.

The 1819 explosion ended the first era of Henshaw Cave. In the decades before the Confederacy again pressed saltpeter caves into service, adventuresome hill people occasionally ventured into Higginbotham Cave. Discounting some wag's "1942" inscription, dates as early as 1855 appear in the old Big Room—a full mile into the cave.

Soon the local explorers ducked under a low ledge at the south end of the Ten Acre Room and emerged into the 250-foot Volcano Room. From that spacious chamber only an ascent of a great pile of huge fallen slabs separated them from the stupendous Mountain Room or Hall of the Mountain King. Eighteen sixty-nine inscriptions show that they ascended that touchy slope and looked onward into the great flat darkness beyond,

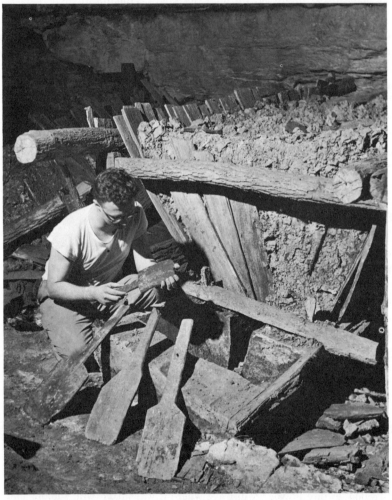

Saltpeter earth was leached in such vats in southeastern caves. Wooden paddles were used to work the mass. *Photo by T. C. Barr, Jr.*

Calcite-encrusted wooden drinking trough left behind by saltpeter miners more than a century ago. *Photo by T. C. Barr, Jr.*

where 18-foot columns rise gracefully to the sloping ceiling in an awesome chamber 150 feet wide and 600 feet long. Here dim lanterns could have shown them only a little patch of smooth ceiling and a circle of gray, slab-strewn, ice-floe-like floor. In the light of only a single lantern, neither wall was visible, nor the greater part of the vast chamber lying ahead. Indeed, they could easily have become lost right in the middle of the room. Yet they pushed on and on. Only failure to crawl through the Keyhole denied them the Great Extension.

The first modern cavers came to Higginbotham Cave in 1945. By 1948 some had crossed the broad, barren flats above the Volcano Room and descended into the echoing grandeur of the Hall of the Mountain King. In its immensity, their carbide lamps seemed hardly more than a pinprick. Even the superbly fluted Pipe Organ, where today's visitors thrill to an inspiring pageant of creation in this cavers' heaven, seemed gray and sullen in the dimness.

The cavers turned to the pleasanter frontier available in abundance elsewhere in the cave. At the far end of the Ten Acre Room they promptly found another chamber almost as large as the Hall of the Mountain King. "You could roll the old Big Room up into a little ball in the center of either," Roy Davis insists, grinning. Freight-car-sized hunks of fallen limestone bestrew this Devil's Quarry. Nearby stands the Monument Pillar—a magnificent column half encircled by an emerald pool. Around its base, calcite crystals radiate reflected light along the corridor. Old Tom

The Monument Pillar of Cumberland Cavern. *Photo by Louisville* Courier-Journal

Barnes in one of his last trips in his beloved cave claimed it as "the world's most beautiful cave formation." Old Tom hadn't seen every cave formation in the world, but more than one far-ranging caver concurs.

More and more spelunkers and speleologists crawled through the narrow portal of Higginbotham Cave. Happily they paced its level corridors, madly they rock-hopped through its slab-strewn auditoriums on their way to unknown darkness. A year of intermittent effort brought Charles Fort to the daylight of a brand-new entrance, but as Fort blinked in the glare he began to guffaw. The proud explorers were less than 250 yards from the "old" entrance of the convoluted cave.

By 1952 almost every weekend saw cavers at work in Higginbotham Cave. Five or six miles of corridors were now known. November was perhaps the most significant month of 1952: harum-scarum Indiana caver Roy Davis moved to Nashville, then set foot in Higginbotham Cave. Neither he nor the cave was ever again the same. Back in daylight, Roy continuously bubbled with enthusiasm. For several months he returned almost weekly, learning more and more of its tortuous ways.

In April 1953 Tom Barr, Tank Gorin, and Dale Smith re-entered Henshaw Cave, seeking a bypass for the exhausting six-hour round trip to the heart of Higginbotham Cave.

Henshaw Cave ends abruptly, a fast one-minute walk from the entrance. At the rear, a jumbled area of narrow cracks and layers of collapsed rock replaces the spaciousness of its lofty Waterfall Chamber. As had many before them, Tom, Tank, and Dale peered into every crevice, wedged themselves into unlikely openings, tested every rock which conceivably might be movable. It seemed hopeless. Furthermore, Tom's map had not progressed far enough to suggest how distant Higginbotham Cave might be.

With the fury of baffled desperation, Dale forced his skinny torso into one more impossible hole at floor level. Helmetless, he turned on one side and reduced his body area by extending one arm. His other arm useless at his side, legs driving like slipping pistons, his chest slowly edged into the jagged nine-inch hole. Muffled ripping noise told the fate of his clothing. Half-stifled gasps and choking breaths indicated some torn skin too. Yet in a series of convulsively irregular jerks, more and more of Dale's body entered the wall.

Crouching anxiously in the little "back end passage," Tom and Tank became more and more concerned. The hole seemed smaller than Dale's body. At that point they could still drag him out at the negligible cost of some additional skin. If he continued, however . . . This was far tighter than Floyd Collins' trap, tighter than any crawlway Tank had ever faced. And Tom was a trifle chunky.

With some misgivings, they queried Dale. Gasping and muffled, his voice filtered back through nearby cracks.

"I can see up ahead over some rocks now. It's a little better, I think."

With tremendous effort, his driving legs jerkily disappeared into the wall, twisting, churning. Down the far side of the rocks he battled, then back up a short, back-breaking angle to an even smaller hole.

"I don't know if I can make it," he panted back. "But it opens up beyond. I'm going to give it a try."

Try Dale did, advancing inch by inch with superhuman effort. Gone were the remnants of his shirt, his temper, and additional skin. But his outstretched arm, then his head painfully emerged into the bottom of a well-like pit, his flashlight pointing to a dark ceiling somewhere high overhead.

Drawing on some last unsuspected reserve of energy and now able to push with his free hand, Dale drove his shoulders free. Additional struggles freed his other hand; then he could push himself out to collapse thankfully on the floor. Panting for long moments in a little rocky chamber, he found himself too exhausted even to call to his anxious companions just a few feet away. Later the oval hole through which he emerged was found to be 9 inches high and 15 inches wide—mute evidence of the compressibility of a lithe young caver.

As strength ebbed back into Dale's battered body, he wobbled to his feet and took stock of his surroundings. His friends reassured, he began to work his way up a steep, rocky slope. A narrow fissure opened upward about 20 feet. Arms and feet braced against the walls, he chimneyed to the top. Falling upward out of the fissure, he flashed his light about him. Its beam showed very little.

For a panicky moment Dale thought of dying batteries, then realization seeped into his exhausted frame: the flashlight beam was merely lost in the gloom of a great stygian chamber. A systematic examination showed him a broadly rounded alcove behind him, perhaps five feet high. Ahead the floor sloped gradually into pockety shadows—unlike the contour of any room he had seen in Henshaw or Higginbotham caves.

But others had been here before. Smoked names were faintly visible on the flat ceiling above his head. As he laboriously rose to scout the surroundings, he found well-marked trails leading to a tiny pool of water in the shadows. A big, barren room, wholly unfamiliar . . . maybe Tank would recognize it—if Tank could get through that horrible Meatgrinder.

Dale descended to the crawlway to shout the news to his worried companions. His words brought Tank through the Meatgrinder after an equally frenzied struggle. Tom simply didn't fit. Climbing out of the fissure after a much-needed rest, Tank blinked perhaps twice, then panted: "It's the Oasis Room. We're in Higginbotham!"

As his breath returned, Tank waxed exuberantly loquacious. "Roy Davis checked that hole you came out of," he told Dale. "Won't he be surprised! But I'm not going back through it. We can get out in three hours the regular way, a lot easier."

Driving homeward, the ebullient trio planned a properly fiendish practical joke at Roy's expense. Tom and Tank prepared a diligent report in utter secrecy. "We wanted Roy to read it in the N.S.S. *News*," Tom insists, gleefully, "but [editor] Bill Hill inadvertently spoiled that."

With pretended reluctance ("You're ruining it!"), Roy joined the others in enlarging the Meatgrinder. Inch by inch, they chiseled the jagged burrow into a caver-sized crawlway.

"We used Tom Barr for scale," Roy laughs. "We figured when Tom could get through, anybody could!"

But Roy miscalculated. One of the first to try the enlarged passage was an older, portly caver. He panicked.

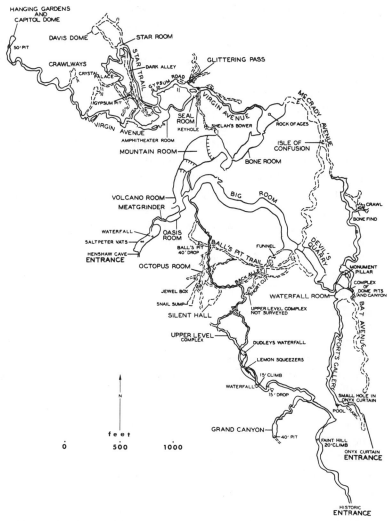

Map of Cumberland Caverns by T. C. Barr, Jr. Courtesy Roy Davis

"I could see him swell up right in my face," Roy recalls. "No wonder he stuck. He frothed at the mouth and went limp. Then we pulled him through with no trouble at all. He went out the long way."

The Meatgrinder slashed six hours and 6,000 feet off the explorations of Cumberland Caverns, as the new Higginbotham-Henshaw combination began to be termed. But, before the new entrance was fully exploited, the gods of mirth intervened. Tank Gorin's teasing led the eager youths to the Great Extension.

The subsequent story of the Great Extension and the Crystal Palace is a glowing tribute to the maturity and vision of its four young discoverers —and of the crowds of responsible cavers who have followed them. More than a thousand persons have viewed its glory. Yet, of all its brittle delicacy, the only loss is the stony pistil of a perfect replica of a tiger lily.

"The guy who broke it felt worse than I did," Roy muses. "His camera brushed it when he was shifting for a better picture. Everything else is just the way it was when we found it. Somebody patted the angel hair, but even that's almost perfect again."

After much deliberation, Roy and Tank signed a contract with the owner of the property and began their labor to commercialize the great cave under the name Cumberland Caverns.

A walkway-sized hole was blasted alongside the Meatgrinder, trails laid out, and superb illumination installed. The change from dim headlamps to electric floodlights wrought miracles. When the murky gloom of the great hall was dispersed, the despised Pipe Organ was found to glow with color. "We weren't even going to include it on the tour," Roy relates. "I don't really know why we dragged the wire down there to see what would happen. I almost went out of my mind when we turned the switch and I saw how magnificent it is."

The Meatgrinder is now only a fond memory, almost obliterated by trail building. During the blasting, careful search revealed a small hole a few yards away. Though impossible for humans, it was ideal for the electric wiring, passed through by a small dog with twine tied to his collar.

One early necessity was a foolproof gate on the historic Higginbotham entrance. Three startled cavers who had camped in the Ten Acre Room in blissful ignorance of recent happenings awoke befuddled in the midst of a floodlit party of high-heeled lady tourists.

From the Oasis Room, the visitor proceeds into the Volcano Room, home of Roy's cherished, hand-carried organ. To perform properly where no self-respecting organ should be, it demands its own electric blanket. Fussing over its crotchety ways, Roy felt personally insulted when four pack rats discovered the electric blanket and cozily set up housekeeping. Regretfully the management was forced to resort to rat traps.

At present the Cumberland Caverns tour extends only to the Pageant of Creation in the vast Hall of the Mountain King. Someday, perhaps, it will continue to the Crystal Palace. But Roy is strangely hesitant. Trails in that area will be a grave problem, and protection of its irreplaceable beauties even more difficult.

Perhaps, too, there is another factor. Roy's favorite practical joke lies in that direction, still occasionally resurrected for unwary visiting cavers.

Some months after the discovery of the Great Extension, someone spotted a short cut by way of a hole that he could hide by sitting on it. Returning from the distant splendors, Roy creates a distraction by pontificating that total darkness upsets man's balancing system. Experienced

Effervescent Roy Davis trims a stony Christmas tree in Cumberland Cavern. *Photo courtesy Cumberland Caverns*

cavers know better and disagree violently. Roy sits down to argue the point and suggests a trial of total darkness. Once blacked out, Roy vanishes down the hole. A partner in crime slides atop the hole and immediately strikes his light, denying that Roy was ever with the party. "Are you nuts? Where is he, then?"

So time marches on at Cumberland Caverns: new trails, new stairs, new lights, new jokes, new stories. Blasting goes on late at night and in the winter doldrums, mapping whenever time permits, exploration whenever someone can make some time free. When loop trails are complete, visitors will admire still more splendors beneath Cardwell Mountain— quite likely some still untouched by any gleam of light.

In the past decade, another six miles has turned up in Cumberland Cavern. Its total length of mapped passages now stands at 22.6 miles, "with an easy five more miles to add," Roy insists. Here and throughout the southeastern United States, caving is a continued, continuing story. Bill Torode rediscovered Long Island Saltpeter Cave only in 1972. More than 600 of Tennessee's 2,100 known caves were first listed in 1973 or 1974.

Despite man's long intimacy here with the netherworld, the story of southeastern caves is only beginning.

10

Follow the Wind

The Story of Breathing, Blowing, and Windy Caves

Before the startled gaze of Burton Faust the wind from the low crawlway died away, then reversed its current. As he continued his spellbound vigil, the wind again slowed, then resumed its original direction. Over and over, every eight minutes, his cigar smoke followed the bright finger of his test candle into the black corridors of the cavern beyond. Just as regularly, the cool breeze returned from the unknown depths four minutes later.

What could produce such a phenomenon? It seemed a very ordinary cave, but the observant Faust had immediately noticed an odd diminution of the airflow.

"You go on. I'll stay here and test the air currents," he suggested to his fellow cavers, fumbling in his pack. What emerged might be termed his secret weapons: an ordinary candle and an extraordinarily foul cigar.

As the faraway clumping and scraping noises of the crawlers dwindled, Burton lit up happily. Settling himself amidst the rocks and mud some fifty yards from the bright summer sunshine, he exhaled a cloud of blue smoke at the crawlway entrance. It blew the smoke back at him.

Strong spelean air currents are not at all uncommon, deep underground as well as at cavern entrances. In 1870 Negro guide William Garvin—later of Colossal Cave fame—followed such a current through the Corkscrew, now one of the main routes of Mammoth Cave. His discovery eliminated almost two hours from journeys to its rivers and the miles of major passages beyond. Although the name of Colorado's famous Cave of the Winds does not pertain to underground winds—as detailed in my 1959 *Adventure Is Underground*—such a breeze also was all-important in its history.

Blowing phenomena were known at the mouths of certain American caves long before anyone paid attention to air currents inside. Hopi Indians long ago wove a blowing hole into their marvelous legendry. Thomas Jefferson recorded that the blast of air from Blowing Cave, Virginia, was "of such force, as to keep the weeds prostrate to the distance of twenty yards from it." Three generations later, naturalist John Burroughs described the gentler outpouring from wide-mouthed Mammoth Cave in matchlessly flowing words:

. . . the cool air which welled up out of the mouth of the cave . . . simulated exactly a fountain of water. It rose up to a certain level, or until it filled the depression immediately about the mouth of the cave, and then flowing over at the lowest point, ran down the hill toward Green River, along a little water-course, exactly as if it had been a liquid. I amused myself by wading down into it as into a fountain. The air above was muggy and hot, the thermometer standing at about 86 degrees, and this cooler air of the cave, which was at a temperature of about 52 degrees, was separated in the little pool or lakelet which is formed from the hotter air above it by a perfectly horizontal line. As I stepped down into it I could feel it close over my feet, then it was at my knees, then I was immersed to my hips, then to my waist, then I stood neck-deep in it, my body almost chilled, while my face and head were bathed by a sultry, oppressive air. Where the two bodies of air came in contact, a slight film of vapor was formed by condensation; I waded in till I could look under this as under a ceiling. It was as level and as well defined as a sheet of ice on a pond. . . . At the depression in the rim of the basin, one had but to put his hand down to feel the cold air flowing over like water. 50 yards below you could still wade into it as into a creek; it had begun to lose some of its coolness and to mingle with the general air; all the plants growing on the margin of the water-course were in motion.

At times the Mammoth Cave–born summertime river of cool air wells up from the stygian depths with a mystic cap observed by few. On warm, still summer nights, a rippling shroud of starlit fog crowns this silent torrent of outpouring air. Almost hypnotic in its perfection, this unsung wonder must be counted one of America's greatest spelean spectacles.

Such caves reverse their basic airflow with the seasons. Large caves with small, single entrances like South Dakota's enormous Wind Cave "breathe" with changes of the barometer. "Five days and nights is the longest time the wind has been known to move in one direction without ceasing," geologist Luella Owen wrote of Wind Cave more than a half century ago. All caves "exhale" somewhat when the barometer falls and "inhale" when it rises, but these ordinarily are slow, obvious matters.

Burton Faust knew all this and more. He had thought little of the air current in this Burnsville Saltpeter Cave until it stopped. That was peculiar.

By the time the cavers' commotion had subsided, the crawlway was blowing again. As Burton maintained his observant gaze, the flickering of his candle lessened, then again stopped. A fume of cigar smoke hung motionless in the air, dispersed only by its own convection.

Suddenly Burton sat bolt upright. As the flame began to flicker again harder and harder, it pointed into the crawlway instead of entranceward. A new cloud of stogie smoke vanished into the black cavity. For many heightened heartbeats, the unknown cave inhaled.

The rate of flow slowed. All was still. Then a barely perceptible wisp of cool, stogie-tainted air touched Burton's excited face. The preposterous cave had begun to exhale once more.

Again and again, Burton tested this eerie breathing. While he awaited the return of his friends, he timed the endless cycles. No doubt about it. The cave was inhaling and exhaling in four-minute periods.

Hours raced by like minutes. Finally Burton's keen ears caught the scuffling noises of his friends' return. Soon they grunted through the little crawlway, babbling about interconnected corridors and flowstone canyons. It was a big cave—how big they could not guess.

Half expecting some tall tale from their comrade, they showed all the skepticism Burton anticipated—for eight minutes.

"Maybe there's a big intermittent siphon spring way back in there somewhere," someone suggested. Obviously a tremendous pond would have to empty and refill with remarkable speed to cause such an effect at the other end of the cave. Besides, intermittent springs are rarer than breathing caves.

"Maybe the wind works it like when you blow across the mouth of a Coke bottle, only lots slower," someone else suggested hopefully if obscurely. "Or maybe it's a kind of natural oscillation after a change in the atmospheric pressure." After profound mathematical analysis, caver-physicist Bill Plummer gave this theory considerable support and a technical identification: infrasonic resonance. Many speleologists nodded sagely. Maybe that was it.

Maybe. That was the key word. Only one conclusion was possible. Many caving hours would soon be devoted to exploring the depths of this mysterious Breathing Cave.

Before his initial report on Breathing Cave reached print, Burton Faust had returned five times to study its mysterious depths. Each time the cave was breathing, at three to seven and one-half cycles per hour. Each time he found no reasonable explanation for the uncanny "Burton phenomenon."

Attracted alike by the remarkable phenomenon and the ever-enlarging cave, more and more cavers flocked to Breathing Cave. Geologists found it of unusual interest, for it continued deeper and deeper where certain theories required it to level off. Barometric studies with instruments specially devised by Don Cournoyer yielded much information—confusing to all concerned.

Before a need for open heart surgery ended his active spelunking in 1971, Cournoyer logged 209 trips into the mysteriously breathing darkness. But it was a quite ordinary cave, even though it unrolled into a lesser, tilted Anvil Cave: an intricate network with 4½ miles of passage, plunging down and down the dipping beds on the west wall of Sinking Creek Valley to a depth of 370 feet. Its explorers eventually met ordinary cave fills plugging the bottom of each branch not stoppered by a pool. In the grudging way of fills far back in difficult caves, they yielded to digging, but at the cost of so much energy that all lost heart. Neither Cournoyer nor anyone else ever did discover an intermittent spring or any other natural bellows that would explain the weird breathing.

This was no way for a self-respecting cave to act. Much more reasonable is Blowing Cave on the bank of the Cowpasture River a few miles farther south. Yet even here spelunkers have found mystery.

Blowing Cave was one of the first caves known in America. Located alongside the old stage trail from Augusta Courthouse (now Staunton) to the Hot Springs Valley and the Indian West, George Washington passed it many times. Old accounts indicate that the cave was a customary stage stop. Ladies particularly seem to have been entranced by the cold blast of air which could suspend their handkerchiefs. Before its orifice was widened by road operations, the force of its wind was no tall tale. The late Bill Foster once found a similar outblast from West Virginia's Cave Hollow Cave. The air current was so strong it shook weeds two hundred yards from the cave's inconspicuous entrance.

All these accounts of Blowing Cave were given by summer travelers. Had they been forced to travel in uncomfortable Virginia mountain winters, they might have been even more amazed. In winter the strong outblast of air is replaced by an equal inward wind.

This is how a cave should behave if it has two entrances at different elevations. In summer, relatively cool cavern air is denser and thus heavier than the summer warmth outside. It thus flows down and out wherever it can find an exit, and is replaced by air sucked inward through the upper entrance. In winter, cave air is warmer and thus lighter than that outside. It escapes upward wherever it can and is replaced by frigid air creeping inward, lower in the cave. The upper entrance of Blowing Cave has not been discovered. Legend says that a dog lost in Clark's Cave, five miles away, came out of Blowing Cave a week later, limping and emaciated. Half the caves of America, however, seem to share this apocryphal tale.

Blowing Cave was the subject of the first *Report of Caves Surveyed by the Speleological Society of the District of Columbia.* Explorations eventually reached a stream flowing away from the Cowpasture River— and thirteen feet below its level. Even at this depth the blowing of Blowing Cave still indicates an upper level somewhere ahead.

Then the United States Army Corps of Engineers took a hand in the story of Blowing Cave. Conquered makeshift-style by Yankee ingenuity, the fortress caves of Okinawa had cost the lives of far too many young Americans. For the bloody assault planned upon the home islands of Japan, the Army wanted new, tested techniques. Blowing Cave was an ideal mock target.

The explosives tested at Blowing Cave, however, were not mock TNT. Their experiments in June 1945 left the outer regions of Blowing Cave a shattered mass of rubble. It took cave-digging teams a generation to break through. Even they were unable to find the source of the mysterious river, nor where it leads.

Yet Blowing Cave and Breathing Cave are mere minor curiosities when compared to the windy caves of South Dakota's Black Hills. For a long time these caves were held in low repute by most of American speleology. "Nothing there but Wind Cave, and that isn't much!" some over-scornful spelunkers long remarked. The 1962 convention of the National Speleological Society was held in the heart of that delightful region, but it was less well attended than any other in recent years. We should have known better.

Unusual among American mountains, the Black Hills comprise an eroded dome. Its heart is of exposed granitic rocks where overlying bedded rocks have been weathered and washed away. On the flanks of the dome, laid out like an off-center bull's-eye, successive rings of ever-younger rocks are exposed. Here in a thick ring of limestone lie dozens of caves. Nestled on the southern flanks of this unusual range are Wind Cave and Jewel Cave, incredibly vast networks which breathe more dramatically with changes in the barometer than any other known cave.

A wonderful story is related that a certain Tom Bingham had his hat blown off when he stopped to look at a foot-wide whistling hole. Returning to "show" his skeptical brother, Bingham lost his hat, sucked into the hole. A storm had passed, and the barometer had risen with the clearing skies. They had to enlarge the hole to recover his hat, legend recounts, and thus discovered Wind Cave.

Authentic details are lacking, for the discovery probably occurred only five years after Custer's nearby last stand. There is still plenty of wind in

Wind Cave, however, even in the artificial entrance tunnel, which is many times the diameter of the original opening. Today reaching a velocity of 18 m.p.h., the air jet through the tiny original entrance must have been truly ferocious whenever the barometer fell.

Earliest explorations revealed Wind Cave as a fantastic honeycomb of big passages, little passages, and horrible crawlways on level after level. By 1904 the noted French speleologist E.-A. Martel could tell the Eighth International Geographic Congress: "It would be interesting to know if Wind Cave in South Dakota possesses, as has been stated, 2,500 rooms, 97 miles of avenues and a depth over 1,000 feet, and if it is really an old extinct geyser. . . ."

It isn't, and it doesn't. "In going to the Fair Grounds," a turn-of-the-century guide told Luella Owen with a straight face, "we travel about three miles [actually less than one mile—W.R.H.]. In each fissure there are eight levels, which makes twenty-four miles of cave from the entrance to the Fair Grounds."

Even without this intriguing multiplication by twenty-four, however, Wind Cave was not to be ignored. In place of stalactites and stalagmites

Original entrance of Wind Cave, South Dakota, where Tom Bingham is said to have lost his hat.

Some of the intricate boxwork for which Wind Cave is famous. *United States Department of Interior National Park Service photo*

(surprisingly uncommon in Black Hills caves) it possesses remarkable accumulations of boxwork. This peculiar petromorph resembles gigantic ice-cube tray dividers gone wild. Unlike stalactites, flowstone, and most other cavern decorations, the boxwork was formed before the cave, as veins in fissured limestone. When the less-resistant bedrock was dissolved around it, the boxwork remained in place.

A thirty-man National Speleological Society expedition spent a delightful but frustrating week in Wind Cave in 1959. Lightheartedly but with serious goals, groups of geologists, biologists, surveyors, meteorologists, and other specialists fanned out in the cave. Specially selected teams of explorers were channeled to groups requiring their services. At preposterous hours, increasingly bearded cavers surfaced to eat. Seemingly they slept not at all.

This was no easy cave. The sharp-edged boxwork discouraged even the most avid cave crawler. So did the very nature of the sprawling cavern. All too often the explorers found themselves merely progressing in tangled circles.

Seeking sense from a thousand man-hours in this bewildering maze, the team drove itself to frustrated exhaustion. Much information was collected but no vast new volumes of cavern were unearthed. Because of the interlocking levels, no satisfactory map evolved, and the formal expedition report was five years late. When the 1966 edition of this book went to press, a "mere" 4 miles was said to have been mapped in Wind Cave. Dividing the old guides' "97 miles" by their factor of 24, it seemed that they had a shrewd knowledge of the honeycomb cavern. Besides, the big caving news was over on the western flank of the Black Hills dome. There, the entrance of Jewel Cave was long posted "THIS IS A SMALL BUT BEAUTIFUL CAVE." Through more than coincidence, this "small but beautiful" was out of bounds to the 1962 conventioneers. Too much was happening.

For many years everyone had considered Jewel Cave a nice little crystal-coated cave in Hell Canyon. True, in 1900 the Michaud brothers who discovered it claimed it to be the north end of Wind Cave. But Wind

Helictite bush in Wind Cave. *Photo by Charlie and Jo Larson*

Cave lies 20 miles to the southeast. More of Jewel Cave was known to exist beyond 900-foot "primitive" lantern-lit tours provided by the National Park Service. Inquisitive rangers, CCC personnel, and curious explorers, however, had reported only a frustrating little maze of small, dirty passages and treacherous climbs leading nowhere.

Late in 1957 ranger Dick Hart and his enthusiastic young seasonal staff poked into a slimy crawlway—Milk River—beyond the Heavenly Room at the south end of the tour. Noting a breeze through a mass of breakdown, they cleared a passable hole and crawled painfully into a canyon-like room: the first sizable chamber of Jewel Cave. Ahead was an area "where the medium-sized rooms would hold a locomotive engine and tender." For squeezeway-rich 2,131-foot Jewel Cave, this was truly exciting. Ranger Bill Eibert and Delmar Brown mapped the new rooms and rocky crawls and totaled the cave's known length at almost one mile. Jewel Cave wasn't so small after all.

Enthusiastic young geologist-caver Dwight Deal obtained permission to study the cave in detail. For help he turned to Dick Hart, Dave Schnute, and Barbara Tihen—and most important, Herb and Jan Conn.

In their late thirties, the Conns were of the venturesome breed which produces exceptional cavers. A love of climbing led this talented pair of easterners to the University of Colorado, then to the spire-peaked Black Hills. Herb's engineering talent and their puckish humor soon marked the history and map of Jewel Cave indelibly.

At first the Conns lacked enthusiasm. Schoolhouse Cave had left them cold. They preferred the clean, free wind of the pine divide. But Jewel Cave soon laid its own special spell upon them. With their rock-climbing experience, its most difficult pits and pitches were child's play. Using new techniques engendered by two decades of intensive spelunking, they began to unroll Jewel Cave. Perhaps it would be more appropriate to say that they turned it inside out.

To his dismay and potentially that of a well-advertised brewery, new ranger Keith Miller found that a miserable but strategic crawlway he had discovered was soon listed as Miller's Low Life. When Dwight missed a few trips, his next discovery in a spacious, seemingly virgin passage was a carefully carved wooden sign, EASY STREET. Remote Helen's Room was named for the mythical Helen Gone; it seemed that far in. Grief is the pit you should stop before you come to. On the enlarging map appeared such cognomens as the Average-Sized Room, the Atom Smasher, the Short Cut to Oblivion, the Snare and Delusion, Sunburn Haven, Monotonous Passage, Black and Blue Grotto, and the Kittycombs. Instead of the inevitable Fat Man's Misery, Jewel Cave boasts a Thin Man's Misery, which no fat man should even try. The Drydock is a chamber containing a large, boat-shaped rock. A 1910 tobacco tin was found in

Tobacco Road. The Dog House is the room at survey station K-9. Far back in the cave, Gyp Joint marks a large passage dazzling with small gypsum flowers.

The first few expeditions under Dwight's direction demarcated the Loft, a domed upper level fretted with aragonite frostwork. In another region Dwight followed a strong air current perhaps a trifle incautiously, dreaming of a second entrance. A half-swung slab of stone opened Fibber Magee's Closet. Out poured an avalanche of small boulders. Splendidly if bumpily, Dwight rode them down a natural slide. Herb and Jan swear that, in turn, a large boulder was riding Dwight's head, fortunately protected by a hard hat.

In time each alcove of the Loft was checked, each pit plumbed, every lead exhausted without any truly extraordinary discovery. Old suppositions that there were interconnecting rooms off to the north and closer to the watery mud of Milk River were recalled and investigated. The winter of 1959-1960 passed pleasantly in near-weekly trips to a new labyrinth beyond Miller's Low Life. Included was a room much larger than those found earlier—the Gear Box. "By the time we crawled out into it, we sure were traveling in low gear," says Jan impishly. Nearby, a 1908 Sears Roebuck catalogue featured Kissle horse-drawn plows. A rotten rope, candle stubs, the cryptic initials J. M., and other artifacts showed that others had passed this way, but not much farther. Dead ends lay close beyond.

The exuberant explorers were far from discouraged. "Just beyond the rotten rope was an intriguing hole in the floor," the Conns reminisce. "We called it the Rat Hole. Rocks poised overhead threatened to block the hole and turn it into a rat *trap,* but after gingerly testing their security, we ventured into the hole. We were in virgin cave again. No one had disturbed the soft mud floor. Even better, there were more large passages on to the north, including Penn Station. About 75 feet wide, 150 feet long, and 50 feet high, it dwarfed the Gear Box. We had found the most extensive area yet, with large passages heading northeast into the unknown."

Through 1960 the Jewel Cave team followed these leads. Track Nine stretched invitingly beyond Penn Station toward the Car Barn, the Monotonous Passage, and the spacious Hippodrome. Devious zigzagging

Simplified sketch of portions of Wind Cave mapped through 1972. Based on National Park Service maps by Windy City Grotto and others. (1) Entrance; (2) Elevator; (3) Fairy Palace; (4) Postoffice Room; (5) Rainbow Falls; (6) Northwest Area; (7) Guides' Discovery; (8) Graveyard; (9) Wind River Chamber; (10) Crossroads; (11) Elks Room; (12) Blue Grotto; (13) Monte Cristo Palace; (14) Fairgrounds; (15) Pearly Gates; (16) Garden of Eden; (17) Rome; (18) Independence Hall; (19) Mammoth Canyon; (20) Spar-tooth Chamber; (21) Red Crystal Canyon; (22) approximate boundary of "new" portion of cave; (23) Land of 1,000 Lakes; (24) Windy City Lake; (25) Calcite Jungle; (26) Base Camp #1; (27) Base Camp #2; (28) Club Room; (29) Half-mile Hall; (30) Chimera Room; (31) Figure Eight Room; (32) Mountain Room

through a labyrinth of small, agonizing passages led to the Breezeway, Carnegie Hall, and finally the crucial Beeline. More than 1,000 feet long and over half a mile "in a beeline" from the entrance, the Beeline was the farthest point reached for many months.

Weird and wonderful things lay in this vast new complex of levels. Via corkscrewing tubes, two corridors connected to overlooked holes in earlier discoveries. One of these obscure routes traverses Einstein's Tube. That impossible-looking pit ignores the obvious corridor 25 feet below in favor of the Fourth Dimension, a corridor 75 feet down. To pass this way requires doffing clothing and snaking the vulnerable body through a vertical slot.

December 1, 1960, was a particularly unforgettable date. On that date, descending a pit near the Hippodrome, George Marks managed to set afire his luxurious beard. He no longer carries carbide lights with his teeth.

Such trips were worthwhile. Beyond Einstein's Tube, the awed explorers came upon a frosty, glistening passage. Lined with sparkling coral-like accretions, the floor of this Treasure Aisle was piled deep with glittering, snow-like crystals. In such areas the Conns swung ape-like, straddling along the passage walls, suspended by their arms and legs, determined to cause no damage to their new fairyland.

By January 1, 1961, Dwight Deal, the Conns, and other dedicated teams had mapped a total of 19,116 feet in Jewel Cave. By chance, Jewel Cave then was a momentous crossroads. Considerable skepticism had long existed in the collective mind of the National Park Service about the national significance of "little" Jewel Cave. Early in 1961 official doubt was expressed as to whether the national monument should be retained.

Much of beauty and scientific interest lay in the new regions, but they were far beyond the reach of the public and of most scientists. As the unknown receded more and more, the endurance barrier loomed increasingly. A new entrance began to assume more importance. Unfortunately, the enthralling new complex lay deep under a hill. The nearest canyon slope—Lithograph Canyon—lay thousands of feet farther southeast. Through the entire year of 1961, only one triumph emerged: with six miles now mapped, "pretty little Jewel Cave" surpassed "immense Wind Cave."

By 1962 Dwight Deal had retired to prepare a thesis on the unrolling cave, but the others redoubled their efforts. In March the Conns followed the wind of spacious Eerie Boulevard eastward to a small opening just beneath the ceiling. Returning on March 24 with Fred Devenport and new ranger Pete Robinson, the Conns took crowbars to some exasperating fallen blocks that obstructed progress. Beyond stretched the interminable crawlway they dubbed the Long Winded Passage.

"It was a real breakthrough," the Conns relate. "Pete, on his second

A particularly beautiful nest of crystals in Jewel Cave, South Dakota. *Photo by Charles H. Anderson, Jr.*

try at spelunking, stepped forth in the awesome blackness of King Kong's Kage, a room that dwarfed Penn Station. It proved to be but the beginning of a series of huge rooms and passages leading northeast. Just beyond King Kong's Kage is the Crystal Display Room where—in a cave literally filled with calcite crystals—we felt they reached the ultimate in beauty. Eastward were gypsum formations, needle crystals, wads of cave cotton, and long silken beards that waved in the heat from our lamps."

Additional progress in that rewarding area had to wait. The cavers had made a giant stride eastward, but now they were angling away from Lithograph Canyon. Would the uncomfortable Long Winded Passage extend farther canyonward? Tedious crawling proved that it did.

Some of the wind had gone from the Long Winded Passage, but enough remained to lure the tired explorers on and on. Curiosities multiplied as they advanced. Pink "popcorn" and glorious displays of many another superb speleothem gleamed in their advancing lights. On April 14, 1962, they entered a glittering chamber, deep red in color. On its walls were reddish, wormlike crystalline growths coated with sparkling quartz. Dwight Deal came, looked, and shook his head. So did expert speleomineralogist Will White. These were not helictites. "Scintillites" they named these extraordinary deposits.

The Long Winded Passage finally ended as a small pit, too narrow to be negotiated. The wind roars on through. Dropped stones echo as if they land in a very large cavity. There were other, easier pits near the

Crystal Display. Lower in the cave than man had ever gone, they led the Conns to bypasses toward Lithograph Canyon. Scores of inviting passages opened in other directions.

In May 1962 the first water was found in the cave. Not far away were dripping stalactites and a 15-foot wall of flowstone—signs that the long-sought canyon side should not be far distant.

The avid cavers pressed southward, only to halt at a difficult pit topped by a blank wall of red clay. Another venture located an alternate route down pink dirt slides to the 250-foot level, where a spacious passage led due south. Ahead gleamed more pools, colorful draperies, thickets of soda-straw stalactites. Pockets of unattached concretions and massive flowstone mixed dramatically with the crystal undercoating of the far-flung cave. On walls coated with squeezable "moon milk," inch-long natural balloons with walls three or four thousandths of an inch thick glistened like soap bubbles. The blackness of an unexpectedly large chamber —the Target Room—swallowed up the lights of the dazzled explorers. Within easy tunnel range of Lithograph Canyon, this area demanded public adulation.

Beyond the moon-milk area, Al Howard felt another breeze on his face. An entrance or vast extension of this great honeycomb? Even the Conns could hardly keep pace with Al as he charged onward, nose to the wind. Up a chimney, then another, the trio raced, sure that they were about to glimpse the telltale blue glimmer of daylight.

"We traced the air flow to a tiny hole which we feared was too tight," Jan recalls. "Here the wind whipped along at a prodigious rate, flapping our clothes and blowing out our carbide lights. The spot is now known as Hurricane Corner. We estimate winds up to twenty-five miles an hour."

The crack was slightly wider than the standard Jewel Cave measure—Herb Conn's head. With difficulty, the team squeezed through. A tortuous obstacle course led to a lake chamber precisely beneath Lithograph Canyon. Automatically it became the Pool Room. But at the next corner, a nasty fifteen-foot drop required a return trip with rappel gear.

Below the rappel, the wind was lost in a large chamber until someone traced it to an inconspicuous hole amid broken rock. The eager explorers met and conquered another obstacle course, but the route curved back beneath Lithograph Canyon. There the wind whistled through a hopeless six-inch crevice.

Every cranny was probed minutely. No bypasses could be found. Dejectedly the crew squiggled homeward, suddenly very tired.

A few dozen yards toward the entrance, another blowing hole turned up. Just barely squeezable-through, it led southward, farther and farther. Beyond Lithograph Canyon, Jewel Cave began to enlarge again. The revitalized cavers mapped several hundred feet, stopping only at a climb requiring pitons and rope.

Beyond Lithograph Canyon is another, shallower gulch, and another, and another. Beyond the horizon lies Wind Cave. Suddenly the ancient fancies of the Michaud brothers seemed less humorous. Whence came this mysterious underground wind so deep in the cave? Would it finally lead to the long-sought second entrance, or to undreamed volumes of cavern heading onward toward Wind Cave?

Initial investigations suggested that the latter was more likely. As at Wind Cave, this air flow alternates with barometric change. Basic investigations revealed that as much as twelve hours were needed for the cavern to inhale or exhale half the colossal volume of air needed to equalize changes in barometric pressure.

Many incalculables are inherent in any attempt to figure cavern size from determinations of air flow. Nor can anyone be certain what percentage of a complex cavern is traversable by humans. Yet, if nothing more, such calculations are always interesting. "Preliminary figuring" by the Conns indicate that the volume of Jewel Cave is about one billion cubic feet *if* no complicating factors exist. And that's an extremely large *if*. A billion cubic feet is the equivalent of a thousand miles of passages 10 feet high and 20 feet wide.

With so much to be done, in 1962 and 1963 the Conns chose to investigate leads closer to the entrance. They brought their total of Jewel Cave trips to 179, with 225 days underground. Each trip increased the cave's importance. The Rambling Loft held superb frostwork and massive coralloidal decorations. A small chamber tucked away between levels near the scintillites boasted hollow stalagmites of popcorn-like accretions 8 to 12 feet high. Through each center and deep into the floor are narrow natural tubes. No one has explained why.

In February 1963 the regional office of the National Park Service formally approved the development of the magnificent hinterlands of Jewel Cave, with a new entrance and elevator shaft leading directly into its innermost splendors, now open to the public.

So the Conns went out and moved the innermost part of the cave.

It seemed as if almost the entire triangle where the two canyons met was hollow. New names that could only have stemmed from the Conns dotted the ever-expanding map: Dire Straits, Exhaust Pipe, The Separator, Brain Drain, Road to Perdition. Beyond Hurricane Corner was Cloudy Sky Room and Storm Shelter. Beyond Lithograph Canyon—quite a bit.

The new entrance brought Point Number 19—its number on the map in the first edition of this book—much closer to the ingress, but not until July 1966 did the Conns return there. With Missourians Earl Biffle and Father Paul Wightman they surmounted the climb but lost the wind and found little. Two years later, however, they sniffed out another wind somewhere in the enormous ant nest filling the map between

Hell and Lithograph canyons. For a year it led them zigzagging westward beneath and beyond Hell Canyon. Incredible dreams arose unbidden: could a single cave ring the entire Black Hills?

That particular dream was shortlived. Instead of continuing into the thick limestone plateaus west of Hell Canyon, "the wind switchbacked abruptly around The Horn, where it has been measured at 32 m.p.h." Often they found themselves forced to siphon off considerable water to be able to pass without getting soaked (at 32 m.p.h. wind chill is serious within seconds, even with the temperature at 47 degrees). On and on the wind led them, back under Hell Canyon, across the V between the canyons "in big cave," crossing anew beneath Lithograph Canyon in a remarkably tight underground strait, then through walking passage to a point just 300 feet south of Number 19 with no end in sight.

The Horn was distinctly the long way around—an extra four hours, to be precise. So the Conns returned to Number 19 and started digging south. Sixty feet onward and 240 feet short of their target, they were distracted. The maze of seemingly interminable crawlways they struck soon bore the proper name: The Miseries.

"Eventually The Miseries shut down into a number of windy holes, all too tight for us," the Conns happily aver today. Yet through The Miseries Jewel Cave surpassed the triumphal 50-mile mark. Somehow the Conns discovered and conned two "minicavers," who they swear can flatten through crawlways just six inches high: Sandy Ramp and Deon Simon. On July 22, 1973, Sandy broke through The Calorie Counter. In retribution, she conned the Conns into burrowing their way after her, into what turned out to be 700 feet of Mini-Miseries. "The wind moved a good deal faster than we did, but eventually it led to the brink of bigger things," they reported. "We climbed down into Metrecal Cavern feeling a bit like a tiny vitamin capsule emerging into the aching blackness of a weight-watcher's empty stomach." Beyond were 3 miles of continuing rooms and passages with some of this extraordinary cave's most spectacular displays: "At Wildflower Walk the gypsum flowers are really wild!"

So was the unique celebration on December 3, 1973, at the magic 50-mile mark. With Mark Stock, Dennis Knuckles, and Karl Martin, the Conns ceremoniously ignited fireworks ("one sputtering sparkler left from the Fourth of July"), and sacrificed Dennis' red nylon left shoelace for a mock-formal ribbon cutting. Tragically, no one had a camera at the unique scene. "If they had, probably we wouldn't have made the necessary survey mileage," Herb asserts.

As I write, the total mapped in Jewel Cave is 54.4 miles. Out beyond Metrecal Cavern and the 1,500-foot Mind Blower, "things are still wide open and the wind is as strong as ever." Perhaps with tongue in cheek the Conns write that they would like to turn the next 50 miles over to younger blood (Herb was 54 and Jan 50 at the end of the first 50 miles). "But we find to our sorrow that younger people find The Miseries every bit as obnoxious as we do so perhaps we don't have a legitimate excuse to retire for a few years yet!"

Most cavers are willing to wager that they will never find a legitimate excuse. Jewel Cave is simply too exciting. As Herb recently wrote me, "The Wildflower Walk gets wilder each trip. We just found a single hair-like filament of gypsum 63 inches long hanging from the ceiling. How many such have we walked by without seeing?"

Probably they'll never have the leisure time to find out. The wind still calls.

Over on the other southern flank of the Black Hills, eleven years separated the 1959 N.S.S. Wind Cave Expedition from the next. The gap was not wholly a vacuum. Members of South Dakota and Colorado grottos of the society periodically investigated and mapped hundreds or thousands of feet of complex cave. So did spelunking rangers, groups from the South Dakota School of Mines, and visiting Jewel cavers. It was the Conns who in 1962 compiled the first modern map of Wind Cave. They returned, periodically exploring and mapping anew, but here somehow without their Jewel Cave ebullience. Yet in December 1968 they and Dave Schnute found an underground lake far to the southeast.

Unfortunately most of these discoveries led the explorers south and southeast: the wrong direction for anyone heading toward Jewel Cave. To outflank the granite core of the Black Hills, a southwestward breakthrough was needed. Sure enough, little by little the big excitement angled southwest. Rooms unexpectedly large for the Black Hills caves added spice to thousands upon thousands of feet of zigzagging passages. With 11.3 miles on the map, unexplored orifices increasingly opened in all directions.

Various cavers suggested different approaches to the new challenge. Sponsors, perhaps forgetting some of the 1959 problems, formed another expedition. For three weeks in August, 1970, eighteen cavers from Chicago's Windy City Grotto of the N.S.S. utilized a base camp at the entrance of Gateway Hall, in the heart of the new discoveries.

Stocking and unstocking any base camp inevitably consumes time and energy. So do repetition of routes and training sessions. Nevertheless, the progress of 1970 was greater than that of 1959. They successfully reached the water table, found at a lake farther down in the southeasterly fringes of the enlarging network than anyone had previously pushed. There and in a remote southern region between base camp and the huge new Club Room they mapped almost two miles. Theirs were such unique discoveries as a "bush" of intertwined helictites six feet high.

It was obvious that they had merely begun. So back for two more weeks a year later. This time the Windy City Grotto planned three

Sketch of Jewel Cave, South Dakota, based on 1973 National Park Service map by Herb and Jan Conn and others. (1) Original entrance; (2) Tunnel; (3) Elevator; (4) Kitty-combs; (5) Penn Station; (6) Track Nine; (7) Easy Street; (8) Mighty Tight Street; (9) Beeline; (10) Eerie Boulevard; (11) King Kong's Kage; (12) Blessed Relief; (13) Esophagus; (14) The Horn, Exhaust Pipe, and High Water; (15) Sinner's Reward; (16) Pool Room, Hurricane Corner; Cloudy Sky Room, and Storm Shelter; (17) The Miseries; (18) Metrecal Cavern; (19) end of exploration in 1963; (20) many leads continue. HC: Hell Canyon; LC: Lithograph Canyon

survey teams, working in widely separated areas. The first didn't produce much: merely "a few thousand feet of good-sized walking passage including two large rooms." Its effort soon halted, and the third team never had a chance to get organized. As the second team worked south from the Club Room complex, Chris Hill stuck his head up a small dome the Conns had bypassed during an earlier survey. Beyond was far more cave than previously had been known in all of Wind Cave combined, climaxed by a chamber 2,800 feet long: Half-Mile Hall.

As the Chicago cavers became more and more familiar with the route from the elevator shaft to their base camp, the fastest began making the trip in less than an hour. Out on the ever-expanding periphery, however, things were different. Fighting jagged boxwork in tight crawlways and chimneys was not the world's easiest going. After about 4 miles of mapping, time and energy again ran out for the Windy City cavers.

Their once-remote base camp now barely an hour from the elevator, the question arose whether this inherently inefficient concept should be abandoned, or the base moved much deeper into the cave. The latter alternative was chosen for 1972, not without some misgivings (especially by conservationists, who pointed out inevitable damage to caves by such camps). Perhaps either alternative would have been wrong, in part because 1972 enthusiasms got out of control. The expedition grew to an unwieldy 65 persons, many of them volunteers unfamiliar with the Black Hills and the idiosyncrasies of its caves. Yet, in six weeks, Wind Cave jumped several notches on the list of the world's longest. Both the passages and the map began to look surprisingly like a Jewel Cave with larger rooms. Their annual report wailed: "There are, as a *conservative* estimate, more than 2,000 virgin leads yet to be checked!" The new, ultra-remote Base Camp II, moreover, again had been found to be within an hour of the elevator shaft. That was the end of the base camp concept at Wind Cave. More routine operations in 1973 and 1974 boosted its total to nearly 29 miles bringing the total to more than 80 miles mapped in the two windy networks. Even though the old-timers never saw any of these new discoveries, the old guides' "97 miles" in Wind Cave no longer seems so laughable. Nor the Conns' "thousand miles of passage."

Nevertheless, in these two exciting decades, the cavers have advanced only about 1 mile from each entrance toward the other. Approximately 18 miles still intervene. Officially, no attempt is being made to connect the two vast caverns, with exploration basically following the most promising, most important leads. American cavers grin knowingly at such official statements. We well recall the similar official pronouncements of the Cave Research Foundation.

Obstacles here within the Black Hills are more formidable than in Flint Ridge. Yet even the conservative reports of the National Park Service point out that "The lateral extent of the cave [around the ring of Pahasapa limestone] is almost unrestricted and in one sense could extend completely around the Black Hills." Within 2 miles of Wind Cave National Park headquarters, the National Park Service considers it "reasonable to assume 50 to 100 miles of passages."

For the Conns' beloved wind to guide them much further along this unique limestone ring, some wholly new breakthrough will be necessary, or new entrances, strategically placed, one after the other. Otherwise, at the present rate of zigzag progress, the irresistible years inevitably will close out their matchless chapter in American caving, long before we know if the two sprawling caves are but one.

Yet the wind they know so well may guide their heirs to Wind Cave itself.

11

Davies Didn't Crawl

The Story of Recent Progress under the Virginias

To geologist Bill Davies, the whole shape of this pitted limestone land shouted real caves somewhere near at hand. Thirty yards into the West Virginia mountainside, a low streamway at the head of Swago Creek blew hard and chilly: Overholt Blowing Cave. Nearby, Bill traversed 1,500 feet of Cave Creek Cave, which collapsed soon after this 1948 reconnaissance. Not far away, he recorded huge Tub Cave, the largest cavern chamber in West Virginia. But neither went much of anywhere. Perhaps that blowing waterway where no man had ever gone . . .

But Davies had come to Swago Creek merely for enumeration and description—part of his statewide survey for the West Virginia Geological Survey. After looking at four smaller caverns nearby he moved on without wallowing through the windy crawl.

It was a mistake. Today that low, uninviting hundred-foot waterway where Bill Davies stopped boasts a strange and wonderful name: the Davies Didn't Crawl.

A stodgy noncaver, misconstruing cavers' boisterous humor, might deem this unfair. Davies accomplished his mission admirably. Gleeful cavers, however, delight in immortalizing each other's frequent discomfitures. I once led a touchy climb to an almost inaccessible, tantalizing cave orifice deep in the most superb section of Hells Canyon. The cave proved just ten feet long and hardly deeper. Though I once was able to expunge a similarly deflating name from California records, this inglorious Idaho cave officially bears the title Halliday's Hole.

In 1953 the old Charleston, West Virginia, Grotto of the National Speleological Society stumbled upon the Swago Creek area. Bulwarked with experienced cavers, the Explorers Club of Pittsburgh and Pittsburgh grottoites soon followed. Four years' efforts yielded thirty-six caves and pits where Davies found eight. Several were of exceptional importance. In the unrolling system which connects Swago Pit and Carpenter's Pit, a mile of passage was soon mapped.

In October 1956 Will White, Rita Battistoli, and Ralph Doerzbacher found themselves outside still-unexplored Overholt Blowing Cave without anything to do. For the sheer joy of caving they bellied along in the rippling underground brooklet. Just 100 feet onward, they rose, irresistibly called by a spacious corridor extending into virgin blackness. Not far beyond, Swago Creek welled up from a siphon, but a short, dry gallery bypassed the obstruction. A stoopway thigh-deep in water was merely a nuisance. A thousand chill feet from the entrance, the cavers found themselves sloshing along, chest-deep, in wall-to-wall water. Only after 2,000 feet of shallower stream passage did the trio retreat, blue-lipped and tremulous but ecstatic. Even in the Swago area, so extensive a virgin cave is found but rarely.

Few but cold-inured Pittsburgh grottoites find such caves fun. A few hundred feet beyond their first exploration, breakdown forced the next explorers to half crawl, half swim a three-foot lead half filled by the chill underground creek. Beyond this Dardanelles, a seeming mile from the entrance, however, beckoned a large dry passage, then a wonderfully spacious room. From it opened superbly embellished side corridors—

Roswell Jones negotiating the Dardanelles of Overholt Blowing Cave. *Photo by Vic Schmidt*

and an infuriating 20-foot pit that momentarily blocked progress. It was a rather nice cave—extensive and fairly challenging, with much to be investigated at leisure.

Other important Swago area projects slowed return to Overholt Blowing Cave. In June 1957, however, six cavers pushed onward along the main stream course. Even the Pittsburghers began to wonder if this was truly fun. Watery crawls, squeezes, and short wet walkways drained energy and body warmth minute by minute. Even before the team reached a superbly miserable sewerlike "narrow, mud-filled tub," three turned back.

Perhaps the cave ended just beyond and they could be done with all this misery. Despite chattering teeth, George Beck squirmed unencumbered into a crack beyond the natural sewer. A thousand feet onward no end was in sight, and his carbide lamp was burning dim. His reserve supplies back at the sewer, George retreated with the cryptic message that the cave was easier past the siphon. They would have to return.

George's tidings were not uniformly popular. Hermine Zotter vividly recalls the return portion of this 15½-hour venture. Staggering with exhaustion, cold, and disappointment, yet buoyed by excitement, she and Beverly Frederick looked like

any two drunks on Sunday morning at precisely 3:25 A.M. outside a tavern; picking each other up as one or the other lost balance and headed for another ducking in the icy water. Bev's torn tennis shoe sole flip-flopping with every step amid stupid laughter and joy. Even through "Davies Didn't Crawl" every grunt produced more wisecracks and more hysterical laughter as shredded clothes exposed bleeding knees and elbows . . .

Fun, Pittsburgh grottoites call it.

But things were looking up. Beyond Beck's Stop, the going was delightfully easy. The next year found new ventures extending the known passages of Overholt Blowing Cave to almost 2½ miles—a really first-class cave. More and more, its stream corridor became the prime target of the Swago Creek area.

A 15-foot waterfall presented no particular problems. Ascending through breakdown above the falls, however, a tired quartet gasped as they broke out into a jagged 80-foot waterfall vault of coalescing fluted shafts: the Cathedral Room. Near the endurance limit of even Pittsburgh cavers, this was an unexpectedly compelling challenge.

Still the total resources of the Pittsburghers could not be channeled to Overholt Blowing Cave alone. The Carpenter-Swago Pit system also was growing to 2½ miles despite difficulties scarcely less than those of Overholt Blowing Cave. Much cried to be done on the surface: new pits to plumb, new springs to trace underground. An entire party in Overholt

Blowing Cave emerged exuberantly splashing brilliantly green water when Hermine Zotter tested a new sinking creek.

To push onward, some kind of breakthrough was necessary. Besides, it would be nice to return some fun to this energy-sucking Overholt Blowing Cave.

The Pittsburgh cavers were equal to the dual challenge. Rubber scuba divers' "dry suits" shattered the endurance barrier. Protected by coveralls and several layers of underwear, vigorous rubber-clad explorers reached the Cathedral Room in comfort, their vitality no longer sapped by insidious hours of cold. Hardy spelunkers vowed that the greatest discomfort of the cave now was donning and doffing the clammy suits. One of the grotto's leading experts on women swore that girdles were nothing to dry-suit contortions.

May 1959 saw the roaring Cathedral Room attacked. Often bathed in sheets of spray, Bob Dunn, Jerry Frederick, and the inexhaustible Roswell Jones emplaced expansion bolts and belayed each other up black cliffs and domes. A new stream corridor beckoned them along an unknown level 60 feet up. Several hundred feet upstream, a third waterfall was easily bypassed. Only past the dry Turnpike, 300 delightful feet farther upstream, did another stream crawl halt progress.

Bob Dunn, Jim Fisher, and Ken Acklin began a "to the end or perish" push after three months' preparation. While other teams labored elsewhere in the enlarging cave, they proudly drove onward 1,000 difficult feet.

An air current strong on the faces of the trio brought unspoken hopes. The cave grew higher. Though a new waterfall whispered from darkness somewhere far ahead, the air was perceptibly warmer. At 8:30 P.M. the walls receded into nightlike blackness.

"We're out!" exclaimed Bob Dunn, hopefully seeking starlight.

But it was not to be. Close carbide inspection revealed merely a very large, warm room—Disappointment Dome. On the far wall the fourth waterfall arched 60 feet along a smoothly fluted, spray-showered wall. Perhaps there was another stream corridor far above, but to the suddenly disheartened crew the murky chamber seemed Ultima Thule. Even with dry suits, 19 continuous hours of exploration of Overholt Blowing Cave is close to anyone's endurance barrier.

To Pittsburgh cavers, however, unclimbable walls do not exist. An unscaled waterfall exists merely to be conquered. But perhaps there was an easy way. Seeking an upstream entrance, the grottoites hopefully reconnoitered the surface. No bypass was evident. A direct escalade of the spray-whipped, ominously shadowed dome was inescapable.

This formidable ascent could not be accomplished by one-day ventures. Rather reluctantly, the cavers planned a mass expedition.

The sand-floored Turnpike, an easy half mile from the dome, was selected as base camp. A twenty-man team set the grand assault for the 1959–1960 New Year's weekend. At 3:30 Friday afternoon a lightly loaded section of the assault team reached the Cathedral Room. The Turnpike was gained by 9 P.M., with every sleeping bag splendidly dry.

Though not quite comfortably warm, all hands slept late next sunless morning, 2½ miles inside Overholt Blowing Cave. The crawl back into chill, clammy rubber suits evoked moans, but a hot breakfast worked wonders.

At Disappointment Dome the delightful August warmth was replaced by January chill. Higher water flung fierce spray everywhere.

After a moment's involuntary pause, Allen McCrady, Oliver Wells, and Guy Wallace attacked the spray-swept chamber while Bob Dunn, Fred Kissel, and Vic Schmidt mapped the region. Endlessly, regularly, hammered pings of metal against limestone cut through the roar of the waterfall. Occasional pauses told of insertion of expansion bolts into the newly drilled holes. Climbers' hardware and stirrups were attached and the spray-chilled driller ascended three feet to begin again the hammered ping . . . ping . . . ping.

Seven hours' exposure to the waterfall saw six expansion bolts emplaced at otherwise unclimbable pitches. Thence a ledge permitted Guy Wallace a ticklish climb to a final overhanging cascade. There he drove the last bolt for a ladder, then retreated: time for the mapping crew to take over the assault. By 10 P.M. the exhausted advance trio were snoring.

To literate Vic Schmidt fell the unenviable task of belayer-from-below:

The ladder hung directly beneath the worst part of the falls and Bob seemed about to be swept from the rungs. We lighted his way as he climbed up and then traversed a ledge . . . and belayed Fred up to his ledge in the dark. . . . Just when Fred got to the lip of the first cascade he turned and took the full force of the water in his face. My light from below was all Fred had to climb by. Everything was completely soaked including my "waterproof" matches and it would have been difficult to start a fire by flint anywhere in the room. Bob sent his lamp down tied onto the belay line and I recharged and lit it and sent it back up, maneuvering it just over the waterfall without losing the flame. The climbers then proceeded up that ledge that Guy hadn't quite been able to climb and soon I heard the Ping! Ping! of steel on rock as new bolts were set. . . . I was alone.

Shivering on his wet rock, for six interminable hours Vic observed the foaming frothy water of Disappointment Dome. Even Vic doesn't claim he enjoyed it. Finally, gearing his extrasensory perception, he readied an alcohol stove for hot chocolate for his unseen friends. It was not as easy as it sounds: "With the waterfall running off immediately through breakdown, I had no choice but to hold the pan directly beneath the cascade to collect several cupsful. I was drenched in a moment and thoroughly chilled, and only then did I realize what Bob and Fred were going through up there."

Vic's clairvoyance was superb. A tiny pinpoint of light appeared high overhead just as the chocolate simmered. But then everything went wrong. Ladders snagged, ropes tangled. Fred had to cut his pack loose. Bob's light went out in mid-snarl. On belay, he finally slid down a freed rope.

Drilling a hole for an expansion bolt in the spray of the Fourth Waterfall. *Photo by Vic Schmidt*

As he arrived, Vic's light too was blown out by the swirling spray. But by now Vic knew every rock in Disappointment Dome by its first name.

Over wonderfully scalding 2 A.M. cocoa, the shivering pair sadly broke the news to Vic. The tremendous three-day expedition had gained just 30 feet of cave. A fifth waterfall lay just beyond. Only 30 feet high, it was perhaps not too difficult. But no one was in condition to attempt it that night. "We were constantly under water up there," Bob Dunn said, shivering.

As the sagging trio sought rest at 4 A.M., their plastic windscreen snapped loudly into the opposite direction. Before resuming its usual wind-pouched billow, it flopped aimlessly for some moments. Exposure-drugged, the weary cavers barely noted these signs of a ferocious storm somewhere overhead. Seconds later they were asleep.

When their watches told them it was midmorning, Wallace, Wells, and McCrady arose and started toward the entrance. Groggily, the others soon followed. Overnight the underground creek had risen considerably, but the inner waterways presented no problem.

Eager to pass the Dardanelles and get home, the still-tired second trio hastened onward. Unexpectedly a light loomed up ahead. "Hello," boomed Allen McCrady with false cheeriness. For five hours his team had been sitting disconsolately wrapped in whatever promised a little insulation. Carbide lights whipped out by a whistling gale through the remaining air space, noses scraping the ceiling, they had found the water too high to pass even the easier section of the Dardanelles. Well-trapped by the sudden storm, they could do nothing but wait.

The furious activity outside Overholt Blowing Cave might have encouraged the disconsolate sextet. Floodwaters of Swago Creek made obvious the diagnosis of the blocked Dardanelles. The supply team rapidly routed Baltimore grottoites out of nearby comfort. The Baltimore cavers, in turn, descended upon the merchants of nearby Marlinton, purchasing great quantities of canned food plus two bicycle pumps and 225 feet of garden hose for an emergency air tube (it worked surprisingly well).

The strong current constantly threatened to sweep the rescuers off their feet as they dragged the ungainly paraphernalia upstream toward the Dardanelles. But, as Roland Kleinfeld prepared to breast the turbulent crawlway, one of the trapped explorers unexpectedly popped through. Bob Dunn had caught a glimmer of Roland's light above the receding flood. The others soon followed, happy even in the icy January night of West Virginia.

Such an experience hardly daunts a Pittsburgh caver. If three days aren't adequate for meeting a challenge, let's go back for four! In August 1960 two supply teams cached packs in the Turnpike. Three assault teams entered the cave Thanksgiving Day. Now familiar with its every peculiarity, they rapidly re-scaled the fourth waterfall. As-

The plastic windbreak in the Turnpike of Overholt Blowing Cave. *Photo by Vic Schmidt*

sembling scaling poles, they hoisted a ladder to the top of the fifth. Bob Dunn hauled himself wearily to its chill apex, looked into a large canyon passage, drove an expansion bolt for the ladder and headed for a sleeping bag. One thousand feet onward next morning, breakdown seemed to block the way. An hour's search revealed no promising leads. The end of Overholt Blowing Cave at last?

As a final resort, Bob Dunn and Jerry Frederick forced their way into a rockbound hole that obviously went nowhere. A half hour later, the pair was still out of sight and hearing. The others followed their squirmy trail, wriggling on and on. Only after a full hour's crawling did they reach a slot where they could sit upright to stretch cramped muscles.

Beyond, the crawlway was even tighter. A little rationalization indicated that this was a good place to wait. After still another half hour the missing pair inched into sight, their news typical of this tantalizing Overholt Blowing Cave. Far ahead, past extraordinarily difficult crawling, they had come upon a little room marked by a cave rat's nest, a woman's garter, and a chunk of asphalt paving.

But the only holes that continued were just large enough for the cave rat.

Such a cave exacts an imperceptible toll of even Pittsburgh grottoites. Hopeful digging soon fizzled out. After mapping to the bottom of the fourth waterfall, the Pittsburghers moved on to other challenges.

The lack of an adequate map, however, rankled a few perfectionists, American and Canadian alike. Among the latter was George Tracey. While caving nearby he chanced to utter a few vigorous words to speleogeologist Tom Wolfe. Tom's retort directed him to go find some new passage in Overholt Blowing Cave. To everyone's surprise, George took three novices into the Penn State section at the end of Anne's Avenue (which most cavers thought merely led back toward the entrance from the Dardanelles) and did exactly that: pits, two fine rooms, and 1,700 feet of narrow, winding passage as much as 80 feet high. And he mapped it, although he needed a 29-hour marathon to finish the twisting, jagged corridors.

The gung-ho caver group at Ontario's McMaster University became interested. A six-man team and several helpers set out in all due grandeur, prepared to camp in the cave for four days and map it in proper style. At the cave, however, their style rapidly became something less than grand. When they laboriously hoisted a scaling pole at the first waterfall, it was three feet too short. For hours they struggled, wracking their brains. Even by standing on tiptoe on the top rung of a ladder hooked to the pole, they got nowhere.

As the frustrated hours dragged on, the inactive part of the group inevitably became bored, restless. Wandering around, a sextet eventually raced back with the titillating information that their fellows were trying to climb the wrong first waterfall. The real one was easy, but no time remained for mapping.

Such setbacks are hardly daunting to McMaster cavers, however. Two months later they were back, grandly disorganized but with three teams working in relays. This time George Tracey was along. In a 26-hour push he and another Canadian, plus John Fish of Texas and points south and north, and Irishman Eoin Finn, surveyed more than 3,000 feet of passage near the back end. Certain difficulties had to be brushed aside in the inimitable Canadian manner. "We had 2½ wet suits for the four of us on the first survey team," George subsequently confessed. "John Mort was nackered by his long drive even before he got to the cave. We had to quit mapping at the first waterfall. By the time we got back to the Dardanelles we were all just about out on our feet. There we all flopped down onto a rock and accidentally went to sleep. Only John Fish's snoring woke us up. You know John and what that means!" I certainly did. "Mr. Indefatigable," John is a legend in his own time. "A trifle heavy," he snorted when he read this, but it's true. A few weeks later the teams brought the survey out to the Mountain Room, and the rest was easy.

With almost four miles mapped, the Canadians' surveys showed that this uncompromising cave rises more than 600 feet above the unprepossessing entrance where Davies didn't crawl. To admiring Yankees, even their finishing flourishes seemed a glaze of glory.

The Canadians, however, remained dissatisfied. Early in 1975, George told me that they hadn't even been able to find the Pittsburghers' Rat

Route. Virgin cave still existed in the Penn State Section, he insisted, and they were excavating two surface pits that fed dye into the cave.

Then, with the map still incomplete, some boulders shifted in the sinkhole entrance of Barnes Pit in Pocahontas County on July 1, 1975. Merely preparing to hang a short ladder, George was crushed to death against a rock wall. Some would say it was the death he sought, subconsciously matching implacable drive against unending, invincible challenge. Few cavers would agree. Few of us see a death wish in caving, especially those who best know the wild joy of exploration of merciless, demanding caves like Overholt Blowing. In caves we seek to live, not die. Yet equally few will attempt to explain their innermost feelings about such caves. Perhaps more than in most caves, spelunkers' motivations here are complex, unclear, unspoken.

"Primarily we do it because we enjoy it," Vic Schmidt has long insisted. "I guess we know it really isn't fun," admitted Hermine Zotter a decade ago, "but it *is* exciting!" Semi-standard is the parry: "Because we want to!"

Clearly the lure of such all-out caving is more than these, more than the often-sublimated "ever-onward" drive of even the ultimate spelean sophisticate. George Tracey's explorations will not be the last discoveries in Overholt Blowing Cave, and some may find death there. Those who push beyond his footsteps will consider the risk worthwhile. Already his stunned compatriots are rallying to a proper cavers' memorial. The completion of "his" map will be their ultimate tribute to George Tracey.

On the other side of the Virginia–West Virginia state line, Kennedy (Ike) Nicholson found a blowing cave of a different type: one which "breathes" with readily detectable changes of the barometer. This particular cave is at the head of Burnsville Cove or Sinking Creek Valley, about two miles from the gaping mouth of Breathing Cave.

Sinking Creek Valley drains underground, beneath the north end of Chestnut Ridge, which forms the west flank of the valley. On the far side of the ridge, Mill Run Spring transmits large volumes of water to the nearby Bullpasture River. In 1956 that clear, trout-thick spring was found to be the mouth of what became known as Aqua Cave. After a 35-foot underwater swim with scuba gear, Bevin Hewitt found a subway-sized corridor leading on and on. For eighteen months the Nicholson family led expeditions to its black ponds and breakdown chambers. Through crevices and along stream passages they advanced a half mile into Chestnut Ridge. The first 1,500 feet of this new-found cave was delightful. Beyond a watery crawlhole, however, 300 feet of a 450-foot passage was neck-deep in frigid water, including one short stretch without sufficient air space to allow the explorers' noses to remain above water. Finally came a deep pond which ended the main corridor.

Belayed with an 80-foot rope, Mike Nicholson sought another

shallow, easy duckunder like the one at the entrance. Unhappily, the wearying transportation of heavy scuba gear proved in vain. At the limit of his safety line the cave still continued downward indefinitely—a broad, waterfilled avenue sloping far into the unknown.

No expert speleologist was needed to announce that Aqua Cave had transmitted enormous volumes of water in the recent geologic past. Such volumes in turn bespoke much more cave somewhere nearby. So did the local geology.

Burnsville Cove is what geologists call a structural valley. Unlike many valleys, whose contours are unaffected by the underlying rock structures, Burnsville Cove follows their pattern rather precisely. If viewed from the south the rock formation is aligned rather like a great pile of paper curved up on each side, with the left side higher than the right. The present ground surface is like a slightly loose top sheet of paper. At the far end of this rough model, the slopes of Chestnut Ridge gentle out to the right toward the Bullpasture River and Aqua Cave.

Such a rounded down-dipping of the bedrock is known as a syncline. Breathing Cave's four miles of complexities occupy one small part of one arm of the syncline.

Even before the Siphon Room dive ended hopes of quick progress there, Ike Nicholson had been seeking an easier way into the predicted cavern system. He and his fellows knew Breathing Cave well. That cave lies far up the karstic valley and on the wrong side to connect easily with Aqua Cave. Though a few passages then remained unexplored, they did not delude themselves about entering Breathing Cave and popping out into this new Aqua Cave.

As Ike hiked the slopes of Sinking Creek Valley, sinkholes funneled everywhere but none opened into significant caves. Crevices and cavelets investigated by other hopeful scouts proved equally disappointing.

Months passed with local enthusiasms largely directed to Breathing Cave's fabulous maze. Then, on May 30, 1958, Ike rechecked a limestone ledge on the ridge at the upper end of the valley. To his surprise, the afternoon sun revealed a grotto he had previously overlooked: Butler Cave. Six short feet inside, a 4-foot pit dropped 35 feet—a pleasant scramble for the renowned Ike Nicholson, supposedly fueled by a single candy bar for 16 hours of strenuous caving.

A 5-foot fissure opened at the bottom of Ike's new pit. It was 30 feet long and it seemed wholly plugged, but there was "a little slot four inches wide and eight inches long." Ike stooped to shine his carbide lamp down the slot. So strong a draft of air was coming up that his light was blown out.

Up the shaft, across Chestnut Ridge, down to his car, back to his nearby summer cabin raced Ike, madly excited. Digging tools soon enlarged the slot so that skinny Don Miller could be eased through.

Down and down went the belay rope as Don found more and more passage below. Finally a call floated back: "I'm not going any farther. There's a big hole in the floor. I can hardly see the bottom, and I'm coming back!"

Two endless weeks dragged by before the excited cavers could mount a full-scale expedition. Two successive weekends vanished as the spelunkers happily probed hundreds of virgin yards of delightful new cave. Interconnecting corridors, squeezeways, shafts, canyons, and mazes of broken rock formed a new Breathing Cave, but passages here were bigger, deeper, and tougher and thus more enjoyable. Down and down went the maze, eventually to a spacious corridor along the axis of the valley. "We paced off 600 feet before the first turn," Ike Nicholson recalls. On and on they ambled, to a lofty junction where their delightful gallery seemed dwarfed. An awesome flat-roofed natural subway led right and left beyond spotlight range.

Now dry, this new throughway once had evidently carried a tremendous volume of water. Downstream, it headed down valley toward Aqua Cave. Would Mill Run lie in a lower level somewhere close below? Was there really a chance of the long-dreamed-of master drainage system cave?

Quick peeks upstream revealed much more cave. So did scouting downstream, where explorers soon came to an underground creek. Sinking Creek they called it, not yet sure of its connection to Aqua Cave.

Despite the stream—and another and another—the cavers pushed on rapidly, "like hounds on a hot scent," to use Ike's own phrase. A few days' exploration brought light to a mile of this new throughway region. Increasing familiarity with the cave and convenient shortcuts repeatedly cut hours off main routes. Soon the pattern of the cave was excitingly clear. In the model mentioned earlier, a soda straw thrust through the axis of the curve in the middle of the pile would suggest the main throughway passage, more than 300 feet below the surface. On both flanks and curving round its upper end was an intricate network of collecting complexes like the entry section of Butler Cave—and of Breathing Cave. Each conducted water down the flanks and head of the valley to the central throughways along its axis.

After two years devoted to mapping Breathing Cave, the Nittany Grotto of the National Speleological Society joined the Nicholsons in an effort which now seems as endless as Butler Cave itself. Repeated explorations lasting as much as a week underground led goggle-eyed spelunkers onward along the axis. Upward too they pushed into additional complexes, ascending both syncline flanks. Within months Butler Cave was larger than its better-known neighbor. Additional side complexes are still coming to light, with several miles mapped in the upper reaches of the syncline alone.

Far down the axis of the syncline, alternate neck-deep wading and belly crawling in frigid water seemed unavailing. Three separate streams disappointingly dwindled into apparent siphons, still more than 300 feet above the stream level in Aqua Cave. Two hundred risky feet into Last Hope Siphon, cave diver Hank Hoover acknowledged defeat in June 1960. The others appear even less promising. Fluorescein dye placed in Sinking Creek took ten days to appear in Aqua Cave. Obviously a human route would not be simple.

Far down in the lightless corridors that channel Sneaky Creek, halfway to its Rats' Doom Siphon, Evasor Gallery leads up the right wall of the syncline almost opposite the notorious Pants Off Crawl where Ike Nicholson once was briefly trapped by his belt. The map in the 1962 *Caves of Virginia* shows an unexplored lead heading southwest from Evasor Gallery.

In time the Butler cavers turned even to that remote, uninviting hole beyond Evasor Gallery. A twisting descent with and in a streamlet was slightly eased by unusually lard-like mud (that 50-foot slide is now dubbed Crisco Way). Only pride drove Joe Faint and Mike Nicholson onward in a tortuous 600-foot squeezeway that ended in a 40-foot shaft. Once more Butler Cave's personality rewarded such exploration in the grand manner. Below a sandstone layer was an entirely new region. The ebullient pair paced off more than a mile of passageway before beginning the tedious return.

"Marlboro Country," Mike Nicholson promptly dubbed the farflung complex below the shaft, borrowing from a certain television commercial. This is a world where few have ventured. It takes exceptional vigor merely to reach its high-ceilinged walking passages, its chill crawls. Even the near-legendary Ike Nicholson is content to leave its mysteries to younger cavers of greater endurance. "I almost didn't make it out the one time I went back in there," he admits cheerfully. Here, beyond the endurance barrier of the average caver, are two more streams, plus a resurgence of Sinking Creek. Here, still more cross corridors lead up the syncline flanks. After ten years, the full potential of Marlboro Country is not known.

Elsewhere in the cave, the past decade has been more fruitful. Bitter effort by Nevin Davis, Fred Wefer, and other members of many eastern grottos of the National Speleological Society has added considerably to its map. The pattern of that map suggests that much of its now-blank areas will be found to have similar drainage caverns. Yet the going is so increasingly tough that the actual size of Butler Cave seems unlikely to be known in its first quarter century. Present figures are impressive enough. As of 1973, 14.3 miles were on the map, and another mile or so has since been added, mostly filling small gaps beneath the upper valley flanks. Connection to Breathing Cave will add 4½ additional miles of drainage complexes. By enlarging a rodent-sized connection to nearby Boundless Cave to human size, another 1,800-foot network can be added whenever convenient.

How imminent is the long-sought connection to Breathing Cave? Not very. Nine hundred fifty feet still separates the two. A decade ago the junction appeared imminent. Sketch maps then suggested that the intervening distance was only 600 feet. Then Corcky McCord and Fred Wefer made a 500-foot advance from the Breathing Cave side, in what seemed exactly the ideal direction. Hopes reached fever pitch, but a disappointing, obstacle-strewn year spent evolving a detailed map proved the original sketches over-hopeful. Much virgin cave turned up in "completely explored" Breathing Cave, but none was much help toward a connection.

Momentarily the cavers were halted by the axis siphons and the ferocious problems of merely reaching crawlways that might bypass them. Attention veered to some small holes near Last Hope Siphon. Through these roars a fierce wind, seasonal in its reversals. Although in the bedrock horizon of Breathing Cave, here about 1,000 feet distant, these

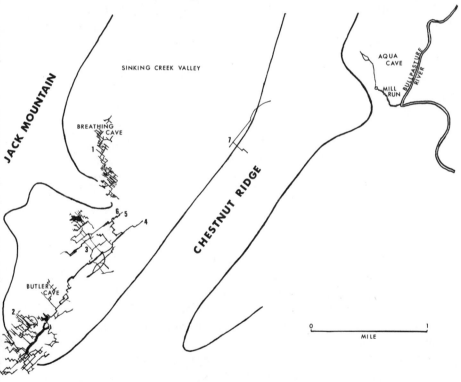

Sketch of the Butler Cave–Sinking Creek area, Virginia. (1) Entrance of Breathing Cave; (2) Entrance of Butler Cave; (3) Marlboro Country; (4) Rats' Dome Siphon; (5) Last Hope Siphon; (6) Fred L. Wefer Memorial Highway; (7) Better Forgotten Cave

holes evidently don't lead to Breathing Cave. The best is infamous as the unlovely portal of the Fred L. Wefer Memorial Highway, ninety feet long. The fastest-known passage through its flowing mud is recalled as 15 minutes. Beyond is human-sized passage, then digging in virgin mud which simultaneously chills the diggers while exciting their hopes. In 1971 Nevin Davis reported an unusual effort here:

A 100-gram flask of ethanediol (10,000 skunks) was broken in the Highway, in an attempt to trace the air. Small groups of human bloodhounds, eager noses close to the ground, discovered not a single skunk in the entire Burnsville Cove. Even Breathing Cave's breath was as sweet as new mown hay. We still haven't successfully traced the air. Perhaps we need more skunks.

Or perhaps a snuffle party much farther away, along the banks of the Bullpasture River. Or those of the Cowpasture River into which it flows, not many miles above Blowing Cave itself.

Even though the pattern of discovery in Butler Cave suggests that this route may be of exceptional importance, nobody has extended the Fred L. Wefer Memorial Highway since mid-1970. Even the Nicholsons and the Nittany Grotto sometimes become discouraged. Predictably, however, they and other members of the Butler Cave Conservation Society—formally in charge here since 1968—will be back, just as soon as they tire of other groady crawls, other digs, expansion-bolt climbs, scaling-pole climbs, blasting, all the tedious routine that keeps adding up the miles in less-challenging sections of Butler Cave. Just as in the Black Hills and many a less-celebrated cave, the wind still calls.

From the outset, Butler cavers have sought easier ways into the unknown parts of the system. Nineteen fifty-nine explorers found a windy little cave in line between Aqua Cave and Butler Cave itself. Seemingly it was located in exactly the proper location to bypass Rats' Doom and Last Hope siphons, but when it had been checked, general acclaim named it Better Forgotten Cave. Three trips advanced cavers to the bottom of a 100-foot pit, 450 feet in and 225 feet down. En route to its lip were "some of the most impossible squeezes ever seen." Mike and Dave Nicholson explored it with Hope Waring, and concluded exhaustedly that the bottom was a dead end. The worst cave they had ever tackled, they reported. They even abandoned 400 feet of rope, unsure that they would be able to regain the entrance if burdened with its weight. It's still there, marking the point where the going changes from extra-tough to near-impossible.

But Ike Nicholson is a stubborn cuss. Nagging gently but persistently, he somehow enveigled Jack Hess and a covey of Davises to try again, a decade later. Their entire first trip consisted of chipping away the tightest spots so that someone besides trim Christine Davis could even get to the 100-foot pit. A week later they plumbed its depths. Following the footsteps of the 1959 party, they found its rivulet disappearing not far beyond its base. Unenthusiastically digging at the point where it disappeared, they suddenly found themselves peering into a black hole that soon became all too well known as the infamous Vertical Crawl. One trip later, vertical crawling led to a more normal 20-foot vertical pitch. Jack and Nevin rigged it routinely, unenthusiastically. As Hess later reported, Nevin went on "to see if it was worth continuing. He came back shouting." Just ahead was their goal: a large streamway corridor that went and went.

Three hundred feet upstream toward Butler Cave the corridor was choked with breakdown. To date no one has made much progress in that direction. Downstream some 750 feet, a neck-deep pool was a totally-damned nuisance, but beyond, another 800 feet of throughway led them onward. Unbelievably big, it had characteristics almost identical to those of Butler Cave. Alas! After a few side complexes, it siphoned 420 feet below the entrance. Dye testing seems to indicate that its stream is a new one, unrelated to any in Butler. The Butler Cave Conservation Society will have to discover and explore many more like it before the full potential of Sinking Creek Valley is known. No longer does it appear likely that a single cave underlies the entire valley and its slopes.

Nevertheless, between the head of Butler Cave and Aqua Cave, there is room for twenty times as much cave as is now known. Or one hundred times. The Butler Cave Conservation Society faces one of American caving's most intriguing challenges. This peaceful little Virginia valley and points beyond still may hide the world's largest cave.

One major function of regional cave surveys is to coordinate knowledge for future assaults on the unknown underworld. This Bill Davies accomplished superbly. Throughout West Virginia modern speleological techniques have built major successes upon his framework. At the last moment Davies gladly revised his accounts of "little" Hedrick's Cave and nearby saltpeter-rich Organ Cave. As Davies declared, "for a number of years, all that was known of Hedrick's Cave was the entrance passage." That was before the coming of Bob Flack and Bob Handley, who casually strolled inside late in 1948. Eight soggy, muddy hours later, they staggered stiffly from the beginnings of the Greenbrier cave system, West Virginia's largest. The first return trip produced a splendid cave, a third trip a second entrance. In May 1949 three Charleston grotto Bobs—Handley, Flack, and Barnes—cocooned in mud, found themselves behind the stalactitic Pipe Organ in Organ Cave. Gleefully they emerged "amidst electric lights, tourists, and consternation." For a quarter of a century, Bob Handley and other cavers have been piling up the miles. By 1962, 15 miles had been mapped—"about half what's explored," Bob guessed. Eight more years passed before the map proved him cautious in his estimates. By then, the cave again appeared only about half mapped. Today the Organ Cave or Greenbrier system is one of the world's half-dozen largest, with almost 35 miles on paper.

Without fanfare, one of the world's most amazing scientific evolutions has come to light just across the Greenbrier River from this exciting system. In the Great Savannah karst around Lewisburg, West Virginia, a decade of compulsive effort by several loosely-linked teams has yielded more than 100 miles of intricate caves. In this mid-zone of Greenbrier County the total is actually well along toward 200 miles. Many caves noted by Davies and others have been integrated into extensive systems. Others like The Hole are new rewards of intensive modern speleological techniques.

In the Great Savannah no single cave currently challenges the length of the Organ Cave system. Instead, a fantastic series of underground networks parallels the line of contact of two tilted geological strata. For many miles along the west side of the Greenbrier River in Greenbrier and Pocahontas counties the ground waters of the Greenbrier limestone

Bill Besse grins as he rounds a ticklish corner deep inside the Organ Cave system and sees a photographer primed for any slip into the 30-foot canyon below.
Photo by Charlie and Jo Larson

come to the surface along its contact with the shaly Maccrady formation which lies below. Only near the southern boundary of Greenbrier County does the river finally cross the contact.

Unless it be little-known Windy Mouth Cave south of the river near the county line, where Charleston cavers led by Don Peters and Gordon Smith have made only a beginning in mapping almost 12 miles, The Hole at 15.7 miles is the longest single cave here. To its northeast lies 4-mile Friar's Hole Cave, specialty of the late Lew Bicking, Ph.D., the legendary hard-charging ultimate in gung-ho caving, for whom the N.S.S. named its annual Lew Bicking Award. Just beyond, and perhaps some-day to be connected to Friar's Hole Cave is the Snedegar-Crookshank system, with 2 more miles crossing beneath the Greenbrier-Pocahontas county line. Still farther north is Overholt Blowing Cave, plus much more. Southward lie Ludington Cave (4.6 miles), McClung Cave (12.6 miles), Maxwelton Sink Cave (9 miles), Benedict's Cave (5.97 miles), and Wade's Caves, neatly lined up along 8 linear miles of the Greenbrier-Maccrady contact. Through the courtesy of 1969 Hurricane Camille, whose floods neatly unplugged Maxwelton Sink Cave, these and The Hole are currently segmented by gaps of only 500 to 2,500 feet. Together the southern quintet contains 31.4 miles of mapped passages. A parallel chain exists about a mile farther west: the Higginbotham system and Coffman Cave, but only about 2 miles are known here to date. Just west of the Higginbotham system, however, is Culverson Creek Cave with 10.6 miles mapped.

As is the case with Overholt Blowing Cave, exploration here may be only in mid-course. Ludington Cave, the Higginbotham system, and the others between The Hole and Windy Mouth Cave drain southward to Davis Spring as parts of a karstic drainage network 15 miles long. South of Coffman Cave is an especially huge gap. Friar's Hole and the Snedegar-Crookshank system are separated from the others by Spring Creek, and Windy Mouth Cave, by the Greenbrier River. As for the others, however, further linkages are quite possible. At the present time The Hole and Culverson Creek Cave occupy separate drainage basins, divided from the Davis Spring Basin by underground watershed divides. In comparatively recent geologic time, however, these also were part of the network which drains at Davis Spring. "Fossil" passages left over from the past may yet lead West Virginia cavers across the stygian divides.

Yet maps and the recording of scientific data are far from the sole lure of these exciting new netherworlds. As in Overholt Blowing Cave, in the vast networks of Greenbrier and Pocahontas counties, and beneath the thin, opaque surface of Sinking Creek Valley, cavers share a supreme comradeship magically woven from the conquest of nature's compelling challenge. Such caves yield man the spacious freedom of great black voids and the delightful poetry of entrancing little grottoes. Here are shared moments of happy banter and of silent communion with nature. Here is the pervading satisfaction of hard-won new knowledge at the happy end of a long, tiring day: knowledge of the cave, of one's fellows and of oneself. Even more than in most caves, man here renews his soul.

12

Determination and Dark Death

The Story of Indiana Caves

At the last homeward bend of sprawling Sullivan Cave, a chance ray of sunlight dazzled the Indiana teen-agers. Like automatons they climbed toward the long-unseen daylight. Amid a half-comprehended crowd of friends and admirers they halted, blinking owlishly. Even through sunglasses the surface world seemed preternaturally bright, vivid, beautiful. Dramatic was the brilliance of the half-forgotten sky, the greenness of each tree, the distinctness of each leaf.

Mike Wischmeyer spotted his family amid the congratulatory throng. After three exhausting weeks in the strength-sapping riverways of Sullivan Cave, the teen-age leader's words were somehow congruous: "Hi. Where's the food?"

Soon to become a Purdue University astrophysics major, Mike Wischmeyer came to Sullivan Cave in 1961 at the age of sixteen. His was the plan for the 1962 First Sullivan Cave Expedition: a two-week sojourn which would be much more than merely a stunt underground.

As Mike envisaged the operation, exploration would yield precedence to mapping teams tediously plotting distant, difficult recent discoveries. Clean-up parties would spend four to eight hours daily restoring and cleansing the vandalized outer part of Sullivan Cave. Soil samples would be collected for biological studies, including search for new antibiotic-producing microorganisms. The physical and psychological effects of the strenuous venture would be recorded and analyzed in detail—all by a bunch of Indiana teen-agers. At seventeen, Mike would be leader. The oldest would be eighteen. Two would be seventeen, one sixteen, and one fourteen. So it happened, and Mike won a Science Talent Search award. Some mere Hoosier teen-agers, it seems, are not so mere.

It was not as easy as it sounds. Preliminary scouting trips were only a part of the Sullivaneers' preparations. Sporting sore arms from tetanus immunizations, they tested the cave's Sullivan River and found it contaminated. Assistance on that and other problems was gleaned from a remarkable variety of sources.

Dietitian-approved foods were selected for nutrition, ease of preparation, variety, cost, and weight: "fourteen types of canned goods, three fresh vegetables, three types of drinks, American cheese and rye bread." Despite the extra weight, canned food was chosen over dehydrated: cans are mouse-proof. Those particular plans worked superbly. "Food is always the best time of day," the expedition recorded. Except for pack-damaged lettuce, the vegetables kept well in the 54-degree atmosphere, though the flavor of the cheese became increasingly stronger: "This, of course, greatly increased the variety of our meals. Fortunately no mold occurred on the rye bread or at least the mold wasn't visible."

The Indiana Bell Telephone Company installed telephones at the cave mouth and donated 6,000 feet of wire. This the Sullivaneers strung 3,000 feet inward to their base camp, thence to the stream level as a precaution against flash floods. To avoid distorting psychological tests, it was used only for weather reports.

On the appointed day, packs, duffel bags, gasoline stoves, and five-gallon cans of gasoline were started on their weary way inward. In bedrolls or in warm, dry "camp clothes" all cavers were wonderfully comfortable. Upon arising, however, each had to re-don "wet, muddy, freezing clothes" for three or four days consecutively. Despite such handicaps, 2,000 feet of virgin passage came to light. "Almost every foot was miserable," Mike recorded. "We surveyed several wet crawlways, muddy crawlways, dry crawlways, gypsum crawlways, domes, and crevices. We surveyed through breakdown, mazes, and a total of five bathtubs"—a total of 5,588 feet. And the conservation teams did their unromantic job so well that Sullivan Cave long remained free of trash and graffiti.

As the cavers' stay stretched on, schedules fell into an irregular pattern based on need rather than plan. Sleep varied from four and a half to twelve hours, averaging nine and a half out of each twenty-four hours. Meals adjusted to an even more bizarre routine—every eleven hours and ten minutes. Perhaps partly because of low consumption of chlorinated, boiled water hauled laboriously from Sullivan River, five of the Sullivaneers lost ten or eleven pounds.

Some curious health problems turned up. Besides the expected head colds there were transient spells of chest congestion, diarrhea, and upset teen-age stomachs. One group began early to suffer from immersion

Teen-agers' Home Sweet Home, 3,000 feet into Sullivan Cave. *Photo by Larry Mullins*

foot—trench foot—after long hours in Sullivan River, but with precaution escaped later complications.

Despite all these hazards, after the arduous expedition Minnesota Multiphasic Personality Inventories and related tests showed decreased anxieties, fewer hysteria traits, and the like. With the single exception of a decreased index of masculinity—and the Sullivaneers included no girls—it seems that at least this highly motivated group adjust better underground than above!

So much remained incomplete, however, that the Sullivaneers found the expedition somewhat disappointing. As plotting of their survey notes elongated the map of Sullivan Cave, exploration teams returned again and again for specific short ventures. Six thousand feet inside, Sullivan River proved only 100 yards from the entrance. A succession of horrible crevices yielded a shortcut, but saved neither time nor energy.

In December, Mike Wischmeyer found that the 1962 expedition had obviously taken a wrong turn. Again seeking the source of Sullivan River, the Sullivaneers sloshed through 5,000 feet of virgin passage and four huge rooms before exhaustion turned them back. The three weekends that followed yielded another 4,000 feet of river passage. At that tantalizing point, Sullivan Cave entered its winter flood stage, halting explorations for six months.

In essence, this new Beyond the Beyond consisted of a single dank river passage roughly seven feet high and twice as wide. Breakdown chambers, flood channels, and tributaries broke the dark monotony. To explore and map this new complex systematically, a second Sullivan Cave expedition seemed essential.

The teen-agers planned two camps for their new venture. The first again would be 3,000 feet inside: Base Camp. From there teams would rotate through a lightly supplied Camp II 9,000 feet up Sullivan River. Their sole task would be the unenviable job of exploring and surveying the Beyond the Beyond, already traced to a point three and one-half miles from the entrance.

Preparation for the new three-week expedition was more complex, yet only last-minute substitute Dick Blenz was over twenty-one. For the sake of morale, twice as many bedrolls were necessary. Each Sullivaneer was to have his own sleeping bag at each camp. Gasoline lanterns were too fragile to survive the trip to Camp II, and its stove gas had to be transported in small containers, requiring the use of flameless battery-powered headlamps. But the Sullivaneers were now a year older and far more experienced. They substituted gunny sacks holding twenty-five pounds of equipment for more elaborate carrying devices. On supply trips lasting fourteen hours, these proved remarkably useful.

In striking contrast to the smooth 1962 operation, however, the 1963 repeat performance seemed trouble-ridden even before it began. Sullivan River flooded mildly for three days. With forty-pound loads, the Sullivaneers' supply trips could conquer swirling floodwaters waist-deep, but when the depth reached seven feet, packs had to be cached. Six thousand feet of telephone line were successfully installed, but not until 9 P.M. on the target date had all the gear been painfully conveyed through the Backbreaker to Base Camp.

Sleeping exhausted until their watches told them it was noon, the Sullivaneers perked up. Body weights and temperatures were recorded. Then everybody ate a fine breakfast of canned wieners, pork and beans, celery, cocoa, and a cheese-on-rye sandwich. Heavily loaded, the first advance team began the 9,000-foot slosh to Camp II.

A second team was due to follow promptly, helping to transport cached equipment to Camp II. At that point, however, the underground telephone rang: there were heavy thunderstorms in the area. Delaying twelve hours for floods which never materialized, the second heavily loaded trip slowly set out at 2 P.M. Tuesday.

This team surveyed en route, regardless of excessive loads. It was an error of judgment. Halfway in, the cavers realized that they were in serious trouble. Mike Wischmeyer recorded that "the surveying was frustrating because Rodney, the note taker, kept falling asleep." Careless with fatigue, fifteen-year-old Darrell Kirby fell into a particularly nasty stretch of Sullivan River. Rodney Grant leaped to the rescue. "He recovered the pack of candy bars, Darrell's pack of clothing, and Darrell himself—in that order," Mike noted. Much of the load had to be re-cached. Not for twenty exhausted hours did the trio stumble into Camp II and collapse thankfully into their sleeping bags.

Recuperation from this tremendous effort seemed to bring better days. Medical studies did not fare well: the Camp II bathroom scale failed to survive the jarring trip. But mapping swung into high gear at both camps. Camp II promptly completed 2,000 feet of survey. As many as five mapping teams were in action simultaneously, scattered almost from end to end of the sprawling cavern. Base Camp frantically plotted the accumulating data—a worthwhile innovation—and the map grew longer and longer. On the eighth day Bob Larson totaled the new surveys and jubilantly telephoned Camp II. Ten thousand feet had been mapped Beyond the Beyond; 17,128½ feet throughout the twisting cavern.

The extraordinary progress of the mappers required a shift to greater emphasis on exploration. How many miles lay beyond the Beyond the Beyond? The Sullivan Cave ridge extends four miles farther north than any caver had penetrated. Its river was hardly smaller at the farthest point reached than at its emergence. The Sullivaneers halfway came to believe that their cave was endless. Would a Camp III someday be necessary to explore the Beyond the Beyond the Beyond?

Explorers splashed on and on in the black natural tunnel. Past Room Six and Room Seven they penetrated the stream channel. In Room Eight they climbed a huge breakdown pile and returned to stream level.

But now they were headed downstream!

Momentarily nonplussed, the cavers trudged back up the breakdown pile in search of an explanation. That explanation proved disappointingly simple. Almost hidden against the somber cavern wall lay a dark siphon pool: their long-sought goal. From it Sullivan River flowed into both branches of an overlooked fork of the cavern. The three-week expedition not even half over, the teen-agers had conquered Sullivan Cave.

Jubilation reigned as the great news flashed along the telephone line. Yet, from that moment of triumph, the expedition began to deteriorate. Suddenly no major goal remained. Gone was the elusive, ever-renewed challenge of increasing miles of cave, of huge black rooms that still awaited the coming of first light. There was no Beyond the Beyond the Beyond. In its place was only "a long, wet, miserable crawlway with a deep, black pool."

A painful redecision kept Mike Wischmeyer, Leigh Lawton, and young Darrell Kirby at Camp II long beyond the normal rotation. Inevitably they spent many hours daily in watery corridors. Dry socks were a half-remembered luxury. Their terrific pace exacted an unavoidable toll. With chronic exhaustion came another insidious enemy—the tingling numbness of immersion feet. Darrell's feet began to bleed, and traces of gangrene appeared around broken blisters. Improved circulation in the relative warmth of Base Camp brought intensified discomfort. Yet none of the teen-agers considered leaving Sullivan Cave. These were mere annoyances, easily dismissed.

Ten days of seemingly unalterable routine were soon behind the Sullivaneers: awaken, complete the "morning" physiological studies, breakfast, survey eight to fourteen hours, "evening" physiological studies, dine, make diary entries, fall asleep, awaken. . . .

On August 21 the Beyond the Beyond survey was complete. Camp II should have been abandoned, but for days cavers sloshed back and forth dragging forty-six pack loads of gear a few hundred feet at a time. One exhausted team was forced to bivouac amid rocky pits, somehow surrounding five cavers with three dry sleeping bags. The early 20-hour nightmare struggle was a pleasant stroll compared to the return to Base Camp. Mike Wischmeyer wrote:

Although I'd gotten plenty of sleep, I was still worn out before [Freshour's group] reached us. My trench foot was worse, my muscles were aching and I was mad at Blenz for not coming back [the older Blenz was in his sleeping bag, totally exhausted—W. R. H.]. The Deep Water [Marianas Trench] drained me of whatever energy I had left. I was completely numb below the waist and I couldn't think clearly. I just dragged one pack after another through . . . I never want to go through the Deep Water again.

At this particular time, the average Sullivaneer had lost 9.7 pounds, was maintaining an average temperature of only 96.4 degrees and was too tired to sleep soundly. Yet not all teen-age spirit was gone. As Mike and "Dixie" Dickson reached Base Camp, Jay Arnold and his sister Pat strolled in from the surface. Mike could still record that the latter "looked extremely attractive even in grubby cave clothes. It could be, though, that after twelve days in Camp II any female would have looked good." (They later concluded that Pat Arnold looks even better by daylight.)

Of necessity, the next day was a day of rest, celebrated by a three-hour, five-hundred-question psychological test. That test, however, was hardly necessary to reveal hypomania and fraying tempers. Base Camp was annoyingly overcrowded. Continuous cookery and other activity hampered sleep and led to factionalism. Yet basic motivation was still intact. Despite aching feet, the little remaining mapping was completed for a total of over 42,000 feet. Without orders, great piles of wet, muddy gear shrank toward the entrance through the Backbreaker. "I'm afraid I'm going to wake up one of these times and find everything gone except for my sleeping bag, cot, and myself," Mike Wischmeyer noted in his invaluable diary. Teen-ager appetites never flagged: the bottomless pits of Sullivan Cave! On the penultimate day, a special call went through to Indianapolis with requests for the great return to daylight:

Luxury in Blue Spring Cavern, Indiana: Ralph Lindberg paddling. *Photo by Arthur N. Palmer*

Fried chicken, barbecued chicken, hamburgers, ten quarts of white milk, one quart of chocolate milk, yeast donuts homemade by Mrs. Mullins, two apple pies homemade by Mrs. Mann, ice cold cokes, a German chocolate cake, butterscotch pie, chocolate dream cookies homemade by Mrs. Dickson and a chocolate cake with thick fudge icing . . . Definitely no celery or rye bread . . . No, we're NOT starving. We just want a little variety!

Sullivan Cave may have been Indiana's toughest. With a little less than 10 miles mapped, however, it is far from its longest. Indiana cavers seem to reach for a compass and tape the moment they lose daylight. As of 1969 they had mapped at least 210 miles under the Hoosier state: a full two-thirds as much as the 1973 figure for the champion Kentucky. No one seems to know today's total.

Until 1964 nobody even thought much of Blue Spring Cave. True, it had a trunk channel 2 miles long and carried one of the largest of the state's underground rivers. But pools as deep as 25 feet made speleoboating mandatory. It suffered from ferocious underground floods, and the stream siphoned 1,600 feet upstream from the spring. Worse, at the upstream end breakdown blocked the corridor with frustrating thoroughness. Furthermore, it was a Johnny-come-lately among Indiana caves. Until the bottom suddenly dropped out of a sinkhole pond on the farm of the Colglazier family in the mid-1940s, no one had even known it existed.

In January 1964, however, Jim Richards, Tony Moore, and Dan

and Dale Chase turned to a maze of unpromising, mud-choked tubes and fissures near the infuriating upstream barrier. After 1,500 feet of intricate passages, they emerged on its far side. Breathlessly they splashed along for a full mile in a gentle, gravelly stream, running pell-mell in a virgin corridor thirty feet in diameter.

This First Discovery soon yielded several side passages and an upper-level chamber with unexpectedly massive dripstone pillars. Even the 7,000-foot Maze, too, was kinda fun, "if one remembers to straddle underwater ledges in Straddle Alley, bridge the pool in the Pool Room, and to follow the path of air movement," according to one N.S.S. report. "Unexpected passage relationships, deep pools, fantastic solution features and hidden calcite decorations make exploration exciting." Just a few weeks later, a muddy crawlway yielded the Second Discovery, blessed with 2,200 feet of another vaulted throughway. Temporarily transplanted Art Palmer undertook the mapping of the exciting new cave as part of his doctoral thesis on karst hydrology. Soon he, his petite wife, Peggy, and other extra-waterproof cave mappers were hard at work. Several trips and two miles onward, at the extreme northeastern corner of the cave, the Second Discovery mainway ended at a pile of surface debris. Hard digging opened a dry entrance, two miles by road and footpath from the Colglazier farm. No longer was boating necessary to get anywhere in Blue Spring Cave. Six and a half miles soon was on the map of the Second Discovery alone. Much of it was large, dry, and beautifully decorated.

Less than luxury in Blue Spring Cavern, Indiana: Peg Palmer in the Pothole Passage. *Photo by Arthur N. Palmer*

The Second Discovery remained pre-eminent only three months. In April, Dan Chase and Tony Moore again broke out of the Maze. This time they found themselves headed north, into the Third Discovery.

This section was less pleasant. Here and there pools were annoyingly deep, and the sand extra-abrasive. Yet here also was an unexpected chamber 330 feet long and as much as 50 feet high. The Pyramid Room, they called it, because of three conical stalagmites, all the while marveling that so large a room could exist in a structurally unstable area without a stabilizing cap of insoluble rock.

On the map, the miles of passage continued to mount. In the first two months 9 miles were surveyed and plotted, and the mapping teams had overtaken the explorers. That was the easy part. Though he still smiles fondly at certain memories, Art grimaces at the three years which followed. The next ten miles, he recalls, included "some of the wettest, most dismal stream passages in Indiana. It was in this latter section of the cave that the fellowship of the mappers who held out until the very last was cemented by a strong bond of friendship and mud."

And, in mid-mapping, Jim Richards and Dale Chase came upon the Fourth Discovery, in a not-so-routine manner. Five hundred feet along a miserable chest-deep waterway in the First Discovery they encountered a pool where they could not touch bottom. Swimming for 50 feet despite their packs, they rounded a corner. Presto! Out went their lights, extinguished by a gale blasting through an air space of just three inches. Noses to ceiling, they broke out just 25 feet onward into what turned out to be 4 additional miles of cave. Inflatable rafts and wet suits were needed to prove it.

Required to return to his northeastern homeland in 1967, Art soon found himself running out of time. Each new discovery became less and less welcome. In the grand manner, however, virtue was rewarded. On the last possible day of his stay in Indiana, the number of unmapped leads plummeted from fourteen to zero. Although it required four final underground marathons during his last week, he and his fellow mappers finished the job at 18.9 miles. Pending the Fifth Discovery, of course, which Indiana cavers are still seeking.

Other challenges beneath the verdant fields of Indiana may prove even greater. Joe Sanders, for example, considers Binkley's Cave "undoubtedly one of the greatest areas for discovery and the study of caves and karst hydrology in the United States today." Although his recent work has largely been one state farther south, Joe's opinion is weighty indeed. A leading member of the Hart Attack Team, he and his teammates recently found a 4-mile river corridor branching out into 7½-mile Grady's Cave under a similar section of Kentucky's Hart County. (Nearby, too, is Crump Spring Cave with 7.9 miles of "underground spaghetti" some consider tougher than that of Flint Ridge.)

Peg Palmer surveying with dipod in Blue Spring Cavern during field work for her husband's thesis. *Photo by Arthur N. Palmer*

Binkley's Cave is in southernmost Indiana, down near the Ohio River, almost in the back yard of Lewis Lamon, grand old man of Indiana caving. Some years ago, teen-age members of the Indiana Speleological Survey became obsessed with its potential. Draining a large sinkhole plain through a 5-mile throughway corridor, it already was considered unusually important. The Bloomington Grotto of the N.S.S.

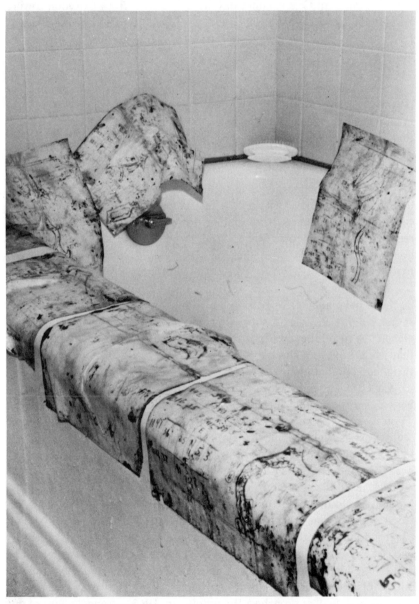

Somehow the map of Blue Spring Cavern emerged from pages of field notes which had to be dried in the Palmers' bathroom. *Photo by Arthur N. Palmer*

had mapped 6½ miles here when asinine grossness by a few visiting slobs caused its once-friendly owner to close it to all comers. Although scarcely older than the Sullivaneers, the I.S.S. members were able to persuade him to make an exception. At last report their map included 17 miles, with another seven to ten miles awaiting compass and tape.

Some years ago Lewis Lamon casually showed me the true face of this verdant land. One summer day we drove west from Corydon toward Wyandotte Cave, away from Binkley's. Even before we left the city Lew had my head whirling as I tried to keep up with his tumbling accounts.

"Right out of town here," he announced as we left the city limits, "there's a bell-type pit that a man and a boy went down on a rope. The boy got out like a monkey, someway, but it took a tractor to pull the man out. They had to put him in the hospital, he lost so much hide. See that junkyard? There's two caves behind it. One of 'em is nice."

As we approached a curve he continued with hardly a break: "Off to the left, there's two pit caves." Second later, another curve, with a pretty green and white house in rolling countryside: "Over on the right, there's a nice pit cave with 1,500 feet of passages. Then there's a nice cave by the next house. And another one over on the left here. See that red barn? Another over there. And three in the woods behind that house on the hill . . . another on the right . . . Barnyard Cave off to the right now . . ."

Silence reigned for a few seconds. I was tired from merely turning my head back and forth, keeping up with the picture, but the let-up was short. We passed a small sink, seemingly too shallow to merit a second glance. "A nice little cave in that one," he mused, almost to himself. "See that white house on the point of the hill? There's a 60-foot pit cave up there, and another along the way."

Caves to our right, caves to our left. In a lifetime of spelunking, Lew hasn't even been able to check all those within 25 miles of his home. "You've got to come back," he insisted. "There's so much you should see!" Lew is right. Someday I must, but no single person ever will know the caves of even this one small part of southern Indiana. Not even the fabulous Lewis Lamon.

Yet, in all the enormous area north of the Ohio River and east of the Mississippi, our target that remarkable day—Wyandotte Cave—has stood alone, unquestionably supreme, since at least 1818. Every serious caver is familiar with its name. As an awestruck teen-ager half a century ago, transcontinental caver George F. Jackson fell in love with its lure and lore. Nicknamed "Wyandotte Jackson," he subsequently became the author of two beautifully illustrated books about this sprawling cavern. Yet, when I first approached Wyandotte in 1963, strangely few of my caving friends had passed through its broad portal. Peculiar helictites, the giant Pillar of the Constitution, awesome chambers—these were common knowledge. So were the commercial claims: "23 miles of underground fairyland," "the largest formation known in any cave in the world," "the second longest cave known"—later amended imaginatively to "one of the three longest caves in the world." Such slogans are always disturbing to cavers. Except for the mammoth Pillar of the Constitution, Wyandotte's stalactitic decorations are overrated even in comparison with nearby Marengo Cave. Almost certainly it is not "one of the three longest caves in the world." Somewhat as in the case of the Kentucky caves, the owners' large-scale map was carefully restricted and failed to show certain known passages. Yet I came away singing the praises of Wyandotte, commercialized yet a wilderness cave *par excellence*. Not soon will I forget its marks of prehistoric Indians: poles carved with flint tools, flint quarries, great quantities of widely strewn hickory bark. "I tried it out once," merry spelunker Lamon told me deep in the spell-weaving cavern. "I cut ten strips a foot long and tied them together with grapevine the way the Indians did. They made a fine, clear light."

Nor will I soon forget the moment when we burst out into a vast chamber at the foot of huge rock-slab Monument Mountain in Rothrock's Cathedral. This awesome underground mountain so fills its enclosing chamber that the only broad view possible is steeply and distortingly upward from a level spot halfway up its flank. Here the viewer finds himself strangely small and unimportant, face to face with inner space. Yet the gloomy Senate Chamber, its steep black mountain superlatively topped by the miraculously red-brown Pillar of the Constitution, is still more awesome.

As at Mammoth Cave, the beginnings of the story of Wyandotte Cave are irretrievably lost in the mists of time. By 1801 white explorers had followed the moccasin prints of ancient flowstone miners to the remote Senate Chamber. For 50 years, that was all that was known of the cave by any human, white or red. Indiana's territorial governor (later President) William Henry Harrison seems to have visited it in 1806, probably to reconnoiter its extensive saltpeter deposits. Soon Dr. Benjamin Adams legally pre-empted the land and began manufacturing that pioneer necessity. Initially Adams called it his "Indiana Saltpeter Cave." Then he found a small but richer saltpeter cave a few hundred feet away, and changed the name of the big cave to "Epsom Salts Cave." It contained plenty of that cathartic, too, with crystals sometimes growing as much as an inch per month. In 1818 he brought it to the attention of the world through a letter to the prestigious American Antiquarian Society. Within a year extensive accounts were in print, describing it as "The Mammoth Cave of Indiana."

Perhaps as at its Kentucky namesake, this was part of an attempt to sell the property after its saltpeter value plummeted at the end of the

War of 1812. In any event, Adams relinquished his claim in 1818. A year later the cave coincidentally was part of 4,000 acres of timberland purchased by Peter Rothrock, a German immigrant. Although he was after the timber, the cave remained in his family until purchased for a state park in 1966. For many years they considered it nothing but a nuisance. Local cavers, it seems, opened its low entrance ever higher for their con-

The Pillar of the Constitution in Wyandotte Cave. Inset shows artist's conception of a century ago. *Photo copyright 1953 by George F. Jackson*

venience. Unfortunately this was also a convenience for the neighborhood cows, which were forever going caving, enthusiastically licking the Epsom salts and coming home with diarrhea. In 1843, the neighbors' patience reached its end. They persuaded the state legislature to enact a special law requiring the Rothrocks to fence off the entrance. The Rothrocks' feelings were not recorded, but the incident was hardly a boon to caver-landowner relations.

The history of the cave soon quickened. In 1850 no living person knew any part of the cave except what is now "The Old Route"—the long, rough route to the Senate Chamber. Then somebody moved a rock. Most of the cave lay beyond, and additional well-beaten paths of pre-Columbian Amerinds.

The Rothrocks reversed their earlier opinion and opened their fortuitous possession to the public as "the second largest cave in the world, second only to Mammoth Cave, Kentucky." By now it was under the name Wyandot Cave, although no one knows why. The Wyandotte tribe had no part in its history or prehistory.

In truth, the Rothrocks' bold claim seemed valid. At that time few caves were known in the United States, and of those, little more than their entrance areas. This was a sprawling cavern of large corridors and huge chambers surprisingly like the historic route of Mammoth Cave. Its lesser passages seemed to have infinite possibilities for discovery. One entire end of the cave became simply "The Unexplored Regions." More easily visited than Mammoth Cave during the heyday of Ohio River steamboats, it long qualified as one of America's truly celebrated caves.

Among those who fell under its spell in the early 1850s was a young Indiana divinity student whose name has appeared earlier in this book: Horace C. Hovey. A generation later he was to vault into prominence as America's foremost speleologist. Almost to his deathbed in 1914 he wrote and lectured widely and compellingly on Wyandotte and Mammoth.

Less than a decade after Hovey's death, harum-scarum young George F. Jackson came to Wyandotte with a group of Boy Scouts from nearby New Albany. When the others went home, George didn't. He had met Wyandotte Cave and the Rothrocks, and his life was forever changed. Charles J. Rothrock then was manager of the cave and little old hotel. Years later George was to marry his daughter and live and work intimately with the entire family.

In 1923, however, young George was primarily fascinated by the stories of Washington and Andrew Rothrock, spry octogenarian sons of the 1819 purchaser of the cave. Making his peace with his own family, George spent much of the summer of 1923 talking with these pioneers about the cave as only they knew it: hazardous pioneer explorations with only homemade candles for illumination, of subterranean windstorms, of discovery of Indian footprints that should have led to a rear entrance, of long-lost passages that eluded their attempts to return. When he became a guide, George built upon their accounts, helping to expand the cave bit by bit through the years. Fifty years later, George is bemused to find himself in a similar role, bridging almost the entire history of Wyandotte Cave.

After the state of Indiana purchased the property in 1966, modern speleology came to Wyandotte Cave. The long-secret, incomplete Rothrock surveys were repeated by Richard L. Powell of the N.S.S.'s Bloomington Grotto. Dick and his friends came up with 6.1 miles of passage instead of the traditional 23 miles. Only a little—part of which inevitably was dubbed The Bloomington Grotto—was virgin.

Which brought a smirk to the face of Wyandotte Jackson. Ever since Powell's maps reached his California home, he has been writing letters and publishing suggestions on where to look for the rest of the cave. Not soon will we know Wyandotte's exact size or its place in the record books. Regardless of exact size and vainglorious comparisons, however, Wyandotte Cave is tremendous.

It may even be Indiana's largest.

Perhaps, however, Indiana's largest cave is still growing somewhere in the Lost River area, where floodtime waters sometimes spout two feet high from farmers' fields. Geologists have counted 1,022 sinkholes in a single square mile of this honeycombed cavers' heaven—or hell.

In desperate moments of heroism and death, in 1961 cavers observed firsthand the prodigiously turbulent forces which enlarge these sewer caves of Indiana. An epic began on the hot, sultry Sunday afternoon of June 16.

On a wooded hillside near Orleans, Indiana, little-visited Show Farm Cave was only a few miles from the main Lost River system. Before beginning a long drive home from a full day's nearby caving, expert speleologists Tom Arnold and Ralph Moreland, Jr., decided upon a quick peek into Show Farm Cave—a mere "twenty- or thirty-minute trip."

Young Alan Lipscomb and Tom's brother-in-law Carl Birky, Jr., stood by, awaiting their return. No one foresaw any problems. Despite frequent thunderstorms, the stoopway entrance was dry. Only a few trickles entered the cave farther inside: practically a bone-dry cave for the Lost River country. A huge thunderhead loomed in the western skies but that pattern had persisted for several days. Adjusting their headlamps, the cavers ducked through the entrance into a comfortably lofty corridor. Delightedly they strode into the unknown.

Slowly the great thunderhead drifted overhead. About the time Birky and Lipscomb began to look for returning carbide flames, sheets of rain deluged the Show Farm. On the hillside behind the cave red-brown rivulets of topsoil sprang to life. Within a few minutes, more than an inch of rain drowned this chance spot. It funneled straight into Show Farm Cave.

The sweeping torrent alarmed Birky and Lipscomb. Dashing inside, Birky found water pouring in from side channels. A hundred yards from the entrance he found himself in difficulties, fighting swift, shoulder-deep currents.

The water was rising perceptibly. Birky barely regained the entrance, now two-thirds submerged. At a run he set out for the nearest farmhouse to gasp the dire news to Bob Nicoll of the Bloomington Grotto and to the Indiana State Police. A state patrol car dashed to the scene, its radio invaluable in summoning and coordinating cavers, skin divers, and a host of other helpers.

Carl's alarm was fully justified. Almost 1,000 feet inside the sewerlike cave, a sudden brown torrent tore at the explorers' legs. In an accompanying underground windstorm their carbide flames flickered ominously. At every straining step, the silty water seemed to gain a higher grip. Boulders rolling along the recently dry stream course, painfully battered their ankles.

Ralph and Tom fought the roaring torrent with thought-blanking desperation. Six hundred feet from the low entrance, a vital crawlway might well mean life or death. The first foam-ridden brown surges had instantly warned them into a high-splashing run. More and more their frenzied dash had altered into an ever-increasing nightmare. The plucking, roaring waters grudgingly yielded each hard-fought step. Each moment increased the threat of being dragged away to eternal black depths.

Silt and debris in tiny pockets against the ceiling told these experienced cavers all they needed to know. As the implacable floodwaters rose higher and higher, superhuman effort drove the men onward, shouldering their way against tremendous buffets, half whirled away at every move. Every knob of rock was a handhold, or a foothold—or a shoulder hold or even a head hold—anything that would gain a few more inches. In their hearts they well knew they were too late, knew that the all-important crawlway was blocked. But perhaps there was a chance.

Then they saw it and knew. For a moment Tom and Ralph gaped numbly at the churning caldron. Just a quarter hour earlier, that crawlway had been virtually dry. Blithely they had twisted through, hardly slowing in their rapid scouting tour. Now a great boil of churning water spurted obliquely ceilingward with a thrumming roar that shook the limestone walls.

It was hopeless. Beyond the crawlway, floodwaters obviously were backed up five or ten feet deep. Unable to shout above the tumultuous din of the hungry waters, the pair gestured downstream, pointed ceilingward, nodded vigorously. A half mile deep in the cave was a single lofty corridor—the only possible safety in this watery trap. Numbly they turned, now floating, grasping the one chance that this would be only a minor flood.

For once, fate seemed against cavers. Two additional tremendous thunderstorms quickly followed the first. All records were broken as more than 5 inches of rain bombarded the countryside within a few hours. By 7 P.M. the entrance of Show Farm Cave lay 16 feet under water.

Fighting 50 miles of blinding rain, Bloomington grottoites began to arrive by 9:30. Sometimes waist-deep in water on flooded back roads they were forced to wade the last 1½ miles carrying their unwieldy equipment. Through the long night other cavers splashed in from several states. The state police continued their outstanding assistance. Portable generators somehow reached the flood-bound scene, and skin divers ferried in by helicopter. Even a bulldozer turned up—no one seems to remember how. But, though the rain had stopped, little could be done with the sucking lake that concealed the entrance. Through the long night, frantic cavers marked the slow retreat of the water.

By dawn the top of the duckunder entrance was exposed. Strapping Dave Howe, Bloomington Grotto chairman, assumed charge of the beginning rescue operation. Because of the exceptional ability and experience of the trapped pair, surprising optimism reigned. If anyone could survive so incredible a deluge, it would be they.

A small rivulet was still running into Show Farm Cave. The bulldozer chugged into action. Unfortunately, the dammed stream promptly poured into a nearby sink and on into the cave. Portable fire pumps were useless for the same reason. But the bulldozed dam permitted cavers to re-enter Show Farm Cave. Often neck-deep in the swirling floodwaters, they sought their lost friends, frantically yet with surprising coordination and self-discipline. Soon the cavers and state police skin divers had penetrated 600 hellish feet to a huge pile of debris, still more than half submerged. Somewhere below lay the key crawlway. Local spelunkers knew of its existence, but no one knew the exact spot. Great air bubbles burst from the bulky mass as the waters ebbed—now so slowly.

As reports flowed back to rescue chief Dave Howe at the low entrance, plans changed almost from minute to minute. Across the flooded land, radio calls went out for peculiar tools—inner tubes, crowbars, posthole diggers. Under almost impossible conditions, a grim hand-dug attack proceeded on the far end of the debris plug. Nerves were taut, emotions strained. These were good, close friends somewhere beyond.

Most of the cavers had gone without sleep for thirty-six hours. The deep, chill water constantly drained their energy. New thunderheads hinted an ever-present risk to every man inside the black, watery tunnel. Tempers frayed as the sopping cavers worked to exhaustion. Humanly snappish comments by a few overtired, near-hysterical cavers helped not at all. A demanding press and overhelpful local citizenry created surface problems. Yet, in the flooded cave, a smooth, coordinated effort took form.

Unexpectedly, a sharp, rather obscene gurgle echoed through the dark passageway. As if someone had pulled a plug, a watery funnel swirled seven or eight feet down into a two-foot orifice at the entrance-ward end of the debris. The desperate hours of work at the opposite end had been for naught.

For a time the water in the 600-foot entry corridor dropped rapidly. New rains came, but these were only ordinary thundershowers which did not affect the water level in the dark cavern. When the weird whirl-pool had dissipated, cavers and skin divers advanced another 600 feet, bobbing along to the end of the air space—in vain.

The receding waters seemed almost to level off for many hours. Late Tuesday night, however, the rescuers were able to push onward 3,000 feet from the entrance. At a major T-junction was a sad discovery. The National Speleological Society *News* reported:

> At 1 A.M. Wednesday the bodies of Moreland and Arnold were found in the right side of the "T," slumped high above the water, apparently drowned. They were a short distance toward the entrance from the highest spot in the cave [ceiling about 25 feet]. Marks on the ceiling seemed to indicate that even the high spot had filled. . . . Moreland's watch had stopped at 7:15.

Caving *is* inherently dangerous. Only if we avoid all caves can we avoid all their hazards. Tom Arnold and Ralph Moreland would be the last to decry caving because of their one-in-a-million deaths.

More, they would have been the first to applaud the teen-agers' triumphs over Sullivan River, planned only weeks after the Show Farm Cave tragedy.

13

Of Rope and Ladder

The Story of the Birth of Vertical Caving

Thirty-five dim feet below the surface, a sloping 4-foot orifice belled out into blackness nearly 100 yards deep. A rope-snarled cable disappeared below a caver dangling in this Mystery Hole.

Hanging spiderlike just above the constriction, Don Black struggled with the messy snarl in a rain of dirt and small stones. Around the mouth of the pit, a leaderless mob crowded close for any new moment of drama. Thrilling indeed, this newest episode in the long history of Chattanooga's Lookout Mountain!

Veteran caver Don Black saw the incident with mixed emotions. To an expert pit plunger, this shocking afternoon should have been merely a tragic, bungled paragraph in a very different drama. But Don was not wholly detached, for the still, limp body far below was that of a family friend. Too, it was no longer merely a matter of the dreadful recovery of the broken body slumped on the rocks. This crowd-enthralling incident of the gaping pit caves of the southeastern mountains now required the rescue of a would-be rescuer.

It had begun so innocently, that pleasant autumn afternoon in 1959. Eighteen-year-old Jimmy Shadden and two young companions had planned to descend to the constriction—the Jumping-off Place—to gawk at the abyss below. Unskilled in spelunking techniques, they planned to lower each other in turn on a short nylon rope. Their frayed old rope was smaller than that used by most cavers—but what of that? "Everybody knows" that even a small nylon rope is so strong that frays were no reason for concern.

Hardly had his friends begun to lower Jimmy Shadden when they were catapulted backward. A free-ravelling rope end whipped from the pit. From the depths came an ominous clattering sound, perhaps a faint

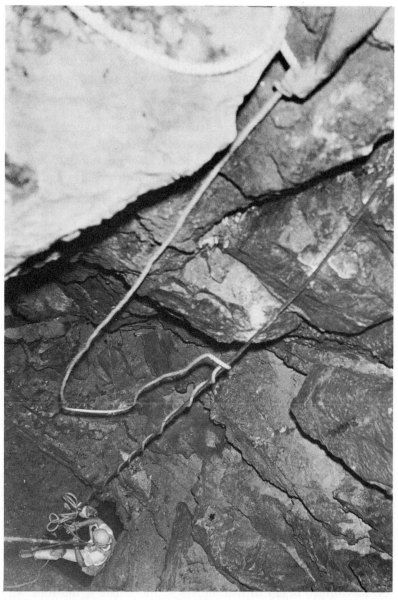

Suspended just above the Jumping-off Place almost 300 feet above the floor of Mystery Hole, Don Black attempts to free snarled ropes. News–Free Press *photo by Delmont Wilson*

faraway scream, then nothing, nothing, nothing but the whisper of a small hidden waterfall. As they rushed to the edge, pale with shock, a faint hollow sound echoed up from somewhere far below. Hoping against hope, they turned their flashlights to the Jumping-off Place. It was bare.

Within a few minutes, local Red Cross authorities contacted Chattanooga caver Don Black. A boy had fallen into a nearby 190-foot pit. Could Don leave work? No equipment was needed except rope.

Don stored 400 feet of rope at his office, but the garbled reports sounded dangerously like Mystery Hole—more properly termed Mystery Falls Cave. With immediate forebodings, Don suggested that he rush home for his climbing gear. The response was insistent—and appalling: the messenger had been sent for Don, not for equipment. No climbing gear was needed. The boy had merely fallen into a hole.

Thoroughly alarmed by the obvious lack of comprehension, Don rushed to the scene. As he had feared, the locality was Mystery Hole, deepest pit known in the United States. Various officials and surface rescue personnel milled about, arguing about procedures and personnel. An impatient crowd was on the verge of taking matters into its own hands.

Just as Don arrived, a local racing car driver was being tied to a stout 1,200-foot rope. Don's protests were ignored. Scores of willing hands began to lower the helmeted volunteer into the abyss.

Don subsided, awaiting the inevitable. Just past the Jumping-off Place, the rope began to spin rapidly. Horribly nauseated, the would-be rescuer shouted to be hauled back to the surface.

The first rash rush ended, Red Cross and Civil Defense personnel showed more inclination to listen to expert logic. Don's three descents to the bottom of Mystery Hole suddenly seemed much more impressive. Sirens announced a hasty 14-mile round trip to his home. There Don grabbed the equipment necessary for a safe descent: Prusik knots, rappel spool, parachute harness, and other gear.

On the full-speed return trip, Don's single hope was that he would be in time to avert other tragic impulses. He was too late. As his speeding car drew up near the cave, the racing car driver was again being lowered —this time by the cable of a tow truck. Perhaps well-meaning officials considered this the least possible evil: muttering bystanders seemed about to start down the cable hand over hand. To a few who would listen, Don predicted a horrible snarl. He got nowhere, and sat down to await developments.

The new descent was smooth. Sharp whistle blasts confirmed the death of the fallen youth. With the body tied to a lowered rope, those on the surface began to take up slack. Immediately, the rope whipped around the cable and snarled. Repeated tugging and flipping increased the tangle.

Don took over with a sigh. No one protested.

Rigging a short rope, he rappelled to a point just above the Jumping-off Place. Hanging free for many minutes, he struggled with the rope in a nasty rain of dirt and debris. Finally he was able to work a little slack into the looping cable. Cutting the rope he passed the free end upward. Holding it away from the cable, he was able to keep it unsnarled as the body was slowly hauled to the surface. Only then could he free the remainder of the snarl so that the trapped rescuer could also be hoisted past Don's twisting aerie.

Members of the National Speleological Society first plumbed Mystery Hole on June 12, 1954. Four hundred feet of lightweight cable ladder was lowered into the pit. Bill Cuddington tied into belay and began the long, bitter descent through the spring-swollen waterfall. Not once could he look up for fear of extinguishing his carbide light. For seven endless minutes, periodic whistle blasts screeched through the roar of the falls as Bill called for slack on the belay rope. To those above it seemed hours until a faint series of whistles told them that Bill had reached the bottom.

After forty sopping minutes of exploration, Bill tied back into belay. Three sharp whistles announced his readiness. The rope tightened in comforting responsiveness, and Bill began the upward struggle. At each rung the pounding torrent threatened to tear him from the gossamer ladder. Soon his lamp was extinguished by the waterfall. Climbing slowly and deliberately in darkness, he was repeatedly forced to hook himself to the ladder to rest his trembling limbs. Under such circumstances, few but Bill Cuddington could have ascended that hellish 316 feet in 11 minutes.

Bill's companions eagerly listened to his gasped report, then looked at each other in dismay. His feat was beyond their capabilities. Some new technique would have to be devised for such stupendous pits.

"Vertical Bill" Cuddington agreed. For many months he had been scaring the hard hats off conventional cavers of the Virginias, developing such a technique from specialized mountaineering practices. In his hands, rappelling down "ordinary" pits and ascending with special loops slid along the rope—Prusik knots—had become easier and quicker than ladder climbing. Experimenting, he found that, in narrow pits, safety ropes tended to foul the rappel rope and often were worse than no belay at all. In their place he substituted a loose Prusik sling looped—just so—around the rope and around his chest. When he slipped, the Prusik knot tightened on the rope and halted his fall. By attaching two additional loops to the main rope, he could stand up, take his weight off this self-belay loop and continue down.

Bill's lack of a standard belay shocked safety-conscious cavers lacking his experience with the new method. "Old-time cavers really lectured me," Bill says. Some urged his expulsion from the National Speleological

Society before his much-predicted death spoiled the society's unblemished safety record. But, as more and more cavers witnessed Bill's extraordinary caution, the prophets of doom began to lose their following. More and more cavers began to realize that Bill had quietly revolutionized vertical caving.

Vertical Bill was well aware of his narrow safety margin. He insisted on tying all knots himself. His rope was never permitted to drag on dirt or rock and he rappelled slowly and cautiously. Some of his imitators have been less careful.

Slightly cocky over his conquest of 119-foot Saltpeter Pit, young Roy Davis was just the man to team up with Cuddington. Gaining valuable experience in multi-pit caverns near Roanoke, Virginia, they soon were ready for bigger game. On October 9, 1953, a superb team of five—Cuddington, Davis, Tank Gorin, Kenneth Bunting, and David Westmoreland—came to Blowing Hole.

Previously published accounts of this first descent bear little resemblance to the facts. Roy tells it in retrospect: "I was just a teen-ager, worried about what others would think of our getting into such a predicament. I wrote the story up as a matter of routine—superior spelunkers conquering all in stride. I was too scared by what came next to admit it then!" What came next was Bill's shout: "Send Roy down!"

Leaping to his gear, Roy found his rappel pad missing. Improvising a pad from a towel, he donned his gloves, got into rappel position and walked backward over the edge of the precipice. As he entered an impressive, smoothly round shaft 15 feet down the narrow upper chimney, his towel slipped. Down, down, down it floated.

The friction of the rope began to scorch Roy's hip like flame. The descent seemed endless. This was more than any 100-foot pit!

The fluted wall belled out into an enormous chamber, its floor still invisible in all-encompassing blackness.

"I thought I was pretty well down," Roy recalls ruefully. "I yelled 'Hello' at Bill. He looked up. I nearly fainted and fell out of the rope. His carbide light still wasn't more than a pinpoint in the black. I still don't know how I got down that last 150 feet."

Perhaps Roy's status saving was justified. Word of the new technique spread. Hole after hole yielded their secrets as this new breed of pit plungers approached them unafraid. More and more southeastern cavers became competent rappellers and prusikers. Several grottoes of the National Speleological Society organized training schools. Graduates of such courses were experts indeed. All over the Southeast, pits less than a hundred feet deep became child's play.

For long free rappels, however, it had quickly become apparent that additional improvements were badly needed. In an ordinary rappel, the caver's descent is controlled and slowed by the friction of the rope around his thigh and body. Friction means heat, as Roy Davis learned the blistery way. In a grotto newsletter, Roy summarized the situation, tongue in cheek:

There are many ways of getting down a pit—the easiest, of course, being to simply jump. This practice is to be discouraged, however, because the jumper might injure someone below. . . . Various cavers have various systems of descent. Some use rope ladders or cable ladders; some prefer to lower or raise by block and tackle fashion; some adhere to rappel and prusik loop techniques, and some prefer to watch the proceedings from a safe distance, without actually participating themselves (smart people).

As the rappel and Prusik loop became more and more accepted by southeastern pit plungers, everyone sought to make the descent more comfortable. Borrowing a mountaineer's refinement, a carabiner was substituted for the painful curve of the rappel rope around the thigh. This modification at first won rapid approval. It permitted the more impatient to edge slowly over the lip of a pit, hang suspended until spinning slowed, then merely relax both hands lightly to descend. With a long zzzzzip, they could descend 100 feet in mere seconds—fairly safely if the rappeller remembered to begin slowing his descent about halfway down. Those who preferred to enjoy the scenery could lower themselves with equal ease and greater safety, "brakes on," taking several minutes for the same trip.

With a Prusik loop self-belay, the carabiner rappel seemed a major advance for skilled cavers. Some of its hazards were slow in appearing. It was obvious, however, that the sharp bend in the rope angling through the carabiner weakened the outer fibers of the rope. For cavers who discard their ropes at frequent intervals, this was no problem. A top-quality 250-foot rappel rope, however, costs upward of $40, and most cavers are young and impoverished.

Mechanical geniuses responded. Again Bill Cuddington was in the forefront. A remarkable assortment of carabiners with brake bars, carefully machined bollardlike spools, sheaves, and similar devices were created to ease the strain on the rope and provide better control.

Only after five years' experience did southeastern pit plungers deem themselves competent to rappel into fearsome Mystery Hole. That huge, echoing chamber was awesome even to experts like Bill Cuddington, Don Black, Roy Davis, and Herb Dodson. As that quartet dared its maw, the great weight of the dangling rope slowed and halted their descent, forcing them to feed it by hand.

By this time, a maturer Roy Davis would admit the sensations which overwhelmed him in the intimacy of the lonely blackness:

Lowering the gasoline lantern to those below did nothing to boost my courage. It spun dizzily and grew smaller and smaller, to become only a flickering speck in space before it was received by the microscopic humans below.

The initial 36-foot ladder climb to the brink of the jumping-off place was shaky enough, in view of the depths below; but snapping into the single strand of ⅝″ manila rope, adjusting my swiss seat and rappeller and mustering last-minute courage were nerve-wracking.

Stepping off into space, the rope began to travel smoothly through the sheve and my descent was in progress. I wished fervently for an instant that I might change my mind and forget the whole thing. The rope twisted and untwisted, turning my light in every direction, and illuminating faintly the vast walls of the shaft.

The symmetry of the well was unbelievable. Not . . . irregular as are most pits, the walls of the Mystery Hole were smooth, unbroken and . . . circular, pushed and polished through the solid rock by the ceaseless abrasive action of the waterfall.

A great drapery hung in majestic folds from the brink of the waterfall and gradually tapered to rock pendants some 40 feet in overall length. Gradually tapering, cone-like, to a maximum width of 150 feet at the bottom, the opening darkness below me became more impressive. The friction of my gloves on the rope generated considerable heat, forcing me to pause and allow them to cool. Solid ground was never more welcome.

This, then, was the pit for which rescue "authorities" thought only a rope would be needed. Just ten weeks later young Jimmy Shadden fell to his death. His death was not wholly useless. In tests, his broken rope was found to snap with a mere 270-pound jerk, one-tenth of its original strength. Widely publicized, the shocking news brought sudden realization of danger to many another untrained would-be caver.

Perhaps equally important, Chattanooga grottoites established a highly successful Cliff, Pit, and Cave Rescue Team with the cooperation of appropriate authorities. All are determined that never again will such a fiasco occur. Next time it may well prove the difference between life and death.

No one should conclude that these new pit-plunging techniques are safe, routine procedures. Disaster always rides with the rappeller, awaiting any miscue. On February 17, 1962, Ed Yarbrough and Nicky Crawford planned a nocturnal carabiner rappel into 110-foot Little Thunderhole, near Cookeville, Tennessee. Seven others stood by on the surface.

With a standard Prusik auto-safety, Nicky took the lead but found his Prusik loop sticking. Immediately below the entrance precipice, he halted to adjust it. Somehow his braking hand lost control of the trailing rope. Nicky plummeted downward, the rope sliding unchecked through his carabiner.

Instantly Nicky locked both hands on the shooting rope in a death grip. The fierce, blazing pain of intense friction shot through his palms, burning them deep into the tendons. But, with the overwhelming terror of the depths below, the pain was hardly a fleabite. With a long whirrrr of the vibrating rope, his frantic clutch perceptibly slowed his fall.

A tremendous blow struck Nicky's back and hips. His helmet flashed into space, and he caromed far across the pit from a slanting ledge. Through pain and shock, a single grim thought rode the brink of unconsciousness: HOLD ON! Long moments later, a lesser blow collapsed him on a talus slope, too shocked to marvel at being alive.

Already alarmed by the flash of Nicky's flying helmet and a weak call, Yarbrough hurried to his side. Nicky was conscious, but in pain from the rope burns and the blow to his back. Fearing a broken back, Ed covered his friend with his extra clothing and called upward for a wire stretcher. Then he sat down to encourage Nicky through the long wait.

"How come your auto-safety didn't work?" he asked at length as color returned to Nicky's face.

"I guess I had one hand clamped on it all the way down!" Nicky admitted with a wan grin. "It couldn't possibly close that way, but all I could think of was holding on!"

Unlike that at Mystery Hole, this operation was controlled by competent cavers. A stretcher arrived within two hours. David Smith rappelled down to help. Within five additional minutes, Nicky was in an ambulance while Ed and David cleaned up the cave. Besides rope burns, his only important injury was a mildly cracked vertebra.

Vertical caving is no procedure for self-taught novices: Bill Cuddingtons are rare indeed. Even the most expert pit plungers must periodically overcome formidable difficulties. Completely exhausted, Mason Sproul once somehow found himself dangling helplessly by his chest loop, feet higher than his body, arms numb from too much strain.

This was more than merely an embarrassing position for the originator of the Virginia Cave Rescue Network. A dangling caver must have some means of shifting his weight from his chest loop. In experiments, strong cavers hanging by a chest loop have blacked out within five minutes. It was suspended thus in an unmaneuverably tight crack beneath an icy New York underground waterfall early in 1965 that James Mitchell chilled and died.

Earl Geil prusiked up Sproul's rope. Burning through a hopelessly jammed knot with his carbide flame, he replaced Sproul's slings. But Mason was too exhausted to advance.

At such a time, no one jokes. Only by extraordinary effort could Geil slide his friend down to a point just eight feet from the floor. There the knots again jammed, this time hopelessly.

Stuart Sprague stood beneath to break Sproul's fall as Earl regretfully cut his helpless friend free. With cavers' usual luck, no one was injured. Had it been eighty feet from the cavern floor rather than eight . . .

I personally had more than my own share of trouble with Prusiks in Neff Canyon Cave, still deepest in the United States despite half a generation's search for a deeper.

The strange setting of this solitary cave seems wholly unlikely for any record. Its well-concealed mouth practically in the back yard of Salt Lake City, Neff Canyon Cave is only two thousand feet up the slopes of the westernmost range of the Rocky Mountains. Parts of the gully in which it lies can be seen from the homes of Salt Lake grottoites. Yet not until 1950 did the world learn of its existence.

In 1949 high school hikers stumbled across this obscure cavern entrance. Able youths on the threshold of young manhood, they had been raised in the Mormon tradition that God will provide for those whose faith is strong. They squeezed into the tiny, jagged orifice. By the dim entrance light they saw that the cave continued downward at a steep angle.

Soon the youths returned with friends their own age. Outside the cave entrance they tied a rope and began to climb down its length. Reaching its end, they climbed back to the surface and returned home to borrow another coil. And another. And another. On March 22, 1950, three of their group tackled a great pit. It led down into the largest chamber they had yet encountered. One by one they slid or were lowered down their rope and continued exploring as far as another pit below.

Then came a problem. They could not slide back up the pit. Slick walls of rotten rock foiled every attempt to climb the rope hand over hand. The audacious trio found themselves trapped.

Fortunately, two others had been sitting patiently between the entrance and this Great Pit. On their first visit to the cave, D. B. (Pete) McDonald and Jerry Hansen had realized the team's unpreparedness for such a pit. Unable to dissuade their companions, they agreed to wait four hours, then summon help if their three friends had not returned.

The time limit passed, with nothing but faraway crashing that suggested a catastrophic landslide. Pete and Jerry rushed to the nearest telephone, an hour's hike from the entrance. The 16-hour rescue operation was a considerable fiasco. So were wild subsequent stories about depths of 2,000 feet, of "a giant room" deep in the earth, and the like.

Not until we formed a Salt Lake Grotto of the National Speleological Society in 1952 did organized speleology come to Neff Canyon Cave. Less than a month later, four of us obtained permission from the U.S. Forest Service—the bureau in charge of the land which contains the cave—to determine what might be behind the stories.

Our scouting party included Dick Woodford and Marvin Melville, later to become an Olympic skier and coach. Dick and Marv were good rockclimbers and had proven their worth a month earlier in a deep new

Nevada cave we had found by following the wrong directions. Equally important was Bob Kennedy, Salt Lake City accountant with experience in Pennsylvania's toughest caves. Others of the Melville party were to act as our support party, waiting at the entrance. Their wait was long.

The actual entrance of Neff Canyon Cave is a vertical slit along the side of a fissure at the bottom of a shallow, sloping sink. Twelve inches

The Great Pit of Neff Canyon Cave. *Photo by A. Y. Owen, copyright 1953 by Time, Inc., courtesy* Life *Magazine*

wide, it is perhaps twice as long. Except for jagged projections of impurities in the limestone, it would have been quite adequate. As we slid into the hole, each of us heard a loud r-r-rip that told us our coveralls would not survive the trip.

We dropped eight feet into a small chamber and lit our carbide lamps. Alton Melville lowered our ropes and other gear through the tight orifice, and we were off, twisting and turning a few yards to the main passage. Looking back, we could see bluish daylight through two small openings. It was the last we saw that day.

Sloping downward more than 50 degrees, the cave passage had the general shape of an inverted V. At floor level, it was 3 to 6 feet wide. At shoulder height, it often was little more than passable.

On the floor were large and small boulders. Gravel and silt also bore witness to the stream which cascades through the cave each winter and spring. The boulders forced us to clamber where we otherwise could have walked. Overhead other boulders were wedged into the narrow crack, but none was loose.

After 200 or 250 feet, we came to a chamber 40 feet high. Its floor was a mass of huge blocks of fallen limestone, securely interlocked. In opposite corners, narrow pits dropped 15 to 20 feet to the original level of the cave. Both were tight, jagged and undercut. Neither looked easy.

Two of us tried each pit. More sounds of ripping clothes were heard, but we passed through and compared notes. Returning, each of us preferred to try the other's route. I still haven't decided which is worse.

Ahead the cave narrowed to a width of about 12 inches. Aside from a few more tatters, this was no special problem until we came to the end of the 25-foot narrows. There the floor dropped away for 20 feet. As we climbed down, we found wet, rotten-feeling shale instead of limestone. What was this? We had seen no shale at the entrance.

We looked back up the irregular, sloping corridor. At the top of the pitch we had just descended, the limestone slanted downward as it did elsewhere in the cave. Below, a thick layer of shale formed the lower walls of the pit. The answer was clear: the cave had extended downward to the very bottom of the thick limestone bed. There the cave stream had eaten into the weaker shale below.

Then we noticed something even more unusual. On one side of the passage, the contact of the limestone and shale was about 3 feet higher than on the other. We were descending along a fault where one block of bedrock had slipped vertically about a yard.

The same situation was apparent at each long dropoff we encountered and dropoffs were increasingly plentiful. First came ledges 12 and 10 feet high, then a more impressive cliff about 35 feet high. For the first time we fixed a rope.

In less than 50 feet, we encountered two additional sharp descents.

Then our voices echoed hollowly, and a great black vault loomed starkly ahead. At ceiling level a narrow crack rose up out of sight. Downward we gazed into nothingness over the edge of the Great Pit.

This was our primary goal. We looped a 120-foot rope around a large, well-wedged boulder. After anchoring it with a bowline knot, we tossed it over the edge. It hung in the stream-cut angle of the pit and did not reach the bottom of the room.

The drop was not quite vertical. The walls of the stream cleft were fairly close together, and an occasional pressure hold appeared possible. From above, the pit did not seem overly vicious. As a precaution, however, we left Bob and Marv shivering at its lip as belayers while Dick and I explored below.

Within a few feet, I learned that the ascent would present problems. I was back on that miserable, rotten shale. It was only a minor nuisance going down, but I found I could dig precarious handholds with my gloved hands—and even carry them along with me. I have never encountered such rotten rock in any other cave.

With relief I saw that the rope touched a steep limestone slope which could be traversed without artificial assistance. I called back the good news, untied the belay rope and gingerly made my way downward another 60 feet to the floor of the vaulted chamber. The small opening of a continuing passage made a good shelter.

The dim loneliness of the cathedral-like chamber surrounded me. At its slanting apex a single tiny glint of yellow light danced encouragingly. Soon came the soft whirrr of falling rock. Dick was beginning his rappel.

With a series of sharp cracks, shale fragments smashed against a limestone slope. Tiny fragments rattled against the cave walls like machine-gun fire. How slowly Dick seemed to move, almost two hundred feet from my protected alcove. Yet in a few minutes he joined me, as delighted as I over the spaciousness of our suddenly majestic chamber.

We called to Bob and Marv that we would be back within two hours, then ducked into the passage beyond. After the vast height and comfortable width of the Pit Room, the usual pattern of the cave suddenly cramped us.

Within 50 feet, I was intrigued by some irregular limestone ridges which permitted us to straddle the passage instead of descending. Passages opened unexpectedly on each side. With Dick close behind, I climbed to the opening on the right. Nestled in a hollow was a small crystal pool with brilliant white shelfstone covering most of its surface. It was the first beauty we had seen in this dark, gloomy cave, and our spirits rose immeasurably.

Mud made the crossing a trifle ticklish, but we stepped over the main passage into the other opening. Here again was a surprise. A short, low passage led to a small chamber floored with soft, powdery dirt, surprisingly dry for this dank cavern.

Back in the main passage, we continued to the lip of another 30-foot dropoff. Our spotbeam flashlights showed the cave continuing, but it was time to return. In four hours we had learned much about the cave and could plan our return trips accordingly.

Coiling our ropes, we returned to the Pit Room. A faraway pinpoint of light, much dimmer than an evening star, told us that we were not abandoned. It was a welcome sight indeed.

Mine was the ultimate responsibility, so Dick was to ascend first. A husky, teen-age mountaineer and strong as an ox, Dick scorned the use of Prusik knots. As he would be on belay, I raised no objection. At first all went well as he chimneyed in the wide V incised by the stream. Our rope provided handholds and he usually was able to maintain a sort of footing by foot pressure on the walls of the V. A hundred-odd feet is a long way to go under such conditions, however. Soon Dick was calling, "Resting!" to his belayer as he struggled upward, yard by hard-earned yard. It was forty minutes before a very tired young man was hoisted over the lip of the ledge by the scruff of his coveralls.

I had planned to do it the easy way—by prusiking. As I began the ascent, however, I advanced eight inches, and no farther. My Prusiks refused to glide.

A second attempt was no more successful. Apparently mud and grime on the rope created just enough friction so the knots would not slide freely when the weight was released.

The details of that ascent are something I prefer to forget. I was able to use one sling as a handgrip on the rope, but that was all. Handholds on the shale cliff came loose in chunks many inches in diameter. A shower of shale shattered far below each time I moved a foot. Twice I slipped and was jolted to a halt by the belay rope. A dozen times I rested to get a little strength back into my overworked arms. My last rest was only a few feet below the ledge. Now well recuperated, Dick gave me the horse laugh. "So that's the easy way!" he crowed. I was too tired even to throw a handhold at him.

It was all over after an hour. While life ran back into my numb arms, the others coiled the ropes. We renewed our headlamps' supplies for the third or fourth time and headed out. At dusk we rejoined the support party, ten and a half hours after we had last seen them, swearing that next time we'd use ladders instead of Cuddington's fancy new tricks.

The following winter saw many pleasant evenings as we planned our next venture and constructed rope ladders. We bought lengths of high-quality rope and acquired a thousand feet of wire for field telephones. We recalled that four cavers ought to be able to stretch out on the soft earth of the small dry chamber we had found. Lists of minimum needs for underground camping were hotly debated. New recruits were trained in easy nearby caves and in night-climbing practice on a cliff just outside the city. The latter nearly led to disaster. An overzealous deputy sheriff drove past during a practice session and almost arrested the lot of us as suspicious characters.

By spring our plans were complete, and we were awaiting drier weather when a bombshell struck. CLIMBER URGES SEALING OF CANYON CAVE announced the Salt Lake *Tribune*. It told of the daring feat of a local climber, wholly lacking in previous cavern experience but well known in the Rocky Mountain area. An expert climber, his mountain exploits had been reported in hair-raising terms by newspapers of the area. Together with several friends equally inexperienced underground, he had penetrated "2,000 feet" into Neff Canyon Cave. The cave was frightfully dangerous and of no scientific or scenic value, he proclaimed. It should be sealed shut so that no one could ever enter it again.

We called an emergency meeting and obtained heartening assurances from the embarrassed staff of the Forest Service. While we were preparing a formal statement, the editorial columns of the *Desert News* took up the cry "Close It Up!" A fire-spitting letter to the editor by Roy Bailey, chairman of the old Utah's Dixie Grotto of the National Speleological Society, saved the day. Nothing more was heard of the closure proposal.

On October 3, 1953, all seemed ideal. Dick Woodford, Marv Melville, Pete McDonald, and I were to spend thirty to thirty-six hours in the cave. We were to advance as far as possible, then retire to the Bedroom to sleep. Support parties led by Bob Keller and Bob Kennedy were to transport equipment and belay us up and down the Great Pit. We packed our gear in duffel bags since the cave was too tight for pack boards. At the last minute, Pete relinquished his place on the advance team to A. Y. Owen, ace Oklahoma spelunker-photographer flown in by *Life* Magazine, which somehow scented something out of the ordinary.

Once inside the cave, we found the duffel bags larger than the cave passages. They snagged on every jagged projection. Tight squeezes which had hardly slowed us now required laborious, lengthy hauling and shoving. Note taking, attempts at photography, and unrolling and unsnarling the telephone wires added up to interminable delays. It took us almost twelve hours merely to reach the edge of the Great Pit.

The frustrating field telephones had stopped working, so we abandoned them at that point. We lashed together three lengths of ladder and Dick rappelled over the brink to straighten them. For the first time we felt better about our planning.

Our bedding and gear were lowered, and the first support party called a distant goodby. We wrestled our impedimenta into the Bedroom. In our exhausted state it looked exceedingly inviting. As we uncurled the sleeping bags, however, we made a dreadful discovery: there was room for only three sleepers. Dick solved the dilemma. He dragged his bedroll onto a half-inch false floor in a little room just beyond the Bedroom, ignoring a formidable pit just beyond his feet.

We munched a cold supper and crawled into our bags. So tightly packed were we that when one of us rolled over the others had to follow. A damp chill soon crept through the linings. The soft dirt floor developed hard, rocky knobs. At intervals we groggily consulted our watches, hopeful that it might be morning.

At 5:30 we gave up, lit our headlamps and rolled our bags. Breakfast was begun by cutting the top off a can of soup and placing our carbide lamps under it. After soup, we heated water in the same can for what looked like cocoa but still tasted like vegetable soup. Canned fruit salad and Boston brown bread followed. It was an odd combination but easily portable, and it put some life back into us.

We dragged our burdensome gear back to the Pit Room and set about what exploration we still could accomplish. It wasn't much. We rigged two more rope ladders and reached the Devil's Slide—450 feet down, we guessed. In our groggy condition, the sound of rocks seemingly rolling forever along its slick surface was too much. Spending our remaining time merely familiarizing ourselves with the intervening passages, we slogged back to the Pit Room. A far-off low growling rumble soon resolved into distinct voices. Part of our scheme had worked: the second support team was precisely on schedule.

Marv took the lead, climbing the ladder on belay as A. Y.'s flashbulbs flared. Then A. Y. ascended, losing a metal tripod leg, which chimed melodiously from ledge to ledge. Next went Dick, carrying the battered piece of tripod, and I was alone.

I have been alone many times in such a place. Shivering, wet, and nearly exhausted, however, I must have been somewhere near the breaking point. Loneliness overwhelmed me. Dick's light and that of someone far above were the last connections I had with anyone on earth. What was I doing here, anyhow? The despairing thought came to me that man has no place in such secret places of the earth. The vastness of the arched chamber surrounded me hypnotically.

A shout recalled me to rationality. "Rope coming down," someone called.

I dragged the last bundle to the rope and retreated to my bomb shelter. The pack moved only a few feet before snagging. Bowing to the inevitable, I called for the belay rope and began to climb the ladder, signaling when the equipment was free enough to be pulled up to the next snag. Even so, I was up in one-fourth the time and with one-tenth the energy of my previous hour-long struggle. A rope ladder can be the most wonderful thing in the world. What a pleasure it was to see the faces of other cavers! No longer was I alone.

But all faces showed strain and overfatigue. We started several helpers toward the surface with part of the equipment, then sought to retrieve our 105 feet of ladder. It jammed repeatedly. Twice we had to send Marv Melville part way down to retrieve it. Though the strongest of the team, he was completely exhausted by the time it was at our feet, ready for rolling and packing.

Then came the nightmare. It had been hard to get the packs down through the narrow crevices. Pushing and hauling them upward was far worse. Two teen-age members of the support party gave out completely, and Marv was in bad shape. When we finally reached the Double Pit Room, we sent them ahead without a load. The rest of us found it difficult to move aside even when a loud clatter and hasty shout "ROCK ROCK ROCK" warned of a swiftly bouncing boulder a foot in diameter, dislodged by someone's dragging foot.

How far was the entrance? We all claimed to see its landmarks at every step, but it always seemed farther ahead, a dimly recollected mirage. When Bob Keller felt the telltale chill of mountain air, he was too tired to pass the good news back to the rest of us. One by one we silently squeezed out into the dusk, dragging our loads behind us. Before the last pack was out, we had been underground more than thirty-three hours. Wet, muddy, tattered, and drawn, we demonstrated amply that ladders weren't exactly ideal here, either.

Hot drinks revived us enough to stumble back through the brush to the jeeps. Everyone concurred with A. Y. Owen, who telephoned his chief: "It's the most miserable cave I ever saw." We were sick and tired of Neff Canyon Cave, we commented to everyone within hearing. Never would we go back. Bob Keller went to the hospital for ten days, and none of us was much good for some time. It was a week before I could adjust to having lost a day underground.

It was six whole weeks before we returned to install a chain and padlock for the Forest Service, and seized the opportunity to do a little mapping. Before that next trip, a touch of comedy entered the story of Neff Canyon Cave, but it was comedy with ominous overtones. Three weeks after our overplagued trip a local newspaper announced CLIMBERS CONQUER DEPTHS OF HAZARDOUS NEFF'S CAVE. We read it and blinked.

The mountaineer who had previously urged the sealing of the cave apparently had changed his mind. With two friends, he had returned to the cave. Now he invited others to look for a marker showing how far they had gone. The bottom of the "4,000-foot-long" cave was at a depth of 2,000 feet, he indicated, almost doubling the American depth record held by Carlsbad Cavern. Using techniques which we considered unduly hazardous, the climbers had reached the bottom and returned to the surface in a mere 14 hours. Taking calculated risks, they had come within a hairsbreadth of being trapped in the Big Room, near the bottom of the cave. Only a successful third attempt at an unbelayed, hand-over-hand 23-foot rope climb saved them. The three climbers were entitled to their triumph, but few who thus approach such a cave will be as fortunate.

Neither their guesses nor ours, of course, were acceptable. Unenthusiastically we acknowledged that we would have to go back to survey the cave. A year rolled by, and again it was the optimum season for an assault on the depths of Neff Canyon Cave. We talked about an expedition spread over three weekends. On the first we would install ladders and ropes. On the second we would explore and map the cave. A third would be necessary to remove the gear. Yet no one seemed seriously interested. Our 33-hour nightmare had left its mark on all of us.

Another year went by, and another. New blood and advancing techniques entered the grotto, and the five of us who had taken the worst beating were no longer in Utah. It was probably just as well. The new generation was eager and we might have discouraged them.

By this time, Caine Alder—one of the trio who had made the controversial descent—had become a good caver and wanted to photograph the cave. Paul Schettler and Dale Green were eager to map it. Painting a glowingly biased picture of the cave, they persuaded Bob Wright and Yves Eriksson to accompany them "to help with technical observations." Little was said about the amount of gear to be carried. Alexis Kelner, an enthusiastic young spelunker and mountaineer, completed the party.

Warned by our dreadful experience, the group planned a combination of ladders and improved Prusik techniques. Wisely they rejected any attempt to carry sleeping bags into the cave. A support party led by Bill Clark and Jim Edwards entered at 6 A.M. October 20, 1956. They carried 250 feet of rope ladder, 600 feet of manila, and 240 feet of nylon rope. By 9 P.M. they had not reported. At the jeep road below the cave, however, a last-minute telephone call by a hastily assembled rescue party relieved the tension. The cavers had just checked in, exhausted, after reaching a depth we now know to be more than 800 feet.

Dale Green tracked Bill Clark to his lair, deliciously half asleep in a steaming bathtub. They had run out of rope at the top of the Big Room, Bill reported. Dale's party would have to take a length off the bottom of the Great Pit. Then he dozed off in the tub, mumbling something about difficulties in getting the gear back to the surface. Dale says he took that remark far too lightly.

With an additional 240 feet of rope, the assault party reached the cave at 6 o'clock next morning. By a last-minute decision, they began to map the cave as they descended rather than on the return. Though accurate mapping is slow and tedious, they moved much more rapidly than on our 33-hour fiasco. Rappelling down each pit on ropes already rigged they reached the top of the Devil's Slide within a few hours.

Dale made some rough calculations and was surprised to find that they were already 650 feet down. In the entire United States, only Carlsbad Cavern was known to be deeper!

While the cavers were peacefully reloading carbide lamps, an ominous rumble echoed through the cave. As one, the group dived for shelter while some 200 pounds of rubble rattled down among them. A large piece of shale struck Paul on the head, knocking him down. His helmet served its purpose, however, and he was uninjured. Another rock landed on Dale's pack, smashing a flashlight and flattening his dinner supply of cheese. A trifle shaken, the sextet inspected the rigging of the next ladder, then descended to the Devil's Slide. Dale says he is still convinced that the rope anchoring the ladder had three ends.

Along the Devil's Slide, there was no problem of narrow, jagged squeezeways. Here the problem was a smooth, slippery shale floor that angled downward at 50 degrees. The few finger- and toeholds were loose and untrustworthy. Mere location of a spot for the tripod of Paul's Brunton compass required a half hour.

Below the Devil's Slide was a small hole dropping 80 feet through the ceiling of the Big Room, largely free of the walls. Bob and Yves remained atop as belayers while Caine and then Dale rappelled down. Dale had no rappel pads, and found the rope burning him in an inconvenient spot. He was descending well, however, when an urgent cry from Caine froze him in space.

"What's wrong?" Dale inquired with mid-air concern.

"Nothing's wrong. I want a picture of that," came the echoing reply. Dale's response, lent force by his painful position, cannot be printed.

With some regret, the little group passed onward without exploring the margins of the Big Room or a maze of breakdown just beyond. At the trickling sound of the cave-bottom stream, a happy shout echoed from Caine and Alexis. Paul and Dale sighed with relief. In 13½ hours they had made 45 measurements and were dog-tired.

At 8 P.M. the six cavers reassembled above the Big Room. Tired but happy, they were distinctly optimistic. Dale later recorded: "We figured that it took about an hour per thousand feet to climb on the surface, and since the cave was 'just a little' harder, we should be out by 10 P.M. Twelve hours later we staggered from the cave with all our equipment still somewhere inside."

An hour of teeth-chattering inactivity was needed to coil the ladders and ropes. At the Devil's Slide, the ladders fell out of the increasingly ripped sacks. Beyond that point, getting the gear up each drop was a formidable undertaking. "In fact," Dale recalls, "getting *anything* up a drop, including ourselves, became a major undertaking." The problems of our 33-hour fiasco again haunted the struggling sextet.

At 6:30 A.M., twenty-four hours after the party had entered the cave, it was battling to haul equipment up the 30-foot dropoff above the Great Pit. The baggage continuously jammed, and it was necessary to belay one of the group as he leaned far out to free the bags.

In the process, an odd noise echoed through the black depths. One of the group was snoring.

The others looked anxiously at the belayer, supposedly in complete control of the safety rope. He, too, had fallen sound asleep!

That settled it. The cavers recognized that they had to get out before their exhaustion caused a serious accident. Abandoning all but emergency

Harry White prusiking in 192-foot Stamps Pit Cave, Tennessee. *Photo by Ed Yarbrough*

gear, they plodded upward. Each step was a distinct effort, each boulder a mountain. Not until 8 A.M., twenty-six hours after entering the cave, did they glimpse daylight. Even then they could not rest. A rescue party was due if they had not reported by 9. Anyone seeing them trotting downhill with glazed eyes and muddy, ragged clothing would have wondered what could reduce man to such a state.

Not until next day did Dale summon the energy to plot the readings they had made. He found the length of the cave, measured along the slope, to be 1,700 feet. Then he computed the depth. Exultantly he telephoned the others.

"The depth of the cave is 1,186 feet!" he announced. "We've got Carlsbad beat!" And so they had.

What of Neff Canyon Cave today—and tomorrow?

Dale Green soon suspected a minor error in the survey data, and insisted on resurveying the entire cave. Rechecking altered the original figure to a depth of 1,170 feet below the spill-over point of the shallow entrance sink—a figure now "official."

Dramatic improvements in vertical caving techniques and increasing familiarity with Neff Canyon Cave have led skilled cavers to many additional ventures into its depths. To date, the cost has been only a broken leg and one case of severe hypothermia. Today it is only a moderate feat to reach the bottom of this deepest cave in the United States. The challenge of vertical caving in the United States has marched onward, to chill waterfall pits in the Montana wilderness, to ice-choked fissure caves in the Idaho desert, to the highlands of California and Arizona, to the hinterlands of Carlsbad Cavern, to the listing of new additions to the 150 major pit caves of Tennessee. . . .

And especially to two surprising, fantastic, incredible caves in Bill Cuddington's back yard. Read on.

14

Surprise! Fantastic! Incredible!

The Story of the Deepest Pits in the United States

Bill Torode discovered Surprise Pit by accident, but he soon made up for that.

The country people of Alabama's Paint Rock Valley had long known of a large, fern-fringed sinkhole high up on Nat Mountain, but it was just one of hundreds back in the hills. Nobody but cavers—speeelunkers, some called them—cared about such things.

Systematically checking each sink and spring shown on U.S. Geological Survey topographic maps, the Huntsville Grotto of the N.S.S. got around to Fern Sink in June 1961. Vertical caving had come a long way in its first decade, and the Huntsville cavers were specifically prepared for the type of mountaintop cave they had been unearthing by the dozen: pleasantly moist, gently-sloping passageways interrupted by a succession of little drops, delightful even though sometimes inhabited by waterfalls. Up the long slopes they puffed for an hour and more, sweating beneath heavy coils of nylon rope and four cable ladders.

The coolness of the deep-shaded sink was welcome. Torode immediately sprawled full length, seeking inspiration from a peanut-butter sandwich. The others went cave hunting, hopeful that this particular sink might be the site of a cave rumored somewhere hereabouts. Immediately the green hid their very sight and sound.

One peanut-butter sandwich later, nobody had come back into sight or hearing. Bill figured he must be missing something. Charging through the brush, he found himself facing an inviting entrance. Taking only time to light his headlamp, he plunged onward. On and on he scurried, trying to catch up: 100 yards, 200, along a small stream, through a nice chamber, a squeezeway . . .

Then, just as was to be expected in such caves, the stream silently vanished into a lightless pit. An impressive one, too.

Obviously the other Huntsville cavers hadn't come this way. Maybe up the chimney back a few paces . . .

Up the chimney Bill slithered, then out above the pit to a step-down where things got a bit airy. Nobody there either. No response to his call, no light in the darkness below. They must be up a side passage somewhere . . . or something . . . better go find them.

But first Bill curiously tossed a rock into the black abyss, counting. One thousand and one, two thousand and two, three . . . Nothing at all . . . the sound probably lost in the whisper of the waterfall. So a bigger rock.

Again nothing but silence, silence, silence. Bill half turned to grasp a real boulder. Then—surprise! From far away reverberated a muffled BOOM!

Torode was impressed. A four-second pit, he guessed: over 200 feet. One of the really big ones. With their limited gear, none of their party was going to get that deep today. Bill turned to find and report to the others. Little did he dream that the challenge of this Surprise Pit soon would make him a cavers' legend in his own time.

As it chanced, Jim Johnston and the others had found a different entrance to Fern Cave, complete with a sparkling waterfall. Enjoying a leisurely lunch in their private wilderness, they kept expecting Bill to join them. By the time they realized that he must have entered the other entrance, they were a trifle annoyed. Meeting him a few dozen yards inside, they were in no mood to accept his smug insistence that all their heavy gear might as well have been left at the foot of the mountain that particular day. A face-to-face view of Surprise Pit somewhat tempered their skepticism, but nobody was willing to acknowledge that Torode's discovery was all that great. At least not until they had something more definite about its actual depth.

With the hectic annual convention of the National Speleological Society in nearby Chattanooga just ten days later, the first attempts to plumb the pit were more confused than usual. Perhaps most inglorious was Jim Johnston's super-effort with magnesium flares, a plumb bob and a digital read-out to calculate the depth automatically. While two other Huntsville cavers watched entranced, Jim magnificently tied a grenade-like flare to a hand line, pulled the pin, tossed the flare into the darkness and waited for sudden brilliance to reveal the unknown depths. Three feet down, it lit on a ledge, bounced twice, and sat.

In sudden consternation Jim scrambled wildly, struggling with the line. As he pendulumed the flare over the lip of the little ledge, it burst into blinding brilliance, just in time to enter the waterfall. Vast clouds of smoke and fog billowed in an awesome inferno. For a fleeting moment, the trio could see that Torode had exaggerated not a whit. It truly was one of the big ones: a huge fluted pit, beautifully sculptured by nature, belling out magnificently as the flare shot downward in the boiling

Fern Cave: the entrance everyone except Bill Torode used. He took this photo on a later trip.

clouds. Then sudden darkness again. With 180 feet of line out, the full force of the waterfall had overcome the flare.

Jim turned to the plumb bob. At the same 180-foot level, the line began to jerk as if taken by the granddaddy of all cave sharks, but it kept going: 200 feet, 250 feet and more, into truly surprising figures: 300 feet, 316 feet—the depth of Mystery Hole. Suddenly there was a new record for free-fall descents in the United States: 325 feet, 350 . . . the string went slack with the plumb bob 351.7 feet below the ledge.

Then another surprise. Jim began to reel the line. Nothing happened. Then suddenly it snapped.

No great problem, Jim thought. He'd get the bob later, when they bottomed the pit. Thirteen years later, it is still there, under the waterfall, caught in a ledge almost 100 feet above the actual bottom of Surprise Pit.

Bill Cuddington came to Surprise Pit ten days later, with Francis McKinney and a fascinated contingent of the Huntsville Grotto. Not even Bill wanted to tackle the waterfall directly, so he and Francis hunted for a good rigging point, well away from the stream.

Twenty feet along the sheer left wall of the pit, a narrow ledge was interrupted by a hunk of nothingness about three feet wide and more than 40 stories deep. The ledge itself wasn't bad as cave ledges go: muddy, sloping a bit toward the black depths perhaps, and rubbly underfoot, but a comfortable two or three feet wide. Except for one complication. About halfway across, a large boulder forced each explorer to the very lip of the ledge. But the boulder provided good handholds, so that seemed all to the good.

To Bill and Francis, the gap was a welcome rigging point, for the drop here was largely sheltered from the thrust of the waterfall. On belay, Bill banged away with his piton hammer for an hour, creating a rigging station: an expansion bolt and two pitons. They anchored a special 368-foot rappel rope and additional belaying tackle. Bill tied a flashlight to the free end of the standing rope began to lower it into the pit. Surprise! When all the rope was dangling, its pinpoint of light still twisted slowly to and fro.

The gear was re-rigged with the belay rope tied to the standing rope, and the latter eased on down until the light stopped twisting. Bill rigged in for rappel while Francis stood by on the ledge. Only after an interminable 35 minutes came four blinks from Bill's light: on the bottom!

In the waterfall spray 80 feet down, Bill had considerable trouble slipping past the knot which linked the two ropes. Correctly anticipating even more difficulty ascending with Prusik knots, he settled for a quick look around, found no obvious passage anywhere, and started back up. One hour and thirty-one minutes later he gasped out the news to Francis:

"That's really a lot of hole! You're against the wall for the first 90 feet or so and then it bells out. Freefall all the way. Francis, it tops anything I've ever seen!"

Bill was right. When the Huntsville cavers got around to measuring the chamber below, the figures were as impressive as the pit: 100 feet long, more than 200 wide, and as much as 450 feet high. Bill Torode had indeed found "one of the big ones."

It was McKinney's turn to descend. But Francis showed the strain

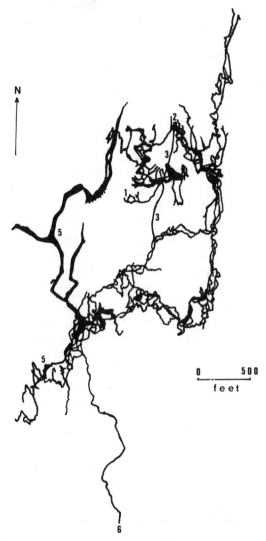

Sketch of Fern Cave System, based on map by Huntsville Grotto of the National Speleological Society, courtesy Bill Varnedoe. (1) Fern Sink entrances leading to Surprise Pit at end of passage curving northeast; (2) Johnston Entrance; (3) West Passage (crosses above eastern part of Surprise Pit complex); (4) Morgue Entrance; (5) Lower Cave; (6) Continues 1,300 feet. Many passages which appear to interconnect on this plan actually cross above or below without connecting.

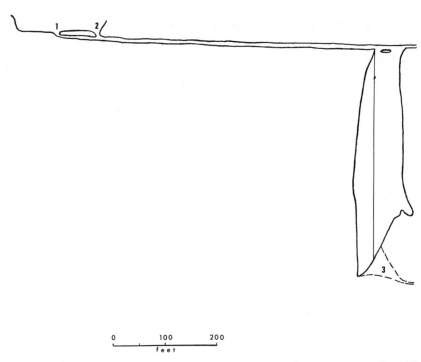

0   100   200

feet

Side view diagram of Fern Cave and Surprise Pit before the connection. (1) Torode's entrance; (2) Waterfall Entrance; (3) Garrison's Grotto and connecting crawlway

of six cramped hours on the narrow ledge. Especially since he would have to be one of two to haul up hundreds of feet of heavy, watersoaked rope, it was not too hard to suspend operations for a week. Then McKinney and Bill Garrison dropped Surprise Pit. Garrison found the nearest thing to a passage at the bottom, but after 50 feet it pinched down into a sand-floored water crawl about 6 inches high and only half filled with air. When the pit was mapped, this was duly entered as Garrison's Grotto, with the formal notation: TOO SMALL.

As months turned to years, vertical caving came of age. American cavers began to measure their totals in millions of man-feet per year. Cuddington's rappel spool and many another hotly-championed descending device gave way to the rappel rack, invented by Huntsville's John Cole. Shaped something like an extra-long, open-ended carabiner with half a dozen metal bars threaded onto its open arm, this new gadget proved the key to near-effortless descents of truly great pits. By changing the number of bars which provide friction against the rope, and adjusting the rope's tension through the bars, virtual fingertip control became routine. Today's latest carefully machined modifications are little if any better.

Ascent techniques changed perhaps even more. While the standard four-coil Prusik knot remains basic, it soon was supplemented by six-coil Prusiks for extra-slick ropes, helical knots that didn't need a rope loop, tandem knots, and semimechanical knots incorporating a carabiner to help loosen the bind. Many began to use mechanical rope-gripping devices instead of knots: Jumar ascenders shaped something like saw handles with precision rope-gripping toothed cams. Or the Gibbs rope walkers, originally designed to be strapped to one foot and the other knee but soon used on a shoulder, or suspended between the shoulder and foot—the "floating Gibbs" technique which is the fastest of all rigs for racing. Today many cavers can ascend 100 feet of standing rope in less than 2 minutes. Some do it in half that time and, in annual championship competition, pursue an elusive 30-second barrier for 100-foot climbs. (Bill Stone was the first to break the barrier, in June 1975.)

Innumerable systems developed as the cavers of half a continent sought ever-improved efficiency and comfort in ascents. High in efficiency is the Inchworm System, born when Charles Townsend fitted a Jumar ascender onto a bar for both feet. When using this device, the caver hangs alternately by a short chest sling while he pulls his feet up, slides the Jumar up the rope, puts his weight on the bar, straightens up full length, slides his chest knot or ascender up the rope, doubles up to raise his feet and their Jumar, and so on. Eventually Bill Cuddington evolved a system with three Jumars and a divided metal frame—Keith Wilson's easy-opening "ascender box"—strapped high against his breastbone, the standing rope running through one side of the box, his long foot sling through the other. Most of the time, his third Jumar (hooked to his seat sling) sits in his pocket, unused but mighty handy for changing ropes in mid-pit or other inconvenient spots.

A new cant developed, too. To see if it "goes," vertical cavers "yo-yo" a pit, especially if they don't do any cave hunting at the bottom. In either case, they "drop" the pit (on a rope, that is), after "rigging" the "drop." Each point of descent is a specific "drop," even if only a few feet distant from another. Usually the wearying part is "derigging" the pit after the fun is over. Nasty pits, ledges, and lips are "hairy" (a lip of more or less suspended breakdown blocks in the Fern Cave system is named the Mustache Pit—extra-hairy). Generally you prusik out whether you use Prusik or other knots, or even ascenders. If the pit isn't quite free-fall, you probably do some wall walking on the way up. To get to openings out of reach across the pit, you pendulum across. Above all, you "pad" your ropes to protect them from friction and sharp edges.

Everyone, it seemed, wanted to drop Surprise Pit: deepest in the United States (according to the Atlanta *Constitution* and several caving publications, Dave Belski found a record-smashing 526-foot pit in the cave-riddled Guadalupe Mountains behind Carlsbad, but if so he's holding out on everyone except close friends). It turned out to be shaped like an upside-down L, with much of its black maw around the corner to the left, out of sight, beyond the three-foot stepover where Cuddington first descended (McKinney was the first to take a deep breath, then the big step, so now it's McKinney's Stepover). Better and better rigging points were developed, some completely out of reach of the bone-chilling waterfall at its awesome worst. Some permitted bottoming the pit with drops of as little at 352 feet.

At first, Bill Torode's enthusiasms lay elsewhere: mapping every cave in Alabama, for example, or at least 1,000 of them and 1,000,000 feet of passage. In 1967 alone he mapped 109, with a combined length of 80,000 feet. His 1970 figure was 121 caves, and as I write he is approaching both goals, which would certainly win him the N.S.S.'s coveted Lew Bicking Award if he hadn't already received it in 1972. For 6½ years he didn't even drop his own Surprise Pit.

But Bill became increasingly concerned about the local traffic problem. Hordes of visiting cavers were yo-yoing the pit. The ledge was a trifle hairy, no doubt about it. And too many visiting brethren were finding themselves stranded at the bottom without light, hungry, shaking with cold, underground much longer than they or their local rescuers wanted. Only one caver was badly hurt here, and she was able to prusik out despite three fractures of the pelvis, three fractured vertebrae, and third-degree hand burns. Yet the stage clearly was set for real tragedy.

Eighteen months after the sobering accident, Bill began what is now fondly termed The Torosion of Fern Cave. Digging away at the wall along the famous ledge, he found its lowest layer more mush than rock. So was the surface of the ledge itself. Soon a half-open crawlway led all the way out to McKinney's Stepover. Just to make things civilized in the face of 437 feet of exposure, he installed a steel cable all the way out to the breakdown bridge, which formed the far wall of the outer section of the pit.

Torosion was contrary to the philosophy of certain purists of vertical caving. Other rumors, too, had somewhat consternated the world of Fern cavers. One group returned especially shaken. When they'd stomped on the chockstone bridge to see how secure it was, part of it had fallen apart with impressive reverberations five seconds later. Moreover, some insisted that Torode's steel cable was rusting rapidly, dangerously.

So Bill happily prepared another surprise at Surprise Pit. Without a word to anyone, he rerigged the cable. The next visitors found a miraculous transmutation from steel to aluminum. Years later the old steel cable, retrieved from temporary storage in a hidden ceiling nook, brought $20 at a sentimental auction benefiting the National Speleological Society.

And what had happened to the friendly big boulder everyone had found themselves hugging as they inched along the ledge? The one that had served as everybody's pet handholds, right on the very edge?

According to legend, Bill snickered when they asked him. While installing the new cable, he accidentally nudged it. Even Bill was surprised that time. Five seconds later, there was one more boulder at the bottom of Surprise Pit. "If everybody who'd used that boulder knew how easy it was to push it into the pit, their hair would turn gray," he quipped. Although he actually had to use a jack to move the huge rock, the heckling subsided.

Bill is still Toroding Fern Cave. Today the top of Surprise Pit and several other touchy ledges are scarcely recognizable by those who knew them only in their original state. Eventually he and Dick Graham dug ledges all the way to the far side. They led to about 900 feet of passage at the far end of the dogleg around the corner, but it didn't amount to much.

Pointing to the location of Surprise Pit, Bill Torode (right) discusses the map of Fern Cave with Don Myrick. *Photo by Jack Ray, courtesy Bill Torode and Don Myrick*

Nobody seemed overly disappointed that they found no passages at the bottom of Surprise Pit. It seemed miracle enough that ordinary little Fern Cave somehow could have produced this freak, this one-of-a-kind, maybe-deepest pit in the United States.

So trim, petite Della McGuffin discovered Fantastic Pit, with a 510-foot free-fall drop, inside Richard Schreiber's new rediscovery: Ellison's Cave. She did it, moreover, in the course of just plain ordinary speleological routine, much as Torode discovered Surprise Pit across the state line.

As every caver would surmise, the stories of Fern's and Ellison's cavers aren't quite that simple. The sequence of discoveries in the spring of 1968 was especially complicated. On March 16, Richard Schreiber and Rick Foote went ridge walking and turned up more than a thousand feet of cave high on the east side of Georgia's Pigeon Mountain. This Ellison's Cave wasn't a totally new discovery but local cavers had largely forgotten about it: "an easy walking-type cave located too high up on the mountain to bother with." Like a good southeastern mountain-top cave should, it dropped about 200 feet in the 1,200 feet they checked. Even though a strong air current called alluringly, that was enough for a start, especially as the next 400 feet was a wet crawlway that soon became infamous as The Agony. A few months later, Schreiber returned with a small mapping crew, entirely unsuspecting the fame the effort soon would bring him.

Meanwhile, in May 1968 two unaffiliated cavers from Huntsville's Redstone Arsenal went hunting for a well-known Alabama cave located on another mountain near Fern Cave. With cavers' usual luck, they had the wrong directions, and ended about a half mile south of Fern Cave. Here they stumbled onto a spectacular sink containing two pits that were beyond their capabilities.

The Huntsville Grotto of the National Speleological Society responded on the run. John Cole dropped the deeper—105 feet—and hollered back: "It goes!"

Indeed it did. In three trips in June 1968 the Huntsville grottoites traced many thousand feet of cave, intertwined on several levels. But it was nobody's favorite cave. A distinct excess of guano—ten feet of it here and there—and overripe skunks and other dead animals led explorers to name it The Morgue. Some of its explorers, it is said, resolved never again to drink cave water. Once the guano was useful: Jay Bradbury slipped, bounced from wall to wall in a chimney and lit flat on his back, speechless but uninjured. Enthusiasm waned. No one returned to The Morgue for seven months.

In October, 1968, Richard Schreiber, Della McGuffin, and Sue Cross—an absolute beginner—got around to mapping Ellison's Cave. The first day was easy: 3,438 feet, through The Agony, skirting a 125-foot pit and onward through the entrance level to the farthest point practical. Next day saw Richard and Della mapping 2,193 feet in another upper-level corridor, which left plenty of time for the 125-foot drop today known as the Warmup Pit. At the bottom was a 100-foot room, then a rather nasty complex of domes more or less fused into something like a stream canyon.

Della led onward. Hunting an easy route to follow a stream that kept disappearing under low ledges, she peeped under just one more ordinary ledge and found another pit. Sixty feet wide, this one was fantastic. Belling

Trim Della McGuffin pauses on its lip during the first descents of Fantastic Pit. *Photo courtesy Marion O. Smith*

out in grooved magnificence, it dropped 6½ seconds entirely free of the wall.

With four vertical-type friends and an 800-foot rope, the elated pair returned to explore Fantastic Pit a month later. Its bottom turned out to be 510 feet down: perhaps a new U.S. record, depending on the mysterious situation in New Mexico. Unlike Surprise Pit, there was cave at the bottom: lots of cave. Downstream, hundreds of feet of waterway extended eastward to a point surprisingly close to the valley floor outside in the warm Georgia sun. Westward toward the heart of Pigeon Mountain

With 510 feet of Fantastic Pit below him, Richard Schreiber settles into his rappel.
Photo by Marion O. Smith

lay thousands of feet of corridors and chambers, intricately superimposed like much of The Morgue.

Explorations of these lower levels required enormous gobs of energy. Until Torosion came to Ellison's Cave, too, perhaps half of it was absorbed by The Agony. In November 1968 Schreiber located a likely spot and began excavating a new entrance down-slope from The Agony. Bill Torode and Jim Turrentine finished the job a week later. Further and further the Georgia cavers pushed in the basement of the mountain. By year's end Fantastic Pit had been dropped 38 times and survey teams had mapped more than two miles of cave.

Early in 1969 five more survey trips added several thousand feet to the Ellison's Cave map, but again returns were diminishing. To push back the new endurance barrier Richard and Della planned an invitational expedition which would camp for a week in the Gypsum Room, a pleasantly dry spot about a thousand feet inward from the bottom of Fantastic Pit. Many vertical cavers sounded interested, but although David Stidham and Charlie Warren helped mightily with the gear, only two underground campers showed up: Richard and Della. Undismayed, they set about their plan.

Each sunless day brought diminishing enthusiasm for cold, clammy cave clothes. So did forcing down "hot meals" eaten cold after their Primus stove incinerated itself the first morning. Bizarre dreams disturbed fitful sleep. Della's included riding camels and surfboards, neither of which previously had seemed of much interest to her.

Day and night, spelean beauty turned to unmitigated nuisance. Once Della woke Richard, shouting in her sleep that "damn old formations keep getting in the way!" Muscles increasingly cramped, limbs bruised, hands grew raw. So did body odors. Yet, in six days, the pair achieved one of the notable triumphs of all American speleology. Passing more than a thousand feet below the cap of Pigeon Mountain, they extended the known limits of the cave by more than two miles. From a point not far from the valley springs on the east side of the mountain, they explored far beneath the western slopes. Along stream corridors they trudged, up into dry middle levels, past waterfall pits, into jewel-like chambers magnificently adorned with the myriad speleothems Della confronted in her nightmares. Here and there they delighted in splendid curlicues of hairlike gypsum and foot-long epsomite needles. Moonmilk, flowstone, and magnificent dripstone, four-inch dogtooth spar, a Hall for the Giants 300 feet long, 80 feet wide, and 50 feet high: Ellison's Cave seemed to have everything.

Including a very special dome-pit Richard met somewhere around midshaft: 150 feet up. Incredible Dome-pit, they called it. The top was beyond their strongest lights, not to mention whatever else might be up there somewhere. Such as another entrance, which was needed very badly at this point.

One reason Incredible Richard and Fantastic Della—to use the overblown names the *Journal and Constitution* pinned to them—had recruiting difficulty lay back in Alabama. January 11, 1969, was the day that Ellison's Cave's sixth survey team reached what they believed was a depth of 981 feet below the highest point in the cave, making it the deepest in the eastern United States (later exploration added 82 feet to this depth). This was also the day that Jim Johnston and Bill Torode went ridge walking about a thousand feet north of Fern Cave and Jim found a crack in a rocky outcrop. This Johnston Entrance was an uninviting little slot but it was blowing strongly. Jim squeezed through and shouted back: "It goes, and it looks like a big cave!"

It was an understatement. Inside were tangled thousands of feet of

Coming over the lip of Fantastic Pit, Richard Schreiber shows the effects of six days underground. *Photo by Allen Padgett, courtesy Marion O. Smith*

corridors, pits, and glorious chambers like The Morgue should have been. New Fern Cave, they called it.

Torode popped through to join Johnston. With traditional southern courtesy—and perhaps also because he had left his headlight outside—he remarked that turnabout was fair play. Since he had discovered Surprise Pit while Jim lolled outside, it now was his turn to loll while Jim ran down this new one, to Surprise Pit. Or maybe another, still deeper.

Jim tried hard, scrambling through a thousand feet of virgin passage, indeed heading toward Surprise Pit. Beautiful walking passage and large rooms contained "unbelievable" helictites. But the only pits were little ones.

Next day with Dick Graham, Torode began mapping this New Fern Cave while Jim arranged a leisurely sightseeing and exploration party. Furiously the duo mapped, charging along hundreds of feet of splendid cave. But something was wrong. "All the time that they were mapping," Don Myrick later recorded, "they were wondering about the whereabouts of all the helictites and rooms Jim had described. They felt that just any old time they would be coming to the far side of Surprise Pit."

By the time they were halted by a 200-foot pit 3,500 feet south of Johnston's Entrance, they knew something was amiss. Without realizing it they had been in a different passage, parallel to Jim's and even closer to Surprise Pit. Much later, when all the surveys were tied together on a master map, it turned out that they had zoomed far past that particular target, just 80 feet away at one point.

But it didn't really matter much. Everywhere the Huntsville cavers turned, New Fern Cave lured them to discoveries seemingly even more unending than those of Ellison's Cave.

April 4 saw New Fern first connected to The Morgue (eventually it happened more times than anyone can remember). It also saw Richard Schreiber ridge walking in his turn. On the far side of Pigeon Mountain, he sought a cave or pit that might lead to the inaccessible summit of Incredible. Incredibly, he succeeded against all odds, although two strenuous weekends were needed to reach his subterranean goal after he had discovered the unlikely entrance. The new route actually presented no great problems. Other than drops of 13, 65, and 83 feet, that is, plus a 35-foot watercrawl aptly named The Misery. And an occasional sink-dwelling copperhead snake.

By Richard's measuring wire, Incredible Dome-pit was 440 feet deep. Free-fall all the way, it was inhabited by a splendid little waterfall. With Marion Smith they dropped it May 3, on their 20th survey trip. Two months later, Della led the first crossover, a 13-hour trip through the mountain. Bernie Jackson added a touch of excitement near the end: his Prusik knots slipped near the top of the muddy rope. Down he

Dwarfed by the immensity of Fantastic Pit, Jane Lampkin and Bonnie McKay tandem prusik near the bottom. This is an especially dangerous technique. Tandeming cavers should not separate more than the distance shown here. *Photo by Richard Schreiber, courtesy Marion O. Smith*

plummeted, squarely onto the head of Ian Drummond, tandem prusiking a few feet below. Ian's knots held while the magnificent midair tangle was being unsnarled, but Bernie suffered a well-scorched seat before he was free.

With Fantastic Pit dropped more than 100 times, Ellison's Cave had grown to 8.08 miles by late 1969. Schreiber wondered if Incredible really was deeper than Surprise, so he took his measuring wire to Alabama. It proved that the Huntsville cavers had underestimated Torode's Surprise, but it was still 3 feet shorter than Incredible's 440-foot depth. Richard and Della were two up on Torode.

To the Huntsville-based crews, this wasn't competition at all. Bill Torode especially was in high gear, turning Nat Mountain inside out. Although bottoming everywhere at a depth of about 450 feet, 15 miles of tangled cave in a dozen different levels was fast becoming one of the world's most important. Fantastic mounds of helictites three feet high climaxed magnificent chambers. Huge natural halls, classically sculptured dome-pits, even short-lived cascades of ice turned up hundreds of feet inside—the latter due to tremendous inward air currents that in winter temporarily vanquish the 60-degree bedrock temperature.

Despite problems with the endurance barrier equal to those of Ellison's Cave (here reduced by building a mile of road), The Morgue and New Fern Cave connected and interconnected over and over again. As in many another cave, mapping proved the key to success after success, providing bypasses, shortcuts, interconnections that smashed the endurance barrier over and over again.

But Fern Cave and its Surprise Pit remained separate, as if this once-deepest pit was its own monument. Even when New Fern virtually surrounded it, on three sides and partially above it, too, solid rock blocked every effort.

As the vastness of New Fern became evident and hopes waned of any connection to Surprise Pit, emphasis shifted. Attempts to achieve a connection gave way to systematic exploration and mapping. The discovery that Survey Station 57 in the West Passage was within 80 feet of Surprise Pit briefly channeled vast quantities of energy to its lowest level. In due time, however, everyone reluctantly agreed that this particular 80 feet of rock was extremely solid.

After shortcuts and bypasses led to a quick route, 3 miles was finally mapped in the "Bottom Cave" section. Here, one large sand-floored passage was found to lead straight toward Surprise Pit. Two trips concentrated on this area, but everything appeared to end heartbreakingly in a large chamber 300 feet short of the base of the pit: Disappointment Room.

Elsewhere in the twisting system, too, even the most determined ventures began to yield diminishing returns. Many other local spelean challenges lured the Huntsville cavers. Interest lagged for many months.

Bill Torode finished the job by leaving Huntsville for a time. At his going-away party, Dick Graham organized a team that proved a recent Torode-Graham discovery an even quicker way to the "Bottom Cave," saving four invaluable hours and 250 feet of rope work. A lighthearted sightseeing tour ensued. Still fresh and eager, Don Myrick rechecked leads in the Disappointment Room area, finally bellying down in what he calls a "gruesome little crawl," following someone else's body tracks. Where the tracks of the previous exhausted slithering stopped, Don didn't. A hundred feet later, he broke out into a large, sand-floored corridor some 1,500 feet long, with a stream passage headed toward the bottom of Surprise Pit.

One week later, a tremendously optimistic group of enthusiasts pushed the stream passage another thousand feet, but to no avail. It seemed as if they had wormed right past Surprise Pit in a miserably mixed-up mess of impossibly low watercrawls.

The stream, however, almost certainly was coming from the general direction of Fern Cave. Dick Graham rappelled Surprise Pit and for the first time attacked Garrison's Grotto, discovered more than 10 years earlier. Belly crawling in 50-degree water, hand-digging his way through stream-packed gravel, Dick gave out after an hour and perhaps 100 feet. Yet it was a real breakthrough. The area was complex and confusing, so, before turning back, Dick dumped a batch of computer chips in the main streamway as a guide. When Jim Johnston and Don Myrick rechecked Myrick's new stream passage from the New Fern side a week later, computer chips were everywhere. After a couple of false leads that consumed time and energy, the two struck a hot lead. Don later described it admirably:

No one else was interested in leading at this point because it entailed belly crawling and digging in about 4–6 inches of water. I wiggled my way through the first puddle which was about 20 feet wide with my hard hat off and my chin in the water. When I got to the gravel bank on the far side, I commenced digging again. I had to dig a trench about four inches deep. I dug this trench for about 10 feet at which time I came to another puddle of water. I could see that the passage continued on and on, but my light was low and I was shaking uncontrollably, so I decided to retreat.

Two weeks later, they were back. This time another team was working from the Fern Cave side. Dick Graham again attacked what is now known as Graham's Grinder. After an hour in its chill, gravelly stream wallow, he again had to retreat, dumping in still more computer chips as a further aid to the New Fern team.

On the New Fern side, Jim Johnston, Don Myrick, and Co. had now arrived. Jim began digging trenches to lower the level of the worst pools. As he worked, more and more computer chips kept piling up: Graham must be somewhere quite nearby.

The team began to shout. A faroff response echoed faintly through the low waterways. Following the voice of Surprise Pit and the computer chips, Don and Sherry Myrick broke through about an hour later. Fern Cave finally had been brought into the Fern Cave system!

That really should end the intermingled stories of Fern and Ellison's caves, but as every caver would predict, it didn't.

Across the state line in Ellison's Cave, recent reports have been especially fantastic. A Virginia Polytechnic Institute caver moved south, Don Davison, Jr., sat one evening with Torode and Reynolds Duncan, discussing really tough caving projects. Bill mused aloud about the smooth, flowstone-coated sheer wall that led upward as well as 510 feet downward in Fantastic Pit. Upward was a lot of blackness that might conceal something important. Shadowed alcoves concealed and distorted, despite the most powerful lights. Quite a few cavers figured that, geologically, a major upper-level passage ought to exist up there somewhere. But for dozens of feet no ledges or other significant projections, no cracks for jam nuts or chocks, no joints for pitons would ease the way. There was only one way to find out: laboriously drilling a chain of expansion bolts into the wall and hanging from each while drilling the next, up and up and up the sheer face.

Southeastern cavers were sufficiently familiar with the technique to be less than eager. As a caver hangs in the climbing slings, each blow with the 18-ounce hammer requires tensing and twisting the body to gain power and control, over and over, endless dozens of times for each bolt. Dust and thirst, sweat and pain build with each stroke. Each reach for a tool or hanger part, each awkward clearing of rock dust from each lengthening hole, each tap testing the rock's soundness brings mounting strain on the cramped body. Rockfall loosened by clumsy reaching for handholds or vibrations of the hammer strokes cannot be avoided, and belaying from below is always a ticklish matter even in the hands of the most skilled. And, if the rock turned bad, as often happened, or if unforeseen obstacles appeared in midclimb, it would all be in vain.

Accepting the challenge, Don prepared a team of fellow V.P.I. grottoites late in 1973. On Thanksgiving morning he trudged up the 2-mile trail with Cheryl Jones, Bob Bartlow, Tom Calhoun, Doug Perkins, and Bob Alderson. Hardly more than unrecognizable lumps, they bore "ropes and personal gear . . . an ammo box, an inner tube, and seven surplus rocket tubes containing over 75 pounds of climbing equipment."

The cave proved to be delightfully dry. Four of the group began to yo-yo Fantastic while Don organized the gear. Bob Bartlow began the laborious hammering, pounding the first two Phillips flush-head self-drilling anchors into the wall. Then Don took over, aiming his line of expansion bolts toward a hopeful-looking ledge 35 feet off the floor on the western edge of what was to become known as Siege Wall. Hanging free, he bang-bang-banged away, inserting the next anchor, then the next. And another and another and another: a good start. Happily the group surfaced after the first eleven hours underground. Its support members went their planned ways. Cheryl remained as Don's belayer.

The Siege of Siege Wall soon turned less than routine, however. Next day saw a rare and odd mischance. With Don angling laterally, high above the floor of the Balcony Room, the stud of his bolt driver broke after just 90 minutes' exercise. Spare studs were about the only climbing hardware Don hadn't brought. One day, one stud, and just one bolt

Don Davison, Jr., nearing the top of Fantastic Pit on Christmas Eve, 1973. The upper part of his route up Siege Wall (at left) is still rigged. Slings ring Destination Rock, at the top of the bolt route. *Photo by Rolf McQueary, courtesy Charlie Larson*

onward, the hammer glanced off its target and found Don's left hand, outstretched against the wall for balance and torque. Deep it bit: an agonizing square dent with a fierce wave of pain that hinted at shattered bones beneath. Numbed with cold water, however, it proved stiffly usable, and the bones unbroken. Onward Don continued, yard by yard, four feet per hour. Twelve bolts, thirteen. Fourteen bolts, fifteen. Almost to the ledge. He could see a slope above it. Would it go? Would it serve as a much-needed belay point?

A crack of sorts permitted Don to place a chock instead of drilling for still another bolt. One more step for a better view . . . then bitter disappointment. The ledge was nothing but the edge of a rubble slope, just waiting to become a landslide. The only way to the top of Fantastic Pit was straight up, out over incomprehensibly vast blackness. More than half the bolts he had inserted so laboriously, angling over toward the ledge, were useless.

One more step onto a second chock for a better look at the route and—WHUMPPP! Jarred breathless, Don found himself revolving quarter turns in midair, shaken, knees skinned, the chock crashing somewhere far below. His anchor techniques and his belay had been tested the hard way.

That was enough. Wearily Don and Cheryl headed out. In these first 10 hours on the wall, the wall seemed to be doing better than Don was. But Christmas was coming, which to Cheryl and Don meant an opportunity to spend 48 or more hours underground, perhaps a fourth of it on the maddening wall. Better organized and prepared, with Don's smashed hand almost healed, and with Tom Calhoun and Rolf Mc-Queary camped in the entrance in support, the Yuletide venture proceeded apace.

At least as smoothly as could be expected, that is, in the face of now-roaring seasonal waterfalls, through which Don, Cheryl, and their support team had to duck, wriggle, and squirm. Dry clothes were part of the plan, but twice water began to rise alarmingly around human plugs in tight slits. Yet no disasters befell, then or in true flood conditions during later derigging.

Cheryl fired up a magnificent underground chow mein dinner. Don scurried back up to anchor number seven on his old route and installed a chain of four more bolts leading straight up, his rhythmic din echoing and re-echoing above the waterfall roar. After supper he set to work in earnest. Only with the 16th bolt firm in the new chain, 551 feet off the floor of Fantastic Pit, did he quit. Perhaps it was too soon. In the deep night of the cave he found himself far too keyed up to sleep soundly.

Eventually Don's watch certified the time as 7:30 A.M. Within half an hour he was back on the wall, only pausing three hours later for what he recalls as "chow mein, gorp, and peas and carrots from the kitchen of Cheryl Jones, renowned cave gourmet." At 12:32 P.M. he was back at the 21st bolt, an almost motionless dot of noise and light suspended in an endless black void.

From below Don's prospects appeared increasingly grim. Human endurance reaches its limit quickly when bolting. From the base of unique Siege Wall he appeared to be merely somewhere around mid-pitch. From Don's level, the prospect was little happier. Overhangs blocked his view, threatened his progress. Each anchor became its own compulsion, the challenging summit an endlessly retreating dream. Possible alternate routes constantly lured him to what he could still recognize as dead ends, but he could feel pressure mounting. No word of despair passed between climber and belayer, but only the most dogged determination now drove onward the Siege of Siege Wall.

At about 1:30 P.M., however, Don's morale leaped several notches. For the first time he was able to spot a promising ledge, not visible from anywhere below. Some 90 minutes later he passed the highest point that could be seen from the balcony room. A flat ceiling came into view only a few yards overhead.

But Don's spirits slumped almost as swiftly as they had risen. Siege Wall ended at the bottom of another rubble-strewn slope just waiting to roar to the lightless bottom 576 feet below. A passage seemed to exist about 15 feet above him, but it seemed suicide to continue.

Off to the east, however, the rubble seemed to thin. Now trembling with tension and fatigue Don tautly traversed its lower rim, at the very top of the solid rock. Forty minutes passed while he cleared a tiny level spot, rock by rock, testing the pressure of each on its neighbors, desperately attaining a safe triangle of bedrock almost a foot wide. A 27th and 28th anchor—the latter on breakdown instead of solid limestone—allowed him to boost himself off the wall, out of Cheryl's sight. Emotionally and physically drained, he nevertheless exulted. Above was the mouth of Assault Passage. Solidly wedged breakdown offered an easy route of stepping stones.

"SUMMIT!" Don bellowed above the waterfall's roar. With only three usable natural projections on its entire face, 28 bolt anchors and 22½ hours against the rock had conquered Siege Wall. Wearily he rigged the full drop of Fantastic Pit from the new summit and prepared a register. Cheryl ascended and joined him in exploring Assault Passage— half-heartedly. "I was just too relieved to be above those problems," Don acknowledges. It was Cheryl who rigged in and—over a knot in the rope —broke the United States free-fall cave record with a 586-foot rappel of Fantastic Pit.

That night in the humid black chamber where he had tossed restlessly 24 hours earlier, Don slept like a corpse. But before leaving the cave he reascended Siege Wall and hauled up some 100 pounds of gear for explorations to come. A few weeks later he dropped a hole Cheryl had discovered little more than 100 feet south of Siege Wall. When measured its depth was found to be 596 feet, almost entirely free-fall, and the greatest of any known cave pitch in the United States. Davison's Drop, Cheryl named it, and properly.

Don's achievement on Siege Wall will long remain a landmark in American caving. Unfortunately it also soon proved to have been a major exercise in futility. Buddy Lane and a team of Chattanooga cavers followed Don's route up the wall and turned to a touchy ledge he and Cheryl had not yet crossed. As they hoped, it opened along the ceiling of a canyon connecting the Warmup Pit and Fantastic, but on this first ticklish venture they were unable to complete the breakthrough.

Don, Cheryl, and Larry Caldwell found a barely workable route two days later. Siege Wall was abandoned, but Ed Yarbrough sums up this new Bypass Passage as a bit hairy: "Several exposed ledges must be traversed and at one point a narrow walkway (clawway?) is negotiated some 580 feet above the floor of Fantastic." At last account, explorations are continuing here, with more than a mile of passages known at the top of Fantastic. For the cave as a whole 10.5 miles are now on the map and Richard's surveying crews are still hard at work.

Yet it was Don's successful Siege of Siege Wall which settled matters, demonstrating that Ellison's Cave contains the deepest pit in the United States. Even if New Mexico cavers really are hoarding that 526-foot hole in the Guadalupe Mountains.

Even less will the world of American cavers soon forget the names of Richard Schreiber and Della McGuffin, who proved Ellison's Cave our second or third deepest cave, barely behind Neff Canyon Cave and at the moment just three feet behind a long-secret California cave: pretty good company for a little Georgia mountain cave nobody had heard of until 1968. Nor will we forget the names of Bill Torode and Jim Johnston of Fern Cave.

Moreover, Pigeon Mountain and Nat Mountain—the habitats of these two surprising, fantastic, incredible caves—are but two of uncounted thousands in the limestone country of the southeastern states. Had no entrances to The Morgue and New Fern been discovered, Garrison's Grotto probably never would have been pushed. The Fern Cave system would probably have remained just Fern Cave through all the foreseeable years. How many lesser pits are overlooked entranceways to equally remarkable caverns?

In the years to come, Fern Cave and Ellison's Cave may well be the guideposts of an entire new age of caving.

15

Aqualungs in the Dark

The Story of Cave Diving

Deep in the warm throat of a sunless vortex, Bill Brown's aqualung clanked noisily against the rocky wall. At once the rhythmic hiss-s-blub-b-bl of his heightened breathing was shut off.

Momentary panic flashed through Bill's mind. Flight was impossible. The nearest aid was 105 feet up the long, slanting limestone slot of this Devil's Hole in the Nevada desert. No one in the sunlit blue-green cavern pool so far above could suspect his sudden heart-clenching fear. Before he could see the first glimmer of the tiny deep-blue window to the world, he would be unconscious.

In a few milliseconds all this and more flashed through Bill's mind. Then the discipline of long, intensive training took hold. In the blackness, he groped backward for the air control of his tanks. It seemed merely knocked out of alignment. With relief he turned the projecting handle to its original position—or so he thought.

Nothing happened. Not even a wisp of tank air reached his clamoring lungs. A few precious seconds were gone.

Slithering backward out of the fissure into the black main cavern, Bill shrugged the harness from his shoulders. Twisting it around, he inspected it with his sealed underwater flashlight. It still looked O.K. Taking no chances with his original tank, he switched on the other. At last he could breathe again, with hardly a mouthful of water to be blown free through the exhaust valve. The weeks of patient practice had not been in vain.

The story of Devil's Hole began long before that dramatic weekend in August 1953. Nearly a century ago, this arid region north of Death Valley Junction supported a sparse population of tenacious miners and grub-staked prospectors. This Devil's Hole with its 93-degree water became the miners' equivalent of the New England cracker barrel. Every Saturday night the lone miners ceremoniously gathered at the hole, luxuriously soaking off the week's grime and reveling in the human company they denied themselves at other times.

In this desert region, water-formed caves of any kind today seem

Complete with diving helmet, members of the N.S.S. Southern California Grotto start man's first descent into Devil's Hole. *Photo by Walter S. Chamberlin, copyright 1955 by the Trustees of Rutgers College in New Jersey*

alien indeed. Yet the towering limestone ranges of the Great Basin are full of caves. In greener, moister times huge lakes filled its sun-burned basins. Here and there storm waves pounded littoral caves out of weak zones in rocky cliffs. Other caverns formed as normal limestone caves, dissolved out by the constant flow of formerly copious ground water from nearby hills and mountains. Some of these desert caverns are of exceptional beauty and interest. A few still trap deep pools and dwindling streams, welcome but strange indeed unless the viewer knows their history. Strangest of all is Devil's Hole, the unknowably deep limestone cave almost filled by a delightfully warm spring.

Pete Neely and I first glimpsed Devil's Hole on a bitter January Saturday in 1950. As we drove onto a plain from a low pass in barren hills, a precipitous sinkhole yawned in the narrow flat between the primitive road and a nearby peak. On its seemingly lifeless slope, wild burros raised their shaggy heads to stare at these intruders.

Pete was out of the car before I could wholly brake it. I was not far behind. We were accustomed to caves and their often-remarkable entrances, but this bizarre pit was amazing.

At our feet a deep gash in the earth dropped almost vertically for 50 feet. It was nearly 100 feet long and 30 or 40 feet wide. At our right, it slanted back into an alcove perhaps 30 feet high and half as wide. Twenty feet of solid limestone separated its arch from the desert floor. Within its mouth lay a narrow pool, sparkling in the winter sun. At the shallow end beneath the arch, it shone crystal-clear and aquamarine. Back against the rear wall, many yards deeper into the alcove, it gleamed with the incredible sapphire of cave pools of great depth.

To our left, ages of surface wash had constructed a sort of giants' stairway to the pool. We climbed down to examine the interior of the exciting grotto. It had the appearance of a limestone cavern carved by the slow, insistent action of subterranean water, and now exposed to alien sunlight. As we peered into the pool, we could see that it slanted far down beneath the rear wall of the grotto.

As we sat lazily in the delightful sun, out of reach of the chill desert wind, we noticed a school of tiny, exquisitely formed fish feeding amid the surface in-wash. Blind blue-gray cave fish?

We leaped to our feet. The fish flashed away. Obviously they were not blind. Even so, their presence here was puzzling. As they returned to their interrupted meal, we could see that they were hardly an inch long.

How came these strange little perchlike fish here, far from any stream? At the moment we had no answer. Weeks later, we learned the curious answer from ichthyologist Carl Hubbs of the famed Scripps Institution of Oceanography—himself a good cave man.

These little fish of Devil's Hole, Carl told us, are found nowhere else in the world. Studying the various fish found in some isolated Mojave Desert springs, ichthyologists had preceded us to Devil's Hole by twenty years. They found its fish descended from small ancestors widespread in the glacial-period lakes of the Mojave Desert. As the great inland lakes shrank into desert, the Devil's Hole school was one of the first to be isolated. Through the millennia it thrived, through extensive evolutionary change.

In commemoration of their unique home, the little fish were given the sonorous scientific name *Cyprinodon diabolis*. Seemingly as friendly as puppies, darting around swimmers, they soon became better known as "pupfish." In recent years cavers roared with laughter over sensational writers' accounts of Devil's Hole. Reader interest is heightened, it seems, if the technical name is translated into "devil fish." Some of these accounts would cause piranhas to turn white with jealousy. In bloody

Scuba-equipped divers in Devil's Hole. Note artificial feeding shelf for pupfish seen above swimmers. *Photo by Alan Heller*

scenes of imaginary drama, our divers had to fight off attacks by ferocious schools of these friendly little fish—our pets!

We sketched and photographed the sink, pool, and alcove, then prepared to leave. I was halfway up a natural stairway when Pete's incredulous yelp spun me around. "The water's warm!" he yelled.

I raced back down the rock slope. I had read of hot spring pools in rare natural caves in Europe, but at that time none had been reported in the United States. Even in Florida and Texas, cavern waters are nippy. Elsewhere they are downright icy.

As I knelt beside Pete, however, I could not argue. The water of the jewel-like pool was gloriously warm. No wonder the miners came here every "Sattidy" night! We spent the next two hours floating luxuriously in their bathtub.

As we studied the pattern of the cave, we became convinced that additional passages lay beyond the barrier rear wall. How deep must we go to bypass it? Near the shallow, aquamarine end of the pool, we dived to a ledge seemingly just below our dangling feet but 25 feet down. Back beneath the slanting rear wall, we again dived to the limit of our endurance. The cave continued down. We could not. As we rested, soaking up the tranquillity of that marvelous winter afternoon, neither of us could have guessed the story to come.

Enthusiastic over our glowing report, others from the Southern California Grotto of the National Speleological Society were soon studying and enjoying this marvelous hole. Nearby desert ranchers offered lurid warnings of boiling springs in the depths of the pool. Some informants swore that the entire pool had been a-boil within their memory. Since neither the pupfish nor the miners seemed overboiled, however, cavers were undismayed. In June 1950 Walter S. Chamberlin of Pasadena descended 75 feet into Devil's Hole by means of a diver's helmet. At that depth, only a faint reflected blue glimmer broke the darkness. Using an improvised light, Walt was able to reconnoiter the openings of several waterfilled passages. The clumsiness of his unwieldy helmet, however, precluded any additional exploration. Rocks dislodged by his trailing lines rolled alarmingly with every motion. Walt discontinued the exploration and returned to the surface to consider the next step. At 75 feet, the hole still seemed bottomless.

In this postwar period, the use of aqualungs was in its mushrooming era. As time progressed, several grotto members became adept in their use. Aqualung exploration of Devil's Hole became a frequent topic.

Study of the area around Devil's Hole had yielded a small nearby pitlike cave 130 feet deep. At the bottom, eager cavers found a pool of similarly 93-degree water. Obviously some connection existed. Would it be traversable with aqualungs?

In the spring of 1953, Pasadenans Bill Brown and Ed Simmons began to prepare for the first aqualung assault on Devil's Hole. Lacking precedents, inventor Simmons developed special underwater lights and other gear deemed necessary for the combination of caving and skin diving. Extra care seemed essential. An ominous newspaper report of the death of a diver in a Georgia cave suggested that minimal carelessness could mean disaster.

At this time, no such dive had ever been attempted in a western cave. Just as the plans were almost complete, 20-year-old Jon Lindbergh made a 150-foot swim to an air-filled chamber in the depths of Bower Cave in the Mother Lode country of California. There the water was unheated, and Lindbergh's dive was shallow, reaching a maximum depth of only about 25 feet. Much later, Bower Cave diving reached greater depths. Even so, this is not directly comparable to the history of the strange thermal spring in the Nevada desert.

For several weeks, the cavers practiced with Simmons' equipment in the swimming pools of long-suffering neighbors. Confidence was established by tedious hours' sitting on the bottom, doffing and donning masks, mouthpieces, and air tanks—more of a trick underwater than it might sound. Only then did they proceed to depth in the nearby Pacific. Soon several of the cavers were able to swim freely in pairs, sharing a single aqualung: a prime necessity for underwater rescue.

A large, tense group pitched camp at the edge of the hole on August 1, 1953. Every imaginable precaution had been taken. Nevertheless, much was still to be learned, and everyone knew it. No report of any previous dive into a warm spring had been unearthed. Would the high temperature predispose the returning divers to the dread "bends"? Consultants in physiology had concluded that if their body temperature did not rise, this would be no added problem. However, no one was quite sure of the effect on the body temperature of prolonged exertion in 93-degree water. Consequently, no one was quite sure how rapidly the divers should ascend.

Under what conditions could a safety line be used? Should the divers operate in groups, in pairs, or singly? Was the brand-new gear too bulky to maneuver through tight cavern passages? Only short scouting dives were planned for this first experimental venture into these doubly hazardous depths.

Lines were rigged and hundreds of pounds of gear transported down to the home of the pupfish. Aqualungs were filled from high-pressure tanks. Expedition leader Bill Brown donned his double-tank aqualung, weight belt, face mask, fins, depth gauge, underwater watch, lamps, and battery cases. Heavily encumbered, he staggered in the shallows. Thankfully in deeper water, he tested the equipment, rocking the little pool with massive bubbles. Waving reassuringly to the anxious cavers grouped around the pool, he fluttered leisurely out of sight.

A grotesque sapphire world of distorted shapes opened ahead of Bill. Cautiously scouting, he swam slowly ever downward into increasing darkness. Soon only his clumsy lights' spot beams broke the eternal night of the warm depths.

As Bill descended, openings led off at several levels, just as Walt Chamberlin had reported. Glorying in his freedom from Walt's trailing lines, Bill mentally catalogued each orifice. As he continued down, his lights showed the cave to open more widely at depth. Downward it continued beyond the range of his lights, almost 200 feet in this incredibly clear water.

Perched on a huge rock wedged between the cavern walls 150 feet below the surface, Bill considered his position. His was to be merely a short scouting dive, yet how easily had he set a new American record! It somehow seemed a trifle ridiculous.

But that was all for today. Deeper lay ever-increasing danger. Too, Bill's friends far above probably were growing concerned at his long disappearance.

With a vigorous shove, he kicked upward. Ever alert for the first excruciating pain of the "bends," he rose slowly, leisurely. A tiny deep-blue window pierced the blackness far above, then became aquamarine as he rose.

On the surface, time dragged interminably. Other divers floated expectantly, prepared for instant submergence with a fresh aqualung. If Bill were forced to resubmerge because of "bends," all was in readiness for an air supply.

The minutes dragged on. The continuing stream of bubbles was reassuring, and watches showed that only a fraction of the seeming hour had really elapsed. Then excited shouts echoed in the grotto. The blue glimmer of Bill's body could be seen, ascending slowly with purposeful strokes. Soon he broke the surface and pushed back his mask. With a spectacular grin, he reported breathlessly to the hushed crew. Everything seemed wonderfully easy!

Sleep came hard that excited night. Early next morning, Bill slipped back into the pool to continue his scouting. This time, the side passages were to be the target.

Within a few seconds of his dive, the eager watchers lost sight of Bill's trail of bubbles. As he entered a shallow passage, his bubbles made little puffs against the ceiling. They danced along the irregular wall like huge upside-down droplets of mercury, then coalesced. Larger entrapped pockets of air reflected his light like flashing mirrors.

This little side passage extended beneath the debris at the outer end of the pool. About 30 feet from the main cave, it ended ignominiously. Bill reversed his course. Turning his attention to the ever-tempting rear wall, he slanted downward in the enlarging tube. The first important opening was at the 80-foot level. It proved merely a small grotto, so narrow that Bill had some difficulty turning around.

Undismayed, he followed the slanting wall to its next orifice. Here the depth gauge on his left wrist read 105 feet. Again the opening was

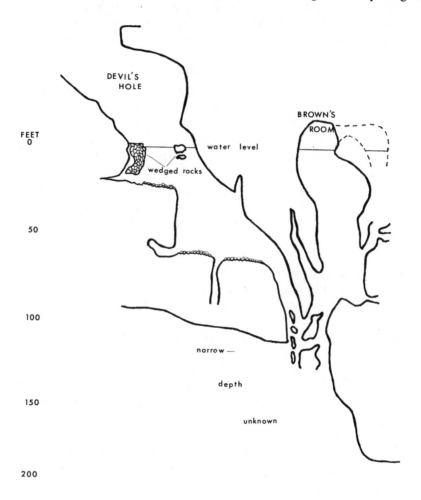

Cross-section of Devil's Hole, Nevada. Data from Southern California Grotto of the National Speleological Society. Courtesy Richard Reardon

narrow. Either his bare abdomen or the tanks scraped the jagged wall as he eased himself slantingly upward. His powerful lights showed a spacious chamber ahead, with a wide tube beyond leading far upward through the crystal water.

Once more, Bill's aqualung clanked against the wall. His air was cut off!

Suddenly Bill understood that, even for scouting, cave diving is not a lone pastime. He reached for the air-control handle, and reaching backward, turned it the wrong way! Only the deep-burned memory of patient practice sessions enabled him to shed his cumbersome gear, inspect it by flashlight, and turn on the life-renewing oxygen of the reserve tank.

Redonning and adjusting his gear, Bill reapproached the 15-foot fissure. Cautiously, speculatively, and with his tanks well clear of the wall, he slithered through without difficulty. In the large channel beyond, he began to slant upward rapidly. The depth registered on his wrist became less and less, but no daylight broke the cavernous night.

The silvery mirror of a free air surface loomed above him, and Bill's head splashed above the surface of a pool. His lights illuminated the walls of a chamber, spacious but stuffy. At its end, the dark opening of at least one corridor led onward.

This was not the nearby pit cave Bill had sought. This room was far too large, and he had traveled only a fraction of the distance. What had he found? Was the air good? It seemed sultry enough to contain noxious gases liberated from the magma far below.

Bill could run no added risks. Clumsily, he waddled from the pool, still breathing from his trusty but burdensome aqualung. It was rough going. Skin divers' gear is not designed for spelunking. After a few yards' exploration, Bill thankfully flopped back into his adopted element for the long return.

Even in an air-filled cavern, it is often difficult to maintain one's landmarks returning from an initial exploration. Everything looks different from the opposite side. Fortunately, however, the few side passages in this new complex were no problem. Without difficulty, Bill relocated the tight fissure. A little anxiously, he swam downward into its constriction. This time he felt only his own heightened heartbeat. Tired, but exultant, he began the long upward strokes.

The excited questioners blanched as Bill casually mentioned his "incident." No more diving today: too much was yet to be done.

Work parties and informal conferences soon developed new plans for Devil's Hole. It was agreed that the cave divers never again should operate singly. The near-deadly air-control handle was replaced. Each swimmer was to be equipped with two or more dependable, independent light sources as well as depth gauges and other skin-diving necessities.

All dives were to be logged, and emergency air was always at hand. Marker lines were to be installed so that a swimmer could feel his way out should all his lights fail. Double air tanks were to be used so that a known reserve would be present at all times.

The priceless training program was expanded. Groups of cave divers soon returned to Devil's Hole, exploring its rear complex both above and below water. Reserve tanks of air were cached in Brown's Room and a telephone line installed. Detailed studies under the guidance of the Scripps Institution greatly enhanced man's knowledge of this unique Devil's Hole, where man first brought light into the eternal night of a spelean warm spring.

By 1960 cave divers were nearing a 200-foot depth, still without sight of a bottom. In February 1961 Brown and Simmons led an eleven-man team, which used more than 100 tankloads of air in a single weekend. Preliminary dives calibrated depth gauges. Spare scuba units were cached 100 feet down. Vital equipment was tested until all was ready for a maximum effort.

Backup teams stationed themselves at various preplanned depths. Four divers fluttered bubblingly to the inky 200-foot level. Their spot beams showed the broad slanting passage undramatic—and seemingly interminable. Despite beginning nitrogen narcosis, two sped to the 240-foot level, their predetermined maximum. Here they dropped a special marker light, known to sink at about 100 feet per minute. The clear water showed it slowly shrinking, farther and farther down.

At this depth, the divers were allotted four minutes. Suddenly the pair gestured wildly at each other. The light had slowed abruptly, but was still shrinking. Perhaps it had encountered a steep slope. Too far away to reveal any details, it seemed to bounce down and down as the divers' precious minutes ticked away. As they turned upward at the last allowable second, it seemed still to be moving slowly—so far away.

Could the divers have been confused by the dread rapture of the deeps? Did bizarre subaqueous optics play a part? Did the light really continue for more than four minutes or stop somewhere on a distant slope? Or did it merely bounce off a chockstone and slowly settle downward for hundreds of feet?

Today anyone's guess is reasonable. In 1961 diver Jim Houtz reached the 315-foot level and returned to report that the cave still appeared bottomless. For the present, the innermost secrets of the Devil's Hole are beyond the glimpse of man. Diving technology has progressed to the point where deeper explorations are now feasible, but the story of Devil's Hole has taken a new direction.

Cave waters as pleasant as those of Devil's Hole are few indeed. An 83-degree stream in Virginia's hazardous Warm River Cave is near-unique in the United States. Missouri cavers once were momentarily elated by a warm subterranean pool fed by a small intermittent spring. Unlike other ebb-and-flow springs, however, this oddity was quickly traced to a washbowl of a drive-in theater overhead. Except in Florida, the average American caver is inclined to consider the icy waters of caves his greatest underground enemy. Diving into them seems a necessity at best.

Only in the great cavernous springs that drain north and central Florida has cave diving truly come of age in the United States. To those familiar with the cavernous regions of other parts of the United States, the drowned karstic region of Florida seems as strange as that of the western deserts. Here only a few caverns are high and dry. Though some are beautifully decorated, most of the dry caverns are small. The great caverns of this subtropical region are water-filled. Theirs are huge corridors which would not be scorned in Mammoth Cave.

Most of the famous springs of Florida emerge from such caves. From these submerged sinkholes or shafts flow crystal rivers which often course in bankless channels through swampy woodland, meandering to the sea or to another sinkhole. Sixty-six such Florida springs each discharge more than 6 million gallons daily. Seventeen of them emit more than 100 cubic feet of water each second. The beds of these strange rivers are often pockmarked with additional tributary pits. Such holes are natural traps for current-borne material, including Indian relics and the remains of extinct mammals.

Even before Bill Brown and Ed Simmons brought aqualungs to the depths of Devil's Hole, venturesome scuba divers were peering into these curious underground channels. Early in 1953, two young divers obtained permission to try their new apparatus in famous Silver Springs, where Navy divers had noted the presence of an underwater cavern.

Like almost everyone else, Charles McNabb and Frank DenBlyker were mere aqualung novices in 1953. It was still high adventure for them to don tanks and face pieces, then float almost effortlessly along the bottom in the clear, brilliant water. Leisurely gliding along, they approached a submerged limestone cliff. It led down to a cavern mouth, 20 feet below the surface, 60 feet wide and 12 feet high in the center. Facing west, great depths of the slanting cavern were bathed in afternoon sun.

With a few lazy kicks of his flipper-clad feet, McNabb approached the cave. To his surprise, he was hurtled backward by a swift emerging current. On his mettle, he tried again at full speed. Again he was hurled back by the outpouring of 5,000 gallons per second. Once more he tried, kicking full speed and pulling himself along the rocky floor of the spring. With determined drive he forced his body through the invisible wall and burst into quieter waters inside.

With an equal effort, Frank joined him. Their eyes adjusting to the dimness, they found themselves in a water-filled chamber 75 feet wide and 30 feet high. It extended beyond twilight.

Returning jet-propelled to the surface, McNabb sought out the owners of the resort. One—Bill Ray—became equally excited. He and McNabb soon returned to the watery cavern. A hundred feet inside it divided into several levels, but the big surprise lay in full rippling sunlight. Nestled in a small depression just inside the dark-framed entrance, lay a huge bone. Digging with their hands, the divers quickly unearthed a weird conglomeration of bones and oversized teeth. Back on the surface, the bones and teeth were readily identified as mastodon, an animal which probably was gone from Florida ten thousand years ago.

Not trained paleontologists, McNabb and Ray unhappily watched much of their trove crumble as it dried. Plunging the remainder back into the spring, they sought expert advice. Then, with a guide line and pre-arranged signals, they again dared the rushing depths. Soon they had located and removed teeth and six-foot tusks of a Columbian elephant and a smaller mastodon. Returning again and again with underwater lights, they found no end to the fascinating cavern. But the operations were interfering with tourist activities and regretfully had to be halted.

Similar bones were spotted in other water-filled Florida caverns. Like their mates in air-filled caves, scuba divers of the great Florida springs promptly sought scientific liaison. Two hundred feet down in the air-clear water of the great cave of Wakulla Spring and nearly 250 feet inside its maw, Florida State University divers encountered the thigh bone of a mastodon in November 1955. In the months that followed, their six-man team made over 100 dives to depths beyond the 200-foot level. The coordinated program in cooperation with the Florida Geological Survey is outstanding in the history of American caving.

Although its flow is only a third as great, the arched cave of Wakulla Spring dwarfs that at Silver Spring. Though occasionally much lower, its height locally reaches 100 feet. Along the floor its width is 70 to 150 feet. At first dimly lit by filtering light, it stretches southeast for some 600 feet, then angles southwest into unknown blackness. No one went more than 1,100 feet from the entrance, where the depth begins to exceed 250 feet and aqualung time is short indeed. Too much was found in the naturally illuminated entrance section.

Floored with pleasant sand, the gaping arched entrance tunnel slopes downward to a depth of about 200 feet. There an equal distance inside the great conduit, limestone rubble covers a gentler slope. Beyond are sandbars, an inner "Grand Canyon" of layered clay—and a treasure trove of anciently strewn bones.

More than a hundred scuba divers have died in underwater Florida caves—many in far less hazardous locations than the Wakulla Spring

Cave. Here the human body tolerates stays of only fifteen minutes without danger of "bends." Even so, coordination is impaired by the dread "rapture of the deeps"—nitrogen narcosis. Here there is no margin for error.

Under the leadership of Garry Salsman, the six young students developed a unique program of effective, safe exploration and study at

Scuba diver approaches proboscidean leg bones in the entrance of Wakulla Springs Cavern. *Photo by D. C. Martin*

these unprecedented depths. Weighted by heavy rocks, they plummeted gracefully to the bottom without significant expenditure of energy. One hundred feet down, they could begin their swim inward, their bubble trail now hidden from watchers far above. If the day was clear, not for 180 feet was it necessary to use underwater lights. It required six minutes to reach the bone area.

Work in the clay matrix around the bones stirs muck and reduces visibility. It became routine for the photographer to approach first. After his underwater flashgun's momentary brilliance, others moved in to measure and record data. In the few minutes remaining at this depth, selected bones were gently freed from their grave. If large, each was lashed to a plastic sack, which was then filled with aqualung bubbles. As such a sack balloons out, the bone stirs in its bed for the first time in thousands of years. With a little more air it rises free of the cavern floor. Fine adjustments readily bring the largest bone to neutral buoyancy, though air must be spilled periodically as the depth decreases ahead of the ascending swimmers. Pushing such a mass of remarkable inertia but no weight is perhaps man's closest earthbound approach to the weightlessness of space.

Each of the 100 dives was carefully timed. Before the calculated time, all divers invariably were free of the mouth of the cave and rising toward safer depths. Ascents were slow, and 10 feet below the surface, all halted 36 minutes for decompression. Largely due to such self-discipline, no accident ever befell the sextet. Scientifically their reward was great. Mastodon, sloth, and deer bones predominated. Occasionally smaller, rarer animal remains came to light.

And in the dim cavern, the scuba team unearthed 600 mysterious spear points. Did ancient man consider Wakulla Spring his butcher shop? No definite clue has turned up.

Nor do we know whether rivals of Mammoth Cave lie beyond the limit of man's physiology in such springs. With the restless cycles of nature, will a great future uplift of the swampy peninsula or a lowering of sea level someday drain these water-filled caves and remove the cave record from Kentucky to Florida? To know, we must wait perhaps a million years until Florida next stands high above sea level.

But neither cavers nor scuba divers are likely to restrain their curiosity for a million years. Just such curiosity recently led Steve Barnett and Dave Jagnow into a 47-degree pond in a miserable little Iowa cave previously noted only as a good place to hide a moonshine still. A quarter mile of scuba work revealed Iowa's largest and most beautiful cave, its name memorializing their triumph: Cold Water Cave. A few carefully trained members of Sheck Exley's Cave Diving Section of the N.S.S. have traversed as much as 6,800 feet of underwater Florida cave and survived depths of as much as 375 feet. With careful training courses now increasingly available, cave diving may be coming of age in many parts of the United States.

For the larger waterfilled caves of Florida and thereabouts, the first underwater speedboats are off the drawing boards. Gas mixtures now in the experimental stage may soon make it possible for cave divers to explore far greater depths than at present. Probably we can never know each hidden crevice of the great waterfilled caverns that underlie Florida until their gaping voids are drained. Yet, before man sets foot on Mars, he may well be caching supplies far back in comparatively shallow sections of these mysterious caves.

Continuing expeditions will lead on and on into caverns now hardly imagined, for here is much more than the thrill of the doubly difficult, often deadly unknown. Here the development of truly great caves is still in progress for the speleologist to decipher.

And, when man has mastered the great submerged caverns of Florida, he will someday revisit the friendly pupfish of Devil's Hole. With luck and determined effort, they will still exist. Today, their survival is gravely threatened. Yet, more and more it appears that future generations will hold this unique cavern and its handsome little denizens in special honor and trust.

Even during the pioneer dives in Devil's Hole, Carl Hubbs, Robert R. Miller, and others in the scientific community had been carefully at work. Few creatures have ever lived quite so precariously as these pupfish, and uninformed or uncaring planners had begun an irrigation project whose pumping was likely to exterminate them. Already there was fear of the effects of a dropping water level—a fear which was to be justified all too soon. Ten other types of desert spring fish have become extinct since the end of World War II.

Incredibly, the pupfish's entire habitat for living, feeding, spawning, and dying is limited to the space of a few inches above the shallow submerged ledge at the south end of the pool, just 18 feet long. So shallow is this vital habitat that, since our visit in 1950, an almost imperceptible fall in the water level has caused several vital square feet of pupfish-depth water to become dry rubble. Only on its eastern half grow the green algae which are their food: the smallest known habitat of any vertebrate species, according to eloquent Edwin Pister of the California Department of Fish and Game, who pinpoints Devil's Hole as "a tiny microcosm which reflects, basically, the same problems that the entire Earth faces, or soon may face."

Inadvertent competition for life is strong here. Nearly two-thirds of the pupfish inevitably die each winter when the algae wane. Should the water level drop just three feet, no more algae and no more pupfish. No one can be sure of the effects of a drop of only another foot or two.

Every species has a critical minimum below which its numbers cannot be restored, and the pupfish are no guppies, seemingly able to repopulate the world from a single pair.

The cave was on government land, so, without fanfare, it and 40 surrounding acres of the ancient Ash Meadows lake bed became a detached unit of Death Valley National Monument (through signing the official proclamation President Harry S. Truman briefly re-entered the story of American caves).

A stout wire fence soon reduced threats from casual pollution. The pupfish were officially recognized as endangered, and intensive studies began to evaluate alternatives of management of endangered fish species.

Our own conservation-minded generation soon learned of the pupfish and their danger. In 1970 the U.S. Department of the Interior created a very special "Pupfish Task Force" to coordinate its preservation efforts. A year later the Desert Fishes Council came into existence. Replete with the names of famous scientists and resource specialists it began to mobilize the conscience of America for the preservation of such habitats as Devil's Hole and the gene pools of their inhabitants. In heartwarming response hundreds of thousands of ecologically-minded Americans demanded protection for the pupfish and the other endangered species of tiny desert springs.

Ichthyologists investigated possible routes of survival for the pupfish should Devil's Hole become uninhabitable. For some other desert spring fish, transplantation to other, seemingly favorable locations seemed satisfactory, but for the pupfish it was a tenuous success at best. To increase algal growth and provide an additional spawning area, an illuminated floating shelf was constructed, but its success also was limited.

Maintenance of an adequate water level was the only safe route. The pumping remained an uncompromising threat. The Department of the Interior and the U.S. Department of Justice obtained a permanent injunction, and pumping for the ill-conceived land development was cut back. At least for the moment. The pumpers are appealing the decision.

The battle for the pupfish is not won, for mankind is not cut from a single cloth and the number of our finny friends has declined to an all-time low. As I write we still lack an Ash Meadows National Wildlife Refuge or a Desert Pupfish National Monument. Even with adequate statutory protection through one of these proposals, just one chance crash of a jet plane, one single chance writhe of international politics might still uncaringly, unknowingly blot out their existence.

Nor are their protectors universal. On April 27, 1973, controversial columnist Jack Anderson dismayed conservationists everywhere. Nearing the height of his pro-impeachment campaign he reported blandly on the admittedly costly efforts of the Nixon administration to save what he incredibly termed "200 ugly, inedible pupfish . . . tiny, wriggly, dis-

Beauty in the eye of the beholder: one of the tiny pupfish Jack Anderson described as ugly and squirmy, shown here several times life size. *Photo by Alan Heller*

agreeable-looking creatures." And the citizens who seize upon every innuendo of the Jack Andersons are legion.

Yet callous Jack Andersons are few, and strong indeed is man's new sense of oneness with even the least of creatures. With today's increasing sanity in geopolitics and with only a little additional luck, the pupfish should outlive the Jack Andersons, indeed until the naturally appointed time of their passing—like that of man himself—from the depths of the earth.

Today Devil's Hole is properly closed to human trespass. Someday science will need to know more about its unknown depths, and man will return to seek answers for which today he lacks even the questions.

He will do so carefully, thoughtfully, affectionately. Devil's Hole belongs to the pupfish, not to man.

16

They Call It Progress

The Story of Onondaga, Nickajack, and
Russell Caves—and Some Others

For a little while yet, Onondaga Cave is still one of America's greatest. But "progress" is coming, they say. If the U.S. Army Corps of Engineers has its way, that political juggernaut will strike down virtually its very name. By the early 1980s, an impressive new dam on the Meramec River may flood most of its age-old magnificence. As an arm of the new reservoir relentlessly rises and falls in its great vault, massive annual clay deposits will coat, then hide ever deeper the glistening splendor which today delights 100,000 tourists each year. Annihilated will be the beauty which a decade ago so impressed the art director of Harper & Row that he selected its Lily Pad Room for the dust jacket of the first edition of this book. Only the tips of a few dripstone mountains will jut from the murk slowly swirling close beneath the ceiling. The clay beneath their bases undermined by the fluctuating reservoir, gigantic natural monuments to timeliness will crash unhonored into the mucky water. This favorite cavern of tourist, spelunker, and speleologist alike—quite possibly the most important of Missouri's 3,100 caves—will be little more than a fond memory. This is progress, the Corps of Engineers assures us, and who dares to disagree?

Every concerned reader of this book should compare the above paragraph with that in its first edition, written in 1965. At that time, the name of the cave and the state were different. The remaining words are virtually identical, and with tragic reason. At that time, they told of a different dam-to-be, and a different victim: Tennessee's once-great Nickajack Cave. No one spoke for Nickajack. Only a little of its once-awesome entrance now remains above water. The murderous crawlway to "Mr. Big" —perhaps the world's largest stalagmite—is only a caver's memory. This would be progress, the government engineers and politicians said, and no mere human then dared contradict them.

Nickajack's gaping mouth lay near the point where Tennessee, Alabama, and Georgia come together. "You play T-A-G in Nickajack," one hearty promoter jested, informing his customers that the huge cavern ran under three states—all in fun and pretty close at that. Nickajack Cave started in Tennessee, sneaked under Alabama, and missed Georgia by just 80 feet.

For a little while yet, Nickajack lives in memory. Mine and that of our children is of a stormy day when I first saw its vast opening. I goggled unashamedly. Half the height of the hill and three times as wide. I had never seen a cave entrance so large.

We turned off onto a dirt road that led to and into the yawning cave. As we approached, the huge squared-off entrance gaped more and more awesomely. Fifty feet high and 140 feet wide, I had been told, and could not disagree. Daylight streamed far back into its spacious vault, so huge that the moment of actual entrance was imperceptible.

Down a muddy slide from the earthy platform on which we stood, the slab-spanned cave stream glowed blue-green. In the deeper twilight it curved away into stygian dark. Delicate ferns and miniature mossy jungles clung to tiny ledges. Cliff swallows flitted to and from globular

Nickajack Cave today. *Photo by Bill Torode and Dick Graham*

nests plastered improbably to the walls and ceiling. From somewhere far back in the huge cavern, faint shouts echoed back from inbound cavers we never met. Nearer at hand, a storm-wrought patter from a dozen leaky roof joints merely accentuated the immense quiet of the tunnel.

A few dozen yards inside the wide entrance, a huge old stalagmite infused an air of unchanging permanence. Little wonder, I told the children, that Nickajack Cave so long enriched the folklore of America. Here countless Indians should have worshiped their Manitou. Here De Soto should have paused on his historic trek three centuries ago, and perhaps he did.

Except for its description of 1819 saltpeter works, the first published account of Nickajack Cave could stand almost unchanged until the dire day of its loss. Much reprinted in later years, this old account provides illuminating glimpses of its early history and near-history:

A few years since, Col. James Ore of Tennessee, commencing early in the morning, followed the course of this creek in a canoe, for three miles. He then came to a fall of water, and was obliged to return, without making any further discovery. Whether he penetrated three miles or not it is a fact he did not return till the evening, having been busily engaged in his subterranean voyage for twelve hours.

Ore's waterfall was never located by those who followed him. Nevertheless, belief in the "three mile river" was current for a century and a quarter.

The story of Nickajack Cave almost began with "Col. James Ore of Tennessee," but it was long before 1819. In September 1794, Ore commanded Tennessee militia which crept stealthily upon two confidently sleeping villages of Chickamauga Cherokees near the cave. Supposedly impregnable, the dual Indian capitals of Running Water and Nokutsegi or Anikusatiyi—Nickajack—were about three miles apart.

In 1777 Tsiyugunsini ("Dragging Canoe"), a fiery young Chickamauga subchief, had bitterly disputed his elders' sale of magnificent hunting lands to the encroaching white men. A disgruntled tribal faction joined him in bloody but unavailing warfare against the immigrant waves. Tsiyugunsini's band was finally forced out of the traditional Cherokee lands near Chattanooga. Friendly Creek Indians granted them permission to settle on their lands. There, deep in the impenetrable wilderness along the Tennessee River, Tsiyugunsini established five new, seemingly impregnable villages of substantial log cabins. Soon they became centers of intrigue against the white man. Long strings of desiccating white scalps embellished the doorways. Warriors of many tribes flocked to Tsiyugunsini's banner. Many were reckless, corrupt, and demoralized, having adopted the worst traits of the white man. It was in this region that Tsiyugunsini lived out his days scheming, raiding, then retiring at leisure to the hidden villages.

One young white man had seen Nickajack and lived. Captured by the Chickamaugas at fifteen, Joseph Brown lived at Nokutsegi as a slave for a year before he was exchanged for a Cherokee chief's daughter.

Even after the death of Tsiyugunsini the Indian raids had not diminished when Brown reached early manhood. Three years after his exchange, Brown guided the militia undetected to Nokutsegi and Running Water. Crossing the broad Tennessee River on logs and newly made rafts of dry cane, they floated rifles and powder in bullboats of steer hide. With simple pioneer vengeance, 268 frontiersmen, including President-to-be Andrew Jackson, fell upon the sleeping settlements. With the loss of only a single man, they utterly destroyed the "impregnable" villages. Never again was there an organized Chickamauga raid on the Wilderness Road settlers.

Much has been written about Nickajack Cave as Tsiyugunsini's headquarters. Those authors must not have seen the cave in a heavy storm: the roof leaked abominably. Huge and ideally located, the cave undoubtedly served as temporary shelter long before Tsiyugunsini's time. Its depths may have harbored his band while the villages were being built. Probably terrified refugees from the 1794 raid found refuge here. Only folklore, however, would have Tsiyugunsini prefer the dank cave to the comfortable village next door.

Similar nonsense has been written about the historic days of this remarkable cavern. Particularly persistent is the tale of Jack Civil, a free Negro captured by the Chickamaugas. Later he was freed by John Rogers, an early trader who is said to have been an ancestor of Will Rogers. "Nigger Jack" enjoyed boasting that the Chickamaugas had named their chief town and the great cave for him! In 1865, abolitionist author David Lathrop took his backwoods boast seriously and antichauvinistically renamed it "the Negro Jack Cave." Other historians were similarly conned at least as recently as 1954.

We have no record of saltpeter mining in the dry side labyrinths of Nickajack Cave until the Mexican War, but it may have begun earlier. During the Civil War its product was particularly important to the Confederacy while it was still far behind the lines. The shifting tides of war, however, soon found it in one of the most bitterly contested of all the war zones. In June 1862 it was shelled by Union artillery and subsequently changed hands several times. Thousands of soldiers viewed its gaping entrance from a nearby main railroad, and hundreds of both blue-clads and Confederate soldiers hiked the half mile from Shellmound Station to enjoy its pleasures. Its final loss reflected the general trend of the war. On February 6, 1864, *Harper's Magazine* reported:

SALTPETER CAVE NEAR CHATTANOOGA

The "Nickajack" Cave near Chattanooga is one of the main sources from which the Confederates have derived the saltpeter required for the manufacture of powder. Its loss is deplored by the rebels as one of the most serious results of our victory at Chattanooga. Six or seven years ago, this cave was visited by

Entrance of Nickajack Cave as developed by Leo Lambert. Visitors are dimly visible beneath overhang. *Mim's Studio photo, courtesy Mrs. Paul Whisler*

"Porte Crayon" [D. H. Strother, of Virginia, lately on the staff of General Banks, of New Orleans], the genial artist-correspondent of Harper's Magazine.

"Porte Crayon" goes on to describe how the cave was formerly the resort of a gang of banditti, whose occupation was plundering and murdering the emigrants and traders who descended the Tennessee River, the cave furnishing a convenient hiding place for their booty. Since he wrote, this cave had fallen into the hands of a worse and more desperate gang, who used it for purposes still more nefarious, extracting from it abundant supplies of the "villainous saltpeter" required for the manufacture of powder for the Confederate States of America. It has now—thanks to Grant and Sherman—fallen into honest hands.

At the close of the war, guide service to this "largest cave and the greatest natural curiosity in the world" was advertised as far away as Chattanooga. At the peak of this short-lived prosperity orchestras were imported and dancing couples whirled in the resonant vault. Soon an epochal biological report depicted the blind, transparent creatures of the cave. By 1901 the scientific world had a surprisingly clear concept of the huge cave—at least as far as the end of the great stream passage. In that year biologist William Perry Hay reduced the traditional "three mile" length to one-half mile; nobody objected. In 1909 a new species of bat was discovered here. Much later, banding experiments demonstrated that Nickajack Cave was equipped with summer bats and winter bats. From October to May, its hibernating denizens included the Indiana bat, *Myotis sodalis,* belatedly recognized as being especially endangered little animals and now given special protection by federal law. These went north each spring. Their place was promptly taken by bats of other species, which snugly slept the day away, then emerged at nightfall to decimate the local bugs. The Corps of Engineers evidently didn't care that some of these, too, were of dwindling species, like the gray bat (*Myotis grisescens*), which congregated here into one of the world's largest colonies. The nearest to a blot on the proud escutcheon of science at Nickajack Cave was a report of Gerald Fowke, then chief of the Bureau of American Ethnology. Fifty-odd years ago, Fowke conducted a hasty appraisal of the impact of saltpeter mining and flooding on the cave's archeological deposits. In 1922 he opined: "There is nothing in the cave to dig for." Unfortunately his influential judgment appeared in the authoritative *Bulletin* of his bureau. As indicated later in this chapter, we of today can only wonder how tragic was this mistake.

No one knows who first poked and peered into the enormous jumble of fallen rock where the Nickajack stream welled from beneath a ledge. Twisting and turning, slipping down great boulders, ascending through narrow slots, someone triumphantly emerged into a broad, dry corridor nearly 1,000 feet long. Leo Lambert gave credit to R. Sageser, of whom we know nothing but the date: 1939. Lambert himself was one of the most knowledgeable and certainly the most tragic figure of the cave's long interaction with man. His was the final and the most perceptive attempt to commercialize it: a splendid development with expensive boat landings, a dredged channel, and half-mile tours in quiet electric boats. In the contentious aftermath of a stupid hoax perpetrated upon him as well as upon a properly untrusting public, he lost everything and soon died, unwilling or no longer able to fight both cancer and fate. Until the coming of today's freeway Nickajack was simply too far from tourist routes for successful commercialization. In 1946 it reverted to "valueless" wilderness. The Corps of Engineers assured the public and Congress that Nickajack Cave and its endangered fauna had no commercial value and therefore might as well be flooded by a reservoir. Its staff evidently did not know that the freeway was on the drawing boards, or else did not care to mention the fact. Today's traffic pattern would have brought wealth to any owner, and immeasurable delight to innumerable entranced visitors. Before the hated dam, this was a friendly, forgiving cave. Despite

TVA surveyors preparing for the demise of Nickajack Cave. *TVA photo*

lurid Civil War tales, no serious trouble seems to have befallen any of its visitors. A few became lost in obscure side passages, but all were rescued easily and promptly. As the grand cave faced inundation, cavers rushed to enjoy its delights "just one more time." Nickajack will remain a fond memory for centuries.

And a shameful monument to unfeeling destruction in the name of progress.

The loss of Nickajack Cave cost America dear. Its great value is sorely missed. Yet its loss is but a single revealing episode. The Chickamauga village sites at its doorstep too were an irreplaceable segment of our cultural heritage. So was the entranceway of Blowing Cave, since its first recognition by Thomas Jefferson two centuries ago. In the Arkansas Ozarks, no man knows how many outstanding caves have fallen victim to similar dams, or are currently threatened, or what has been and may yet be lost in their depths. Perhaps a glimmer of the truth stands revealed at Bull Shoals Dam. Here, cavers aver, its main leak has suddenly created the second-largest spring in the state of Arkansas, its water passing beneath the White River and emerging unwanted in a state park.

Nor is the much-damned Corps of Engineers the only agency gravely at fault. Duplicity by the Bureau of Reclamation and its adherents threatens the structural stability of incomparable Rainbow Bridge, even now being undermined by the ceaseless sapping of an illegal fluctuating reservoir.

That agency quietly also won Congressional approval of the Brantley Dam on the Pecos River near Carlsbad, New Mexico, which would triple the size of the present McMillan Reservoir. It seems unlikely that that bureau told Congress the entire history of this proposal, or the warnings of such noted geologists as Preston McGrain about "unsatisfactory conditions" of karst for "extensive impoundings of surface waters." The Brantley Dam is intended for "extensive impounding of surface waters," but the history of the McMillan Reservoir suggests that it may be less than successful. Built in 1893 and destroyed several times during floods, McMillan Dam is cited in several textbooks as a classic example of poor placement and design. Evidently its builders failed to consider the bedrock underlying the west end and side of its reservoir. Promptly it dissolved into an extensive gypsum karst. Neatly sandwiched between layers of limestone, large caves came to be characterized by complex mazes on several levels. No one knows the entire extent of these caves. This type of gypsum cave is rare in the United States, but in similar Russian deposits is the world's largest gypsum cave, with 67 miles of network passage mapped.

By 1905, large caves and sinkholes were engulfing the entire Pecos River when not in flood. The water development company had gone bankrupt. The Bureau of Reclamation took over, building long levees to keep the reservoir away from its largest leaks. With new caves forming and collapsing, few of the levees were effective for more than a few years. Recently the bureau has turned to large earth-moving equipment, sealing off many caves from cavers and slowing down some of the leaks. With some six miles of caves thought to underlie this area, Coffee Cave's 6,000 feet of passages can still be visited, and others of lesser extent. Inevitably cavers view the future here with mixed emotions. The present plans of the bureau are dependent on Congressional funding. Recurrent rumors hint at its grouting the caves with cement, and wider flooding of the karst plain.

Yet not all would be lost. More likely than not, construction of the Brantley Dam will merely divert the Pecos River underground again, in new channels. Encouraged by the new reservoir, one of the new caves might possibly grow to challenge the Russian giant, if not Mammoth Cave.

And achieve one unchallenged record: the world's most expensive cave.

Today's cavers, however, are unlikely to wait to see if the Brantley Dam actually grows us the world's largest gypsum cave. These are no longer the spineless days when we shamefully gave up Nickajack Cave with hardly a whimper. Painfully we have learned that we cannot trust bland assurances of governmental special interests. Public outrage forced the Corps of Engineers to build a supposedly watertight dike to prevent its flooding of Washington State's Marmes Cave. The dike failed, and that internationally significant home of early man was flooded precisely on schedule. Conservationists permitted enactment of the Upper Colorado Storage Project Act that cost America incomparable Glen Canyon— later symbolized nationwide as "The Place That No One Knew"—only because that act contained an amendment supposedly protecting Rainbow Bridge National Monument for all time. Less than a generation later, that solemn compact by the American Congress lies scandalously broken and Rainbow Bridge is doomed.

But America's counterattack is underway. Terming it "a miscarriage of basic engineering and elementary geology," the N.S.S.'s Virginia Polytechnic Institute Grotto demands a halt to the unfinished Gathright Dam Project. Bitterly they publicize a four-fold increase in the costs of its first half, due in part to caverns that were "apparently unknown and obviously unexplored before the early part of June 1974," according to the contractor working on that part of the project. "Shameful," they call "the idea of attempting to build a dam over an extensive network of caverns in extremely porous limestone." Compared to the *Engineering News-Record,* their language is restrained. That respected periodical termed the site a dambuilder's nightmare, with "rock like Swiss cheese," and the foundation of the dam "as weak as its economic justification." Satiric indeed were the words of *The Potomac Caver:* "It seems odd that the [dam] geologists are just now finding out what almost any novice caver knows, i.e. the speleogenic properties of these limestones."

From California's New Melones Project to the sea caves of Acadia National Park speleologists and their allies are struggling to save caves like Onondaga, and their fragile, defenseless contents. With appalling speed so many irreplaceable resources are falling victim to uncontrolled population pressure and narrow concepts of progress. Our nation is not so poor that we must sacrifice the Nickajacks, the Onondagas, the Blowing Caves, the Rainbow Bridges for the sake of dams of dubious cost/benefit ratios. Such irreplaceable losses reemphasize that technological progress alone is not civilization. Desperately we seek recognition and preservation of unique underground values.

In central Indiana the once-revered Soil Conservation Service is the latest villain. There, that bureau and its political allies recently urged destructive "channelization" and flooding of an internationally famous karstic wonderland: the "Lost River Watershed Project."

And they apparently did it without even pondering the significance of that name, much less the underground stream tracing of S. H. Murdock of their own staff. Seemingly without thought for the biological, wilderness, scenic, recreational and other values of these celebrated Lost River karstlands, they proposed eleven "flood-control structures." Several would flood some of the most famous parts of Lost River itself and seriously damage many classical karst features nearby, including at least thirteen caves. By flooding and thus eliminating present-day natural flood-control capacities of this spongelike terrain, several of these "flood-control structures" would defeat part or all of their stated purpose. In further havoc the bureau planned to poison part of the endangered Lost River animal life. "Rough" or "trash" fish would be annihilated in some areas, including blind cave fish and other innocent cave life, merely to improve the local fishing—in the very region whence blind fish first reached the world of science a century and a half ago!

Lost River and its basin are much more than any ordinary sinking stream, any common karst. Its watershed of barely 150 square miles is a remarkably compact, comprehensive living museum of fascinating geological phenomena, of scenic features, and of varied ecosystems of scientific import. True, some of its features are those common to all karst in such a geographic setting: sinkholes, sinking streams, and the like. Although there are those who love them, its caves are comparatively lacking in recreational and scenic values. Much more is here to be saved or lost: the winding 22-mile channel recently abandoned by Lost River.

A complex system of swallow holes and stormwater rises. Gulfs, karstic hanging valleys, subterranean cutoffs, and stream piracies of notable intricacy.

This is a strange land of unheralded surprises. Travelers driving east on U.S. Highway 150 may question the need for a certain bridge about five miles west of Paoli. On their side of the road is no stream, no gully, no evident excuse for construction of a bridge. Extra-curious motorists turning their cars for a better look may have difficulty identifying the spot. West-bound drivers see Lick Creek placidly flowing northward in what looks like an entirely proper channel. By taking a short subterranean cutoff which chanced to rise precisely where the highway had to go, Lick Creek is saving itself about a mile of meandering surface flow.

To those attuned to ordinary karst forms of the central United States, the sievelike Lost River karst is amazingly shallow, so new that its upper reaches even form a sink-free "island" surrounded by pitted fields of other Indiana karsts. The abashed Soil Conservation Service has circulated a draft of a "Revised Environmental Impact Statement" that is much improved from its earlier debacle, but no such project befits the Lost River country. Long overdue here is an extensive national monument where the National Park Service could protect, interpret, and display the features of this unique juvenile karst for all time to come.

Already a good beginning has been made here through registration as National Landmarks of three famous but endangered features. Except in awesome flood, Orangeville Rise is the least spectacular of these, yet probably the most important. Here water merely rises some 20 feet from a water-filled cavern some 70 feet wide, welling up into a rock-walled basin 110 feet long and half as wide, whence it flows back into Lost River's original bed. This, however, is one of only two resurgences of the main Lost River drainage, and is vital to its interpretation. More than 1,000 feet long and 350 feet wide, Wesley Chapel Gulf is the largest and most dramatic of four such steep-walled gulfs in the basin. A gaping maw of a half gulf, Tolliver Swallow Hole is the spectacular orifice of a dangerous low cave usually blocked by a log jam that renders needless any lecture about the prodigious forces of underground water. Perhaps lacking in exceptional scenic beauty, such features yet exemplify the scientific values which cry out for proper interpretation and preservation here.

In many ways the future of America's caves is brighter than when I wrote about Nickajack a decade ago. Increasing numbers of ordinary people are daring to disagree with the government engineers and to demand meaningful planning. Already we are seeing successes unbelievable a mere generation ago. The 1972 establishment of the Buffalo National River saved Beauty Cave from flooding (or so we fervently hope; a previously authorized dam hasn't been officially de-authorized). Despite the tragedy of Nickajack, the "emerald cameo of life" of Ezell's Cave and the Devil's Hole pupfish are about as safe as can be. So is the rare animal life of Shelta Cave, atop which the N.S.S. headquarters building in Huntsville now squats protectively. It and McFail's Cave were purchased as vital conservation measures in the teeth of a lack of funds for such purposes, and the funds soon were forthcoming.

Even without protective legislation, Onondaga and the hundred-odd other caves threatened by the proposed Meramec Park Dam seem likely

If the proposed Meramec Park Dam is built, Onondaga Cave would be flooded to approximately the level of the three small stalagmites on the ledge in the upper center of this photograph. *Photo courtesy Onondaga Cavern*

to be saved, and with them the threatened, vanishing species of bats and blind Ozark cave salamanders which they house. Despite its customary methodical mobilization of public support—here including such Congressionally condemned measures as illegal patterns of land purchase benefitting "cooperative and speculative owners"—the roof has caved in on the Corps of Engineers here. All over America, outraged public opinion is fast mobilizing through the N.S.S.'s Meramec Valley Conservation Task Force. Caving professor Tom Cravens, Don Rimbach, and their teammates have organized for action a host of other caving and non-caving fellow Missourians. At last count Rimbach alone had delivered 92 lectures. Lester Dill has thrown much of his Meramec Cavern bankroll into the increasingly bitter struggle.

Just about everyone seems against the Corps here: property owners, fed-up taxpayers, plain people who enjoy the beautiful Meramec Valley as it is. The director of the Missouri State Park Board politely but pointedly asked the Corps to respond to well-documented charges that its cave study was obviously inadequate. One current lawsuit avers that construction of this particular dam would be illegal under the 1821 articles of admission of Missouri into the United States. Another by the Sierra Club and local citizens is attacking its Environmental Impact Statement, originally just eight pages long—triple-spaced!—and obviously an erroneous "short piece of promotion trying to sell the dam" according to Professor Cravens. The U.S. Department of the Interior formally requested that its construction be halted "until studies can project on the rare and endangered Indiana bat." "The tone of the letter was impatient, even testy," reported the St. Louis *Post-Dispatch*. Indignantly it pointed out that the Corps had "paid little heed" to the federal Endangered Species Act of 1973, despite "previous attempts of lower-level officials of the U.S. Fish and Wildlife Service." Time will tell—and the support of millions of Americans.

Throughout our land, fast-spreading recognition of the intangible human values of these fragile natural wonders thus is stimulating broad, effective action. The era is long past when members of the Salt Lake Grotto of the N.S.S.—plus one middle-aged truck driver—found ourselves standing alone against the Bureau of Reclamation, thinly disguising ourselves as the Utah Committee for a Glen Canyon National Park. We were eventually steamrollered, of course, and America lost Glen Canyon, now a famous symbol of irreplaceable loss. And, as a result, Rainbow Bridge National Monument is unlawfully invaded by a destructive reservoir.

Other little groups of cavers similarly began to stand alone, and lost. But even our bitterest defeats were progress. We learned to join forces across the nation, around the world. When a similar proposal came for the next canyon below Glen Canyon, our exposure of the heavy-handed tactics of the Bureau of Reclamation had forewarned, forearmed American conservation. Marble Canyon was saved, and with it the green canyon-bottom oasis that is Vasey's Paradise. And the extraordinary cave system which waters it.

Even the monolithic Corps of Engineers may be a trifle nonplussed. Ostensibly because of fear of leaks, it recently moved the proposed site of Meramec Park Dam more than a mile upstream, where its reservoir would spare some of the finest caves in Meramec State Park: a sign of the times unthinkable a mere decade ago.

With much of Indiana deeply disturbed by the Lost River proposal, that battle is far from hopeless. In more than token opposition, conservationists here purchased the famous Orangeville Rise, that would be one of the chief victims of the project. Noted scientists and cavers soon published details of the impact of the proposed project across America. Then they brought suit in Orange Circuit Court to block the proposal. Already the Soil Conservation Service has offered "compromises," extensively revising its infuriating, self-contradictory Environmental Impact Statement that incredibly asserted: "There will be no known adverse effects upon the environment." Only time—and effort—will tell here, but the future looks hopeful.

One state to the south, the story of Wolf Creek Dam is almost beginning to be fun. This $80 million dam was authorized in 1946, long before cavers or conservationists had even heard of Sloan's Valley Cave, much less realized its immense size and importance. Today it stands second in length only to Mammoth Cave in all Kentucky, with 22.4 miles mapped. Although only 14 miles has been surveyed to date, nearby Hyden's or Cave Creek system is little less.

In frustration over the dam's flooding of key connections of these unexpectedly vast systems, some Kentucky cavers began wearing slogan pins, willing the dam to leak. To their surprise, it did. According to one published account, the Corps foolishly plugged caves near the damsite with clay, which promptly washed out. Other accounts lay the problem to defects in the dam itself. (Some may feel that I should have contacted the Corps to obtain "the facts" rather than repeating mere contradictory allegations here. A generation of personal experience with phony fact sheets, deception, and outright lies by these once-idealistic bureaus has led me to believe that such queries are a waste of time. Somebody else can do that for a different book, if they want. I did so at one time. No longer.) In any event, the Corps repeatedly has had to lower the reservoir level thirty to fifty feet, for many wonderful, tremendously productive months.

This new opportunity created new problems, however. On November 17, 1974, the reservoir's level stood 40 feet below its high average elevation. On that date, the two halves of Lou Simpson's team came within

50 feet of linking the Firestone Cave and Goldson's Cave sections of the Cave Creek system for a total of 8.4 miles. The groups could see each other through a 4-inch air space between the ceiling and the reservoir level, and cheered each other's attempts to fight on through the chill water. But the water was 6 feet deep. Lou himself swam on until only one eye and nostril remained above water, scraping the ceiling. At that point, everyone quit, expecting a drop of an inch or two during the following week. Instead, the reservoir rose. Even the passages through which they approached the line-of-sight connection became impassable.

Infuriatingly, it was only a few days later that smoke from Paul Unger's cigar in the 6-mile Hyden's Cave–Humongous Hole section turned up—shades of Burton Faust!—in Firestone Cave, and the connecting passage that would have produced a 13.7-mile cave was cleared three months later. Fortunately, the reservoir level fell a foot still lower in 1975, permitting Paul Unger and Alan Henning to float the tantalizing voice connection. Promptly the Central Ohio Grotto team began to seek new connections to other nearby caves.

Only Wolf Creek Dam thus prevents Kentucky's Pulaski County from placing two caves high on the list of the world's longest. Hopefully that barrier will be only temporary. The Corps of Engineers, however, is now reported to have asked Congress for an additional $64 million to build a 3,000-foot "cutoff wall" to protect the original $80 million "investment," alleging that the entire dam is in danger. Local cavers are suggesting that Congress would do better to tear down the damn dam, for otherwise we can never know what other irreplaceable resources lie submerged here. They still have their slogan pins ready.

Less romantic but equally revolutionary concepts of cave conservation have drawn national interest at Devil's Icebox, virtually in the back yard of University of Missouri cavers. Perhaps even more than that at Nevada's Devil's Hole, this battle shapes up as an exciting prototype of cave protection through political involvement based on scientific principles.

Here the problem was more insidious than any concrete monolith: uncontrolled urbanization and pollution by sewage rather than by dams. In the Lost River karstlands, slowing of sewage dispersal by the "flood control structures" would increase the pollution level drastically. Of the problems there, however, this would be one of the least. At Devil's Icebox, it is the greatest.

Previously, uncontrolled urbanization generally had been overlooked as a threat to caves and their unique values. Yet horrible examples had long been evident. Part of the nauseating pollution of Missouri's Sequiota Cave—once a beautiful tourist attraction—was traced directly to a urinal in the boys' washroom of the local elementary school. Once pleasantly developed as Hidden River Cave, Kentucky's Horse

Cave is a revolting example of what a community should not be allowed to do with sewage. Seeking to prevent the same fate for Devil's Icebox, cavers quickly mated the principles of modern science and conservation.

Prophetically located squarely between the University of Missouri's main Columbia campus and its Environmental Health Research Facilities, this unusual cave was long noted primarily for stirring triumphs of human doggedness against dismaying adversity. An obscure 1873 book proclaimed Devil's Icebox 7 miles long. Years ago Missouri cavers stopped snickering at that "wild" claim. "At the Icebox," a Missourian long ago declaimed with tongue in cheek, "J Harlen Bretz went downstream 260 feet and wrote six pages. It's sure lucky he didn't go upstream where it's low, muddy, and wet—and keeps going for almost four miles: 19,013 feet, if you want the exact figure. Counting side passages, we've got more than 30,000 feet on paper and may reach *eight* miles. I'm not sure it's worth it."

For a long time, most other Missouri cavers were inclined to agree. Deep in the cave are spacious corridors and chambers, with several areas of notable beauty. Most of the cave, however, is dark and dreary. Speleothems are few, and waist-deep wading common. Just 300 feet upstream from the sinkhole entrance the cavern roof dips within two feet of the stream. For almost a half mile, a boat is a necessity. Through "The Low Spot" itself, cavers must lean far back in their craft, pressing

Today virtually an open sewer, Kentucky's Horse Cave once was famous as Hidden River Cave. *Photo courtesy Miss Caroline Withers*

it downward into the water. Farther inside are horrible watercrawls, with as little as 8 inches of air space at low water. High water is at least 17 feet higher. For half a generation, only an inexplicable *compulsion to know* drove on the dogged explorers.

In 1964 Gene Hargrove unintentionally took over leadership here, in part because of cavers' typical luck. On his second trip into the Icebox, he and Tau Smither missed a crucial bypass. Pushing one watercrawl after another, they remained unconvinced for hundreds of feet of virgin streamway.

For two years Gene concentrated on the part of the cave he had been seeking when he missed the bypass. Then he returned to the site of his serendipitous error. Promptly it yielded a wildly branching new section inhabited by numerous domes, a splendid Formation Room, and water often up to his muddy, chill-purpled nose. Before Gene and his teams finished exploration and mapping, a new chapter in American speleology was underway.

At first surface and subsurface geologic and geographic studies proceeded at a routine pace. Topside, however, Jerry Vineyard and fellow speleologists found subdivisions threatening to spread over two sinkhole-dotted square miles which surround and feed water into the cave. None planned the comprehensive waste-treatment facilities needed to prevent pollution by sewage. Surprisingly little acrimony followed, however. Through the state's Water Pollution Board, the Missouri Geological Survey twice was able to discourage would-be developers without great difficulty. The entrance and part of the sinkhole plain were included in Rockbridge Memorial State Park, created in 1967. Everyone seemed to take pride in new recognition that Missouri no longer was a backward state (if, indeed, it had ever been).

Unsuspected problems lay ahead, however. In 1970 Geoff Middaugh sniffed unbelievingly as he strode along the main corridor. No question about it: raw sewage, and not just a little. In near-panic he and other speleologists revisited the 1,200-acre karstic plain. They found no unlawful subdivisions, but individual houses had begun to dot its expanse. Each had its own septic tank and nothing more. Water sampling revealed that cavers and the cave's natural biological values indeed were endangered by sewage bacteria.

Never before had American cavers planned a battle against this type of threat. No one was sure of the best approaches. Geoff began a detailed study of the overall implications of sewage pollution for cave management. To do this he found it necessary to create an entire new method of baseline measurements of the germs occurring naturally in a complex cave of this type. His findings were so impressive that the Missouri State Park Board and the state's School of Forestry co-sponsored publication of his report. The National Speleological Society quickly chartered a special conservation task force to monitor and seek correction of the problem. Among its concerned members were four of America's most famous speleologists. Newly returned from European studies, indefatigable Gene Hargrove was acclaimed its chairman.

This spreading teamwork promptly mobilized further impressive assistance. Civil engineering professor Darrell King soon joined the effort. Professor Dennis Sievers provided the facilities of the Animal Science Research Center's Waste Management Laboratory. Scientific documentation mushroomed. Missouri's departments of Wildlife, Forestry, and Fisheries began long-range studies of the cave's unexpected biological values. The Governor's environmental counsel became concerned, directing the state's Clean Water Commission to expand even these studies.

As data accumulated, it became clear that this particular sinkhole plain already was overpopulated, overpolluted. The National Speleological Society and cooperating organizations formally proposed a widely-hailed four-part plan: expansion of the state park through long-term land acquisition, agricultural zoning for the sinkhole plain, strict controls on household sewage, and interpretive development of outstanding sections of the cave.

The Boone County Planning and Zoning Commission, however, had other plans. Despite the accumulating data, on November 5, 1973, it convened a hearing on whether to zone the karstic plain for residential development. The Missouri cavers duly presented the problem and went back to work. Biological, public health, and other studies were expedited. Special interpretive tours were arranged for influential citizens and administrators. The official N.S.S. recommendations and their documentations were published as a special issue of the respected *Missouri Speleology*. Three weeks after the hearing the data were formally submitted to the Missouri State Park Board. Impressed, that board immediately voted to call for a delay in permanent zoning until it could analyze the report plus Boone County's proposed comprehensive plan. The Boone County court agreed to a delay—of just three weeks, and not in writing. Interestingly, its presiding judge (also a member of the County Planning and Zoning Commission) announced that he would not get involved because of a major financial interest in Pierpoint Enterprises, Inc., a corporation seeking rezoning for residential development.

Then on December 5, the Commission unexpectedly voted to zone the karst for development—supposedly because of some ambiguous statements made by a park board official who was trying to avoid committing himself prematurely.

Recovering from their initial dismay, the conservationists redoubled their efforts. The Sierra Club began to play an increasing role. Wide public support for protection of the Icebox broke into the open.

On December 18, 1973, the Missouri State Park Board concurred with the cavers' recommendations. Two days later, the state's Clean Water Commission adopted essentially the same position. Before formal notification reached the county court, however, it formally zoned the sinkhole plain for subdivisions, with unacceptable sewage regulations.

Missourians clearly retain their traditional character, however. The resulting brouhaha eclipsed the old cave war between the disputed ends of Onondaga Cave. On one hand, outspoken Gene Hargrove found himself the target of a libel suit by the controversial presiding judge. On the other, local cavers handed the state governor an extra-hot political potato one day after the county rezoning: a well-documented request that he direct the state's attorney-general to determine whether the county commission had acted illegally or ineffectively in meeting its legal obligations under the Missouri Clean Water Law. The attorney-general eventually ruled that part but not all of the county court's actions were unconstitutional, further complicating matters.

Neither the libel suit nor the inexplicable drowning of two of their number when their canoe overturned in the chill waters of the Icebox daunted the cavers. Snowballing caver involvement in Boone County politics gained still more momentum. Letter campaigns and other traditional means of mobilizing still further public opinion unearthed numerous enthusiastic volunteers. Additional state agencies found themselves deeply involved. In today's era of accountability, their support of the speleologists' documentation was virtually automatic. By mid-1974 state action had partially circumvented the rezoning. Some residential development seemingly proceeded uncontrolled, but other developers held back. To their credit, some had become concerned about environmental impact. Others may have feared prosecution under the state's Clean Water Act or the Consumer Protection Act. At last report, the Missouri attorney-general was considering requiring developers to warn prospective buyers about possible dangers of sinkhole collapse and other karst-related problems. With the aid of a $250,000 loan from the Nature Conservancy, long-term land purchase is underway for enlargement of the state park. Already several public-minded owners have proven willing to sell to the park and its allies rather than to developers. All but the smallest subdivisions seem effectively halted, and the next election or two may complete that part of the job.

Nevertheless, mid-1974 studies showed additional sewage pollution at three points beneath the cluster of houses that is Pierpoint. A second section of the cave has begun to smell like a sewage-treatment plant. The future of Devil's Icebox increasingly seems to depend on key land purchases.

No matter how successful, however, such emergency last-ditch defenses are hardly the ultimate key to the preservation of our cultural heritage. A far greater hope for the future may lie in another cave not many miles southwest of Nickajack Cave: Russell Cave.

Everyone seems to have become interested in Russell Cave about the same time. Late in 1954 Nashville cavers Tom Barr and Bert Denton investigated northern Alabama's cave country. With Dan Bloxsom and Fritz Whitesell they planned an extensive scouting trip. Studying topographic and geological maps they noted the disappearance of a creek into a Russell Cave.

No one seemed to know anything of this Russell Cave. The local geology seemed promising, however, so they sought Oscar Ridley, its genial owner. Ridley reckoned to speleobiologist Tom Barr that he'd seen lots of them salamanders back in thar.

Tom was beside himself. Under his avid goading, the caving quartet hurried through the hillside brush. Abruptly they halted on the rim of a complex sink 200 feet wide. Ahead loomed an immense black entrance, naturally divided like some fantastic double railroad tunnel.

With hardly a glance around, the little group plunged into a long, twisting stream corridor. Sometimes vast, sometimes narrow, it proved a biologist's paradise.

Tom Barr and his companions were not the first to seek out the secrets of Russell Cave. A year earlier, amateur archeologists Paul H. Brown and Charles H. Peacock marveled at Russell Cave's double arch. Where cavers soon would avidly seek the water passage, they gloried in a great dry grotto. Adjoining the gaping stream passage was a sheltered alcove 107 feet wide and 26 feet high. Dry and attractive, it faced east to the morning sun. High above the stream channel, the 270-foot shelter was naturally warmed in winter and air-cooled in summer. If ever a place existed where ancient man should have lived, this was it.

In November 1953 the amateur archeologists obtained permission for an exploratory trench. Three members of the Tennessee Archeological Society set to work with shovels and screen. From almost the first shovelful, their wildest dreams were exceeded. That group's entire Chattanooga chapter swarmed to the cavern. Arrowheads, spear points, pottery, awls, and other bone tools, shell ornaments, and munched fragments of animal bones came to light in enormous quantities. Even six feet down the ancient relics seemed to have no end. Bands of Indians must have lived here for uncounted centuries.

The excavators halted, awed by the magnificence of their find. This was big—too big for amateurs who might unwittingly do untold damage in their unskilled enthusiasm. Eagerly they sought out Matthew S. Stirling, successor to Gerald Fowke as director of the Bureau of American Ethnology. Dr. Stirling notified Carl F. Miller, noted archeologist of the Smithsonian Institution who was then working in nearby Tennessee. Miller visited the cave with Brown and his friends, and became as enthusiastic as they. Obtaining a generous National Geographic Society grant, Miller hired seven Alabama coal miners and began a full-scale dig May 1, 1956.

The first few trowelfuls from the surface properly yielded modern debris. Next should have come relics of the era of Indian traders—glass beads, copper bracelets, and the like. But none appeared. To Miller's trained eye, it was clear that all the Indian relics were older than A.D. 1650.

Broad squares of the cave floor were excavated by hand, trowelful by trowelful. Each solid object—rock, cracked bone, point, shell—was set aside and identified. Each scoop of dirt was placed in a wheelbarrow. Careful sifting of barrow loads yielded snail-shell beads and many another delicate relic. Gradually the cave floor was lowered in careful layers. Each told its own story of the sequence of vanished ways of life.

The uppermost feet of dirt revealed mostly the cracked bones and broken pottery of a myriad primitive meals. Small, rather crude arrowheads indicated that the occupants were of the Mississippian and Woodland periods, not more than three thousand years old. As the excavation

Archeologist Carl F. Miller painstakingly uncovers a human skeleton in Russell Cave. The white quartz spearpoint which caused death is clearly visible. *Photo copyright National Geographic Society*

deepened, pottery coarsened, then disappeared about five feet below the surface. Arrowheads had vanished a little higher, for the bow and arrow is a relatively modern invention.

As this wealth of cultural material was sorted, a clear-cut picture emerged of the life of Russell Cave's primitive man of over three thousand years ago. He was primarily a hunter employing stone clubs. He had a few stone-tipped darts and spears, often propelled by a spear thrower. Probably he reserved his valuable, tediously chipped spear points for his superlative foe, the bear, whose teeth he wore proudly. He had no tools for cultivation, but gathered nuts, berries, and wild grain in finely woven baskets. He fished with bone hooks. Stone-cracked mussels and crayfish varied his primitive diet of crudely roasted raccoon, opossum, deer, rabbit, snake, turtle, turkey, and bear. He cooked in fire pits in the cave floor or dropped hot rocks in food containers of skin. He built shelters to deflect the drip from the rocky ceiling. The women—and perhaps the men also wore necklaces, rings, and nose plugs of bone, shell, and stone. Probably he painted himself with hematite when he took the war trail. He had a few highly valued dogs, and gave them honored burial. As his garbage rose around the fire pits, he brought in basketfuls of sandy earth to bury it, thus thoughtfully preserving two and one half tons of artifacts for modern archeologists. Occasionally he had to flee

The joint Smithsonian Institution–National Geographic Society expedition in full swing in Russell Cave. *Photo copyright National Geographic Society*

great floods which rose up out of the stream passage into his comfortable grotto—and rotted away the perishable items in his refuse. And here he buried some of his dead, including a tribesman mortally wounded by a spear point of white quartz.

Life at Russell Cave was desperate and marginal. Nevertheless, it may well have been better than that of most of the world in that dim day when shepherd kings still ruled the children of Israel.

Deeper and deeper, farther and farther back in time the excavation continued. Below the eight-foot level, nobody had bothered to bring in sand to bury the garbage. The diggers had to toil in soft, sticky clay that accumulated naturally in the cave. It clung to their shoes and instruments in great sticky gobs. No longer could the workmen sift for artifacts. Instead, each trowelful had to be squeezed through the fingers.

Strange things began to appear in this ancient goo: torches of bear bone, hinged fishhooks like those of modern Eskimos. Nothing like these "new" artifacts had ever been found in the southeastern United States. Northern tribes moving southward along the Appalachian Mountains must have strongly influenced these Russell Cave people of many thousand years ago. A new chapter in man's knowledge of ancient America was written in sticky clay.

As the trench deepened, digging became more difficult. Slabs of rock which fell millennia ago blocked the way and had to be deftly dynamited. Artifacts became less and less plentiful—not surprising since the archaic population was pitifully small. Yet the charcoal of their fires continued down and down. During the first year's excavations, a fireplace at the depth of thirteen feet yielded a radiocarbon date of 6,000 B.C. Three feet lower, a curious spear point tantalizingly suggested the still earlier paleo-Indian culture that hunted ancient beasts with the superb Folsom spear points.

Broken bones, lumps of charcoal, and chips of stone tools continued still farther down. Little else came to light in the greater depths of this remarkable cave fill. Then, at 23 feet, a small pocket of charcoal unexpectedly turned up against the north wall of the chamber. Avidly, Miller collected it for radiocarbon dating. The intolerable wait was amply repaid. That insignificant-looking charcoal was 9,020 years old, give or take some 350 years.

That was the end. Thirty-two feet below the surface of the ancient deposit, the devoted miners reached the water level. Russell Cave had yielded a layer-by-layer story of daily life over a longer period than any other site in the Southeast.

The National Geographic Society soon purchased Oscar Ridley's farm and entrance to the elongate cave. Two excellent articles delighted readers of the *National Geographic Magazine*. On January 7, 1961, that society ceremoniously presented the deed to the Secretary of the Interior for its preservation as a national monument. Melville Bell Grosvenor formally expressed the hope that many Americans would "visit the cave . . . walk in its parklike surroundings and . . . come away with a better understanding of how mankind lived in prehistoric times."

Today that hope is being fulfilled through the agency of the National

The multicolored layers of Russell Cave form a ladder of time. The earliest charcoal was found ten feet deeper than the lowest sign. *National Geographic– National Parks photo, copyright National Geographic Society*

Park Service. Increasing crowds of visitors are poring over exhibits, gaping at the fog-spanned double entrance of the great cave. Inside they peer at multicolored layers of antiquity—gray-brown, red, ocher, brown, charcoal—exposed in the trench wall. With the aid of the expanding interpretive program of the National Park Service, thousands annually are coming to know the rich values our caverns harbor. If emphasis understandably is centered upon archeology at Russell Cave other spelean values are admirably interpreted at other cavern national parks and monuments, scant as is their number. Not every beer-can-strewing tourist at Russell Cave will perceive and champion our cause. Yet such programs are leading Americans increasingly closer to a re-understanding of man's cultural symbiosis with nature. In a proper balance between technology and preservation of our natural heritage lies our hope for true civilization, underground and aboveground alike.

Nickajack Cave, too, with its historic neighbor sites, could and should have been another marvelous living museum interpreting exceptional cultural values for all time. Instead, we had to settle for a belated "salvage" archeological study, quite likely all-too-halfhearted in view of Fowke's 1922 dictum. Cavers well remember that Fowke missed Russell Cave entirely.

But it was far too late to stop the rising dam even if by some freak its relics had been found to equal or surpass Russell Cave.

The coming decade will decide the fate of many of America's great caves. The battle is joined, and the picture is clear: all knowledge is incomplete, and there is so much yet to be learned.

Here the past can help the future as men increasingly open their minds. Let us vow that when I again rewrite this chapter for future editions, never again must I merely replace the name of Nickajack—or Onondaga—or Lost River—or any other great cave with that of the latest victim of an overnarrow concept of progress.

17

Beneath the Ice

The Story of Glacier Caves

Where had the pretty lakes gone?

Alongside the Martin River Glacier, the crisp stillness of the Alaska morning hardly hinted sudden geological catastrophe. Yet *something* had occurred during the midnight dusk. When the 1963 summer geology field group had last viewed the glacier alongside their tent camp, its surface had been mirrored with six circular ponds. Now only a dumb-bell-shaped depression marked the beds of two of them.

Gingerly, professors and students ventured onto the stagnant ice that formed this part of the glacier. At the bottom of the deepest basin gaped a broad natural shaft—a giant moulin—where the lake had drained.

An experienced glaciologist, Professor John Reid was thoroughly familiar with such pits. But this one was remarkable. Three hundred feet down its indigo throat, at the bottom of the glacier, the faraway roar of a vast subglacial torrent rumbled ominously, fearsomely punctuated by occasional explosions of sound.

A natural dam obviously had burst, Dr. Reid told his students: a dam that had caused the hidden river to flood caves and moulins to a height of more than 300 feet. Down the valley at the snout of the glacier, the excited group found a jet of water spouting 20 feet into the air—further impressive evidence of the prodigious forces at work in these unknown depths.

Then chance played an unexpected card. As some of the group

peered back down the jagged moulin, the muted rumble halted as if flicked off by an automatic switch.

"Run for high ground!" someone yelled. Water rose in the moulin at a rate of a million gallons per minute, spilling out into the empty basin, the churning cavern deep below again dammed by some unknowable barrier. Suddenly the missing glacier-top lakes were back.

For a while. Before the glacier's dyspepsia settled down, the disappearing lakes emptied and refilled three times, quite enough to demonstrate that this was no freak occurrence. Four others nearby underwent partial drainage also.

Nor was the Martin River Glacier unique. The Black Rapids Glacier has more sinks than many a karst. At the edge of Yakutat Bay's Malaspina Glacier in 1890, pioneer geologist Israel C. Russell came upon an entire river spurting high from the throbbing mouth of a deep-buried cave. Inhabitants of high valleys of the Alps have long feared flash floods from glacier caves. A torrent which destroyed the village of St. Gervais in 1892 burst from a cave 270 feet long and 100 feet in height and width. Few glaciers of the United States are located where cavern-born floods menace habitations. But this also causes their outbursts to be noticed only rarely, and studied even less often. Washington State's majestic Mount Rainier seems to be almost the only place in the coterminous United States where they have even been recorded.

As the speleologist views Mount Rainier, however, caving and climbing have been interwoven for more than a century. At about 5 P.M. on August 17, 1870, two intrepid but wind-chilled climbers first set foot atop its 14,410-foot summit. For eleven hours they had battled gales, crevasses, altitude sickness, air hunger, and exhaustion. Ill-equipped, marginally supplied with food, and facing a subzero night, General Hazard Stevens and P. B. Van Trump frankly considered their condition desperate:

The wind blew so violently that we were obliged to brace ourselves with our Alpine staffs and use great caution to guard against being swept off the ridge. We threw ourselves behind the pinnacles or into the cracks every seventy steps, for rest and shelter. . . . On every side of the mountain were deep gorges falling off precipitously thousands of feet, and from these the thunderous roar of avalanches would rise occasionally. . . . The wind was now a perfect tempest, and bitterly cold; smoke and mist were flying about the base of the mountain . . . thoroughly fatigued and chilled by the cold, bitter gale, we saw ourselves obliged to pass the night on the summit without shelter or food, except for our meagre lunch. It would have been impossible to descend the mountain before nightfall, and sure destruction to attempt it in darkness. . . . Never was a discovery more welcome! . . . a deep cavern, extending into and under the ice, and formed by the action of heat . . . its roof . . . a dome of brilliant green ice. . . . Forty feet within its mouth we built a wall of stones, inclosing a space 5 by 6 feet around a strong jet of steam and heat.

The heat at the orifice was too great to bear for more than an instant, but the steam wet us, the smell of sulfur was nauseating, and the cold was so severe that our clothes, saturated with the steam, froze stiff when turned away from the heated jet. The wind outside roared and whistled, but it did not much affect us, secure within our cavern. . . . However we passed a most miserable night, freezing on one side, and in a hot, steam-sulfur-bath on the other.

Over the years, many another climber thankfully followed the example of Stevens and Van Trump. Larger, sulfur-free caves in the main crater were especially esteemed, but when benighted atop Mount Rainier, any cave in a storm!

Before Lou and Jim Whittaker became curious, nobody seems to have thought much about these Summit Steam Caves as *caves*. In 1954, while preparing to climb Mt. Everest, these rugged mountaineers panted up and down Mt. Rainier so many times that it was getting boring. One not-so-fine day they decided to carry flashlights and look into the steam caves. The largest sloped downward—a long way downward—between compacted snow and the steep crater wall of rock and pumice. Deep

The author at an entrance of the Paradise Ice Caves. *Photo by Charles H. Anderson, Jr.*

in the steamy blackness, the cave continued. But flashlights don't cut well through steam, and pumice is about the world's most tiring stuff on which to climb (you seem to slide back four steps for every three you climb). The Whittakers developed headaches and knew that these might be the result of poisonous volcanic gases. They didn't push their luck.

Many Seattle climbers came to know about the Whittakers' exploration. Nothing resulted, however, until some unknown Boeing company personnel manager pushed some impersonal computer button. It transferred mountaineer Dick Mitchell from Seattle to Huntsville, where Northwest-style mountaineering just isn't. Dick switched to caving, with spectacular results, some mentioned earlier in this book. The limestone caves of Washington State being nasty little holes, however, he promptly switched back to mountains when he was transferred back home. Yet Dick was marked irrevocably by his years as a caver. What of the Whittakers' steam caves?

Dick and other members of the Cascade Grotto of the N.S.S. talked with the Whittakers. Interest jelled into a team: part cavers, part mountaineers. All were curious about what might be in those odd caves up there, more than 14,000 feet above our Seattle homes. Equipped for both mountaineering and caving, they began their ascent early on August 10, 1968. Twenty-four weary hours later, Dick led the way to a promising cave entrance on the south rim of the summit crater, one breathless step at a time in the thin air. Step, rest. Step. Rest. Step. Then down into the welcome darkness, sliding, rock-hopping downward where the Whittakers had led fourteen years earlier.

Some 200 feet down, the ice met the floor but sizable tunnels led left and right. Rightward they went, through an increasingly cloudy corridor, struggling over loose rocks, fighting sandy pumice. Around them swirled wraith-like vapor. Amid the nearby rocks steam vents hissed eerily.

The passage curved along the crater wall. Two hundred feet, three hundred, more. Step. Rest. Step. Rest.

An ice wall loomed ahead. "The passage continued upward, perhaps to another entrance," they later recorded. But photography and air testing had taken too much time and energy. Lacking altitude conditioning, the lowland cavers had reached their limit. One weary step at a time, they retraced their path to announce to the world of caving an unexpected challenge atop Mount Rainier.

Because of that lack of altitude conditioning and overloading, not much was accomplished on that first venture. None of its exhausted members even sought to return. The spring of 1970, however, saw a lively section of the Cascade Grotto turning repeatedly to unfamiliar habitats: the outside of high northwestern volcanos, in that odd stuff

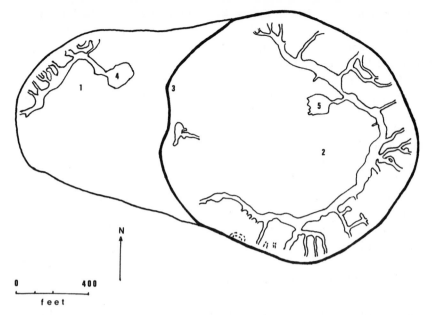

Summit craters of Mount Rainier and their geothermal caves. (1) West Crater and its caves; (2) Main Crater and its caves; (3) Columbia Crest, summit of Mount Rainier, elevation 14,410 feet; (4) Lake Room; (5) large chamber at deepest level of main cave. Data courtesy Dr. Eugene Kiver

called daylight, preparing for an attempt scheduled early in July. For one reason or another, all but three eventually dropped out, but their enthusiasm was contagious.

In late June, Lou Whittaker camped on top of the mountain while leading an expedition seminar. Assisting was a Tacoma fireman, Lee Nelson. Nelson must have had his eye on the caves, for he packed along two Chemox oxygen rebreathing units, "just in case." Not to mention other equipment Lou and Jim had lacked sixteen years earlier.

Donning the oxygen gear, the pair plodded down the steep pumice slope, quickly out of sight of their chattering students. In an hour they pushed upward where the Cascade Grotto party had turned back. Only a few minutes later they popped out into daylight on the far wall of the crater, half a mile from their admiring group. Discarding their picturesque but cumbersome oxygen tanks, they turned around to explore a minor maze of side corridors, then installed mountaineers' wands as a reliable guide through the steam-clouded main corridor. The entire seminar party promptly traversed the crater in a brand-new way.

"It was eerie and lonely in there . . . very dark," the explorers told the Tacoma *News-Tribune* upon their descent. Not so, a week later when our own team—Chuck Coughlin, Ron Pflum, and Greg Thomson— now well conditioned to the high altitude, began mapping and studying these curious caves. Nor in the weeks, months, and years which followed. Very dark, certainly; totally dark, in fact. Eerie, perhaps, with hissing fumaroles and swirling steam. But hardly lonely. For those with limited time and resources, altitude conditioning had opened a new door.

With more than a mile of passage soon mapped, the unique nature of these geothermal caves became evident. Two hundred feet down the north, south, and east walls of the mountaintop crater, a curved line of steam vents has melted a large, near-level "perimeter passage" more than 3,000 feet long and as much as 35 feet in diameter. Innumerable side corridors extend up the crater wall, their number, size, and openings changing with each year's climate. Downward another 100 feet, perhaps halfway toward the center of the crater, one leads to a chamber 120 feet in diameter and 70 feet high.

Elsewhere in the summit craters are other caves of the same type: much smaller, with a total of only about 1,000 feet mapped. Yet, in one of these smaller caves in the older west crater, Eugene Kiver—a limestone caver long before he became professor of geology at Eastern Washington State College—found Mount Rainier's subterranean crater lake. Perhaps calling it a lake is a bit of an overstatement, but at more than 14,000 feet, a pond of 33-degree water even 30 feet wide and 130 feet long is still worth recording.

These Summit Steam Caves have become a key point in the Volcano Watch, created to warn the Pacific Northwest should any dangerous recurrence of volcanic activity begin to flare. But the big news about Mount Rainier caving today is halfway down its southeastern slopes, where the Paradise Glacier and the Paradise Ice Cave or Caves keep giving fits to the Board on Geographic Names.

American cavers and speleologists were slow to enter these and other glacier caves. Flint Ridge's Fred Dickey still remains unconvinced. Some years ago near Juneau, Alaska, Fred came upon the awesome mouth of a large cave in the Mendenhall Glacier. As he gawked, an earsplitting crash shook the ground. "Like an iceberg calving," he recalls: a multi-ton flake of ice collapsed into a hidden underground lake. Seconds later, "a terrific surge of water . . . looked as if a giant were throwing a bucket of water out the entrance!"

Yet the Paradise Ice Caves have been a major tourist attraction of Mount Rainier National Park during much of the twentieth century. Sections suitable for public admiration sometimes cannot be freed of snow for several consecutive years. Nevertheless, tens of thousands of northwesterners watch for newspaper announcements that at least one of their entrances is now accessible.

The early 1920s seem to have been a period when the snow was especially uncooperative. In July 1925 Park Naturalist Floyd Schmoe became impatient. Vainly he dug toward the caves with an ice ax, later explaining: ". . . only rarely, following winters of extremely light snow-fall, are the openings sufficiently exposed to permit safe entrance into their fastnesses." A few weeks later, however, he succeeded, advertising their glories all over America in *Nature Magazine:*

Stretching away for a hundred yards was a low tunnel of crystal ice some twenty feet high and sixty feet wide arching over the mad, white water of the embryo river. Beyond, a large domed room appeared through the roof of which a ray of sunshine fell from a crevasse far above. Into this chamber other tunnels opened, and beyond, up dark passageways, could be seen ice bridges, waterfalls, and light wells. The roof above . . . was vaulted, like the nave of some great cathedral and everywhere was diffused a wonderful light. . . . The whole area was filled—saturated—with the richest blues and greens imaginable and along the surface of the white water hung a haze of old rose and gold as though beams from a mountain sunset had been im-prisoned there. Looking above, there were areas of pearly white with every-where streaks of clear ice painted every imaginable shade of blue and green by the sunlight beyond.

As naturalist Schmoe wrote, however, the snows were lessening. By the end of the Roaring Twenties, the Mount Rainier Guide Service impres-sively conducted guided parties to these mysterious caverns, barely two miles from Paradise Inn. From head to toe they ceremoniously outfitted greenhorns by the hundred: floppy hats, mackinaw jackets, baggy shirts, shoes and socks, goggles and gloves, and archaic alpenstocks. At some forgotten but daring date, ladies' long skirts gave way to the traditional male garb: lumbermen's canvas trousers known as "tin pants." Then everybody could squeeze together into human toboggans, sliding down the steep snow slopes on the way home. Whooping and screaming de-lightedly, the tourists loved it. Forgotten photographs in innumerable family albums nostalgically recall those happy, unsophisticated days.

Yet not until the 1960s did speleology come to the Paradise Ice Caves. By that time, matters were about as mixed up as could be. There was only one of them. It wasn't the same cave or caves which bore the same name in 1925, much less a generation still earlier. Definitely it wasn't in the Paradise Glacier. For a time some of us began to doubt that the Paradise Glacier still existed.

Early in the century, matters were comparatively simple. The Paradise Glacier spread a wide sheet of ice on the slopes of Mount Rainier high above Paradise Valley. Above Stevens Canyon (named for spelunker Hazard Stevens, of course) northeast of Paradise Valley it cascaded down a steep rocky wall into a basin it once had filled to the brim. Gullies in this basin were and are the sources of the Paradise River and of Stevens Creek. Near its center a bulge of ice marked a minor division of the ice mass. The part of the glacier north and east of the bulge drained into Stevens Creek, and some called it the Stevens lobe of the

Paradise Glacier. Others called it the Stevens Glacier. Nobody got very uptight about either name. Around 1908 mountaineers visited some caves in the area of Stevens Creek, and a few photos appeared in obscure writings. But the Paradise Ice Caves were at the head of the Paradise River. Everybody knew that.

Despite the temporarily heavy snows of which Floyd Schmoe wrote, the Pacific Northwest was warming. The glaciers retreated; the Paradise Glacier more than most. Instead of a broad, gently sloping icefield spilling into the basin, its main body shrank to the point where it was little more than a compacted snowfield. Its once-magnificent lower segment wasted away into several independent little glaciers, each in its own basin or perched precariously in a protected canyon-rim gully. The part of the basin where the Paradise Ice Caves long disgorged the Paradise River lost its much-admired hundred-foot wall of ice. About a dozen years ago, each summer saw it become nothing but bare rock and blowing pumice. The caves themselves had shriveled and crumbled away considerably earlier, probably during World War II when nobody was likely to notice.

Somewhere around 1946, hikers and mountaineers started going to some "new" Paradise Ice Caves up at the head of Stevens Creek. Nobody knew what glacier they belonged to. That hunk of ice clearly wasn't the Stevens Glacier any longer, if it ever had been. Several of the little residual glaciers drained into Stevens Creek, and who was to say that the particular one that held the "new" Paradise Ice Caves ought to be singled out? Besides, it was only possible to go a short distance into the new attraction: far enough to admire the beautiful blue light filtering through the body of the glacier and illuminating the cave in a most magnificent show, but that was all. Beyond, a roaring torrent filled the cave from wall to wall.

Memory of the "old" system faded. Somehow, an odd misconception sprang up: glacier caves were transitory little structures that melted back along the stream each summer, then filled up each winter as a result of glacier flow. The Cascade Grotto (including yours truly) had a look in August 1961, nodded wisely, and didn't bother returning for six years. I didn't even bother to go along the second time.

It was a mistake. A new era had begun. The section we had visited in 1961 was gone entirely. Alongside the roaring creek the cave was wide enough to penetrate a hundred yards, to an echoing chamber fifty feet in diameter. A side passage here presented problems: quicksand and tilting boulders yards in diameter. Immediately it was dubbed Suicide Passage. The explorers retreated.

But just inside the gaping entrance, an alluring dry corridor wound hundreds of feet to the west side of the glacier. Along the way were all the beauties which had delighted Floyd Schmoe and the tourists of bygone years. In sheltered alcoves cavers' headlights sparkled back a millionfold from giant frost crystals. A magnificently glassy ten-foot ice stalagmite and several lesser columns reached toward fluted moulins high overhead. From glistening white ice and from odd clear bodies of glare ice sprouted frozen ribbons, entrancing rows of helictites, occasional transparent stalactites. The greatest glories of limestone caves seemed transmuted into an incredible natural palace of scalloped ice in this miraculous Pillar Passage.

The challenges of limestone caves were present, too. Backed by the Cascade Grotto of the N.S.S., Charley Anderson formally proposed a scientific exploration of the entire cave. At first, the local staff of the National Park Service was dubious. Too dangerous, perhaps, and just not worth it. Chief Park Naturalist Norman A. Bishop, however, was impressed by the concept of scientific exploration. He authorized the study with the tacit warning that permission would be revoked if the scientific exploration turned out to be less than scientific. He needn't have worried. An entire new subdivision of speleology stemmed from his decision.

Just as if they had been in limestone, the Cascade grottoites forced themselves into squeezeways, dove into water crawls. Just as in limestone, most pinched out. One on the left wall, hardly 200 feet from the junction, was different. Back toward the glacier snout it curved, then westward, paralleling the lower margin of the glacier. Soon the enthralled

The steamy atmosphere of the geothermal caves in Mount Rainier's summit craters makes photography difficult. *Photo by Curt Black, courtesy Dr. Eugene Kiver*

explorers broke out into an awesome lightless natural auditorium 250 feet long and almost 100 feet wide. Beyond, four separate corridors led upward to daylight along the basin wall. Between this Big Room and the Pillar Passage lay a complex of corridors and small domed rooms. Some of these, too, led to new entrances along the basin wall. This entire part of the glacier seemed hollow. The first return trip alone yielded ¾ mile of mapped cave.

With this writer and other members of the Cascade Grotto, Anderson returned time and time again, mapping, photographing, studying. Soon we all recognized a myriad obscure glacier features, here exposed in ready view. Sandwich-like layering of various years' snow was prominent. Vertical layering resulting from compaction and glacier flow could be traced along the axis of the glacier. Some was found to curve away from the axis to special zones of accumulation. Here and there large rocks and earthy deposits within the ice told their own stories. The bottoms of a few small crevasses stimulated excited discussions of chimneying inside glaciers (so far, no one has found one that seems just right).

Almost at once, we recognized a unique caving hazard: flakefall. Often weighing many tons, flakes are swordlike slabs peeling off from innards of glaciers as a result of natural release of intraglacial pressures. Occasionally they split off in a single long convulsive shudder. Usually the process is gradual, with the flake's narrow attachment increasingly thinned by the cave's atmosphere. Glacier cavers soon learned to scan far down each passage, automatically spotting inconspicuous blades dozens of feet long yet so tenuously attached that a chance touch could collapse tons of ice. A decade later, our most experienced glaciospeleologists still shudder occasionally when they turn around and see what they have missed.

Jumbo-grade flakefall promptly impressed the new-style cavers. It also ended a minor semantic dispute: is (are) the Paradise Ice Caves singular or plural? Oregon Caves is embarassingly singular, and quite a few others from which I have deleted the offending letter *s* in this and other books. Although clearly multiple in the dim past, the Paradise Ice Caves was as singular as Oregon Cave when these studies began. Through flakefall, nature settled that particular problem. As the glacier front shrank backward and ablation further enlarged its internal corridors, the passage connecting the Big Room and the Pillar Passage broke down into an ever-changing complex of collapsing, splintering flakes. Soon that entire part of the glacier was no more.

Prominent landmarks appeared: impressive glacier snout entrances opening into four or five newly-separate caves. The Paradise Ice Caves were plural again.

Mapping and remapping the changing caves took time and effort but enthusiasm remained high. An old friend of Don Cournoyer, Charley Anderson had always been fascinated by Don's record of 209 trips into Breathing Cave. After a few dozen trips into this unrolling system, Charley began to wonder if he could beat that record.

The winter of 1967–1968 brought its expected cold but surprisingly little snow. The new entrances at the snout never closed. Since Stevens Creek is fed largely by seasonal meltwater, cold-weather explorers found its flow stilled to a mere subglacial trickle. Exploration and mapping hastened up the main stream passage, now almost dry. Even Suicide Passage yielded, although cavers still treat its loose, car-sized boulders with particular respect. Winter exploration seemed the key to glacio-speleology.

But outside the stilled caves winter still reigned on the blizzard-swept middle slopes of Mount Rainier. All the explorers were equipped for winter survival and repeatedly undertook additional training in its techniques. Moreover, it was such a mild winter. Sometimes they had to resort to navigation by compass, with visibility reduced to two or three feet, but that was expected and they were prepared. About half the teams tried out the current weather conditions, decided it wasn't worth the effort, and turned back, but the rest had no real difficulty reaching the caves. Once inside, of course, they could wait out the worst storm in something close to comfort.

It was too smooth, too easy. A weekend trip in February 1968 began in extra-balmy weather and the team left part of their survival pack in the car. Halfway back, they learned about Mt. Rainier blizzards. When windchill kept them from further progress, they dug a "snow cave" for shelter and waited—correctly. Mountain Rescue teams soon were near at hand. Visibility, unfortunately was about five feet. Hearing was slightly less. At least one rescue team passed unknowingly within yards of the snow cave and its marker.

Dawn brought a slackening of the blizzard. Dave Mischke was able to stagger down to the ranger station to summon aid for Charley Anderson and his wife Edith. When it arrived minutes later, Edith's down jacket and jeans had not been enough. Hypothermia had claimed another life on Mount Rainier.

Inevitably the reaction lay heavy on all of us. Yet it did not kill the project. The summer of 1969 yielded little, but in November, routine investigation of an ordinary side passage 2,000 feet up the stream passage quickly discovered the Rockslide Room, larger than the Big Room itself. Atop its treacherous slopes lay another entrance, surprisingly close to the upper end of the little basin. Then the snows closed in.

Nineteen seventy was the Centennial Year for Mount Rainier and for its caving, courtesy of Stevens and Van Trump. As part of the celebration, the grotto attempted a special breakthrough on October 3. It was also Charley Anderson's ninety-ninth trip into the system (his hundredth was in dense underground fog next day), and perhaps his most disappointing. Over and over again the cavers found themselves knee-deep in ice water before they even reached the Rockslide Room junction. Charley and Van York both had frostbitten their feet years earlier. Each time they were forced into deep water, increasing pain tormented them. Grim reality pointed out that they would have to face the same agony if the breakthrough was unsuccessful and they had to return almost the entire length of the subglacial stream. Both gave up while still in mapped passage. Ron Pflum and Bill Zarwell grimly splashed on, returning surprisingly soon. After the first known traverse beneath the entire length of a glacier, after nearly a mile under the ice fighting 34-degree rapids, they had burst out of a huge entrance just 200 feet further on.

With more than two miles mapped, it appeared that the Paradise Ice Caves system was going to total about three miles of cave inside this mile-long glacier. Especially from some temporarily northwesternized eastern cavers, however, Anderson and his teammates were the recipients of distinctly barbed comments. Some gung-ho limestone cavers scoffed outright, asserting that this was inexcusable exaggeration, that all of this didn't amount to much. It wasn't in limestone, which was the only stuff that grew caves worth considering. It wasn't even all one cave now. And the Big Room and Pillar Passage were fast over-enlarging and melting away.

At least in public, Charley and his co-workers merely smiled. With soft answers they turned away the bitterest taunts. The main cave indeed had more than two miles of passage. Moreover, as the lower sections collapsed and melted away, once-impenetrable slits off the stream passage were becoming full-scale side passages, some with their own brand-new moulins freshly grown out of intraglacial bodies of glare ice. The little ice caves' glacier might be disappearing rapidly, and the name Paradise Ice Caves might have to be moved again—maybe all the way up to the summit craters—but, for the moment, it was a darn important cave.

As if annoyed by these "limeys"—as some defiant glacier cavers began calling skeptical limestone buffs—nature again took a hand. The 1970 breakthrough had been just in time. Somewhat in the pattern of the "old" Paradise Glacier, record snowfalls began to accumulate and compact. Most of the moulins were plugged. Many recently-spacious chambers and corridors were greatly narrowed or largely filled. Glassy speleothems became scant.

For a time the new snows dismayed the glaciospeleologists. Only through hazardous basin-wall entrances on the far side of the glacier could any part of the system be entered. But much scientific observation and documentation obviously were needed. Though morale sagged, the dogged team plugged ahead.

For the first time, the annual convention of the National Speleological Society was to be in the Pacific Northwest in mid-August 1972. The National Park Service had authorized a special post-convention field trip to the caves. It became obvious, however, that no large group, unskilled in glacier travel and glacier caves, could safely follow the only open route: across the glacier and down between the rock and ice. Gloomily the team contemplated still further criticism if the trip had to be canceled.

On July 1, 1972, a group returned from the upper basin entrances by way of the Paradise River, "mostly for a change of scenery." About a mile upstream from Sluiskin Falls, where the Paradise River tumbles out of the glacier basin, they unexpectedly encountered a large hole in the snow-filled gulch: Surprise Entrance. With an hour to kill, they casually decided to investigate. Upstream they sloshed, expecting the cave to end at any moment. A thousand feet upstream under the snow, they ran out of time, not cave.

This was nothing but a snow cave, however. At first only moderately interested, the cavers returned on each of the next several weekends. Each revealed more and more snow cave, its patterns surprisingly like those in ice. Some odd relics turned up, too: a rotting wood ladder, an old DANGER sign, its paint flaked almost to illegibility. Although in snow rather than in ice, the original Paradise Ice Caves was back! Soon a mile of snow cave was added to the map, not counting a 2,000-foot lower section that lasted just long enough for the post-convention field trip.

But it is hard for anyone to get really excited about even a mile of snow cave. Anderson and his colleagues continued to undergo merciless ribbing. Deadpan "limeys" spoke half-facetiously about northwesterners growing caves with carbide lamps or smuggling flame throwers into the national park. Other skeptics held that the whole thing was merely an exaggerated farce. Even the glaciospeleologists didn't expect much of these new finds. Less than a month after the N.S.S. convention, the last virgin lead became lower and lower. Finally it ended, pinching out against a rock wall.

Not very surprised, the team took a short rest. "Then Mark Vining decided to nose around," Charley now reminisces happily. "I was right behind him. Soon he found a low side passage maybe 300 feet back toward the entrance. When we crawled through, it opened into a real honeycomb. We had to leave 19 passages unchecked. They went up and down, interconnecting all over the ridge."

Returning as soon as possible, the elated explorers found the Honeycomb Area expanding atop the low basin-center ridge that separates Paradise River from Stevens Creek. Soon they announced that a connection from the resurrected Paradise River cave to Stevens Creek appeared possible. Maybe even likely.

To the scoffers, this was manna from heaven. Glacier caves—and snow caves, too—follow stream courses. Thus they couldn't possibly go up over and across a divide between two drainage basins: "More of Charley's big talk!" Charley just grinned again, then went back to work.

While the rest of American caving enjoyed the fruits of its labor, the Oregon Grotto of the National Speleological Society had worked its heart out making a success of the 1972 convention. My wife and I decided to throw a thank-you party for them. On September 30, 1972, cavers from all over the Pacific Northwest converged on our home. Charley and his team were conspicuous by their absence, mapping the dripping black maze of the Honeycomb Area. At about 6:30 P.M., just when our celebration was fast becoming enthusiastic, a weird, almost inaudible vibration increasingly impinged on Charley's awareness. Not for several body-lengths did comprehension click.

"Do you hear a roar?" he breathlessly queried Chris Miller, crawling close behind.

Miller halted, listened, pounded Charley on the back. Muted but unmistakable sounds revealed a torrent not far away. At double speed they slithered onward, fifty feet or a trifle more. Below a black little hole about the size of a coathanger roared Stevens Creek, 25 feet wide.

"Yippeee!" For the first time, Anderson's long-suppressed feelings burst out in pure joy. "Now they can't say I'm exaggerating!"

Nor was he. The new enthusiasm quickly spawned an International Glaciospeleological Survey with—of course—Charley Anderson as director. Various members of the Cascade Grotto of the National Speleological Society and British Columbia Speleo-Research served in initial supporting roles. Besides the Paradise Ice Cave project, which the new survey soon formally inherited with the blessing of all concerned, other studies were soon underway on both sides of the border.

Nor were the glacier cavers exaggerating when they reported that in 1973 they had mapped 60 passages in the ever-enlarging system. Increasingly teaming with young Mark Vining and two British Columbia cavers —Gerrit Van der Laan and Clarence Hronek (father of Vancouver Island caving)—Anderson ran wild. In the course of a two-week mapping expedition, his total number of trips into the cave passed Don Cournoyer's record.

Soon it was evident that the now-reunited cave contained the headwaters of both Stevens Creek and the Paradise River, plus miles of passages between and around them. Everywhere they turned, new networks begged to be explored and mapped. Some were in ice, some in

Water freezes into fantastic stalactites and stalagmites of ice in the Paradise Ice Caves. Note also the banding and other features of the glacier itself.

fresh snow, some in every intermediate stage of compaction. By the end of 1973, the map of Paradise Ice Cave showed more than ten miles of passage.

In 1974, however, further progress seemed hanging in the balance. As late as August, every basin-edge orifice appeared to be still snowed shut. The snout entrances were deeply buried.

The glaciospeleologists turned their recollections to a little warm spring on the basin wall. Barely warm enough to support a distinct plant ecology, its warmth proved sufficient to have opened a minuscule, hidden orifice. Although two tortuous, soppy hours from any corridor walled with ice, it sufficed. As I write, Charley Anderson reports that the cave map has passed fifteen miles, and he himself is well along toward his 300th trip. No end seems in sight for either—unless the globe's weather

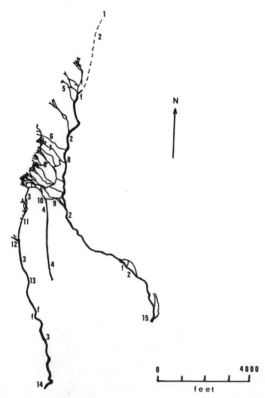

The Paradise Ice Cave system as of December, 1973. (1) Upper Entrance as of 1970; (2) stream passage of Stevens Creek; (3) stream passage of Paradise River; (4) stream passage of unnamed tributary of Paradise River; (5) Rockslide Room; (6) Suicide Passage; (7) Pillar Passage; (8) approximate lower edge of glacier in 1970; (9) Link Passage; (10) old sign; (11) Surprise Entrance; (12) site of possible geothermal activity; (13) Avalanche Alley; (14) resurgence of Paradise River, just upstream from Sluiskin Falls; (15) resurgence of Stevens Creek; (f) major waterfalls. Simplified from surveys of International Glaciospeleological Survey

pattern again shifts and the snows again slacken and fade each summer, the caves resuming their course toward extinction, the name Paradise Ice Caves migrating up the long, tedious slopes to the summit craters. . . .

At Mount Rainier, an entire subdivision of speleology thus was born, complete with its own principles, its own techniques and challenges. The growing files of the International Glaciospeleological Survey reveal that its glacier caves are far from unique. Larger examples probably are present in Alaska and western Canada. Several major rivers burrow beneath the entire ten-mile width of the Malaspina Glacier, where Israel Russell spotted glacier caves almost a century ago. Glaciospeleologists' minds boggle at the potential of Antarctica, home of glacier sinks that dwarf all others presently known.

Yet not even the most expert northwestern glaciospeleologists are rushing to investigate every newly-noted glacier cave. Even in the comparatively friendly Paradise Ice Caves, Charley has seen Stevens Creek suddenly and inexplicably rise more than a foot within seconds, the new level remaining constant for many hours before the stream drifted back to its previous flow.

This is a brand-new sport and science, still imperfectly known. The Malaspina Glacier's Fountain Spring reminds us that we have much to learn before confronting the caves of that vast glacier. "A death trap," John Bridge called it after a swift but effective study. For the moment, he is correct. So are the Martin River Glacier and a thousand others. Glaciers come in different forms. Each has its own characteristics, its own distinct personality. Already we recognize several types of glacier caves, as different as the glaciers which surround and surmount them. Slowly and cautiously we are proceeding from one type of glacier to another, from one kind of cave to others, learning the secrets of each before going on to the next.

Already such caution has paid off handsomely for Gene Kiver in the sulfurous geothermal caves of Mt. Baker, first trodden by Gene's team in August 1974. "Trodden" is perhaps not entirely precise for parts of this daunting cave system, where stays of twenty minutes without gas masks cause dizziness and severe eye and lung irritation. Gene's own words sum up some of the novel problems here:

[In] small passages with restricted air circulation in the lower part of some of the west passages . . . Oxygen is extremely low. . . . Gas masks were useless in these passages. . . . A boiling pool, a very large fumarole, falling ice blocks and slabs (flakes) and sediment that turns into a quicksand when disturbed are some of the other hazards. . . . We reached a point where daylight and the large roaring fumarole were visible, but we were unable to emerge from [this] entrance because of the treacherous quicksand floor. Moving one's foot is sometimes all that is needed to cause the sediment to lose its coherency and a shoe would begin sinking out of sight. We attempted to move along the edge of the passage with shoes removed to keep them dry but turned back when the depth of the quicksand exceeded the length of the ice ax we were using as a probe. Waders and a rope around the leader are recommended for future attempts.

Some of the world's most fascinating caves thus lie beneath glaciers, and its longest may be here rather than in limestone. No glacier caver needs extra incentive to take extra care to survive long enough to find out.

18

The Molten Sewers

The Story of Lava Tube Caverns

This jagged, glisteningly black pit ought not to exist. As the "All clear!" call floated up from young Luurt Nieuwenhuis, I eagerly wrapped a nylon rope around my coverall-clad body. An extra-hard tug by three of us tested the anchor point: all secure. Sliding the rope freely as I backed toward the yawning orifice, I resettled my hard hat and newly filled carbide headlamp. Leaning back on the rope, I waddled wide-legged onto the jagged upper wall of the broad vertical chute.

My balance and confidence established, I could begin the ever-lengthening sliding bounces that delight the rappeller. Ten feet down the shiny black wall, twenty, thirty . . .

Abruptly the wall receded. Lost in the pleasure of the rappel and the unexpectedly broad chamber coming in view, I was paying insufficient attention. Suddenly I was in mid-air, ten feet from the floor with my feet dangerously high. All at once I understood how easy it would be to fall out of my unbelayed body rappel.

Tensely I called to Luurt in the shadows ahead. Sizing up the situation on the run, he calmly talked away my momentary panic as I had done for so many others:

"O.K. Hang on with your left hand. Slide some rope through with your right. Now a coupla feet more. That's it. Now a little more. Now you're O.K. Slide on down."

Unwrapping with a heartfelt sigh of relief, I looked around with delighted amazement. This broad, level-floored "ballroom" was more than 50 feet across. From behind our pit it stretched away into a dark tunnel that called to us irresistibly. Along its margins were superb natural gutters. In one was a phenomenon at which certain experts had scoffed: boils of reddish lava extruded into the cavern.

This fine chamber and its spectacular entry pit would be admired by cavers even in the mammoth limestone caverns of the central United States. Here, formed in lava instead of in limestone, it was the extraordinary climax of an exceptional cave.

Almost where I could have touched it during my rappel, a huge pile of rocks rose up nearly to the overhang. It bore silent witness to a near-tragedy which triggered the naming of this Dynamited Cave. A few years earlier three venturesome boys had slid down into the pit on a rope not much larger than a clothesline. To their inexperienced surprise, they were unable to climb it hand over hand when they were ready to leave. More resourceful than many another trapped novice, they spent many hours building the rock pile. From its summit, the strongest miraculously was able to scale the jagged wall of the pit and summoned aid.

A helpful resident of the Columbia River Gorge rim country wrote us of the new discovery and a team of caver friends from nearby Portland described it in glowing terms. Before we could come from Seattle, however, they sorrowfully mailed us a self-explanatory clipping: RECENTLY DISCOVERED LAVA TUBE DYNAMITED BY VANDALS. As it turned out, the "vandals" were a self-appointed safety expert: not the first time that pomposity has led to jackassery or worse.

The dynamiter was thorough. Only after six hours of backbreaking work on a magnificent June morning of 1961 did we near a breakthrough. Crowbars, shovels, and tackle laid aside, we grubbed with our hands in the cool northwest sunshine, excitedly ignoring dislodged dust blowing strongly into our sweaty faces.

Laboriously but with extreme care, we pried yet another slab of compact gray lava out of the huge rock pile, then peered hopefully into its bed. After tons of rock, we could see the nothingness we sought, seemingly about a yard away.

The hole was tiny. Less than a foot in diameter, it was remarkably jagged. I prodded with a wrecking bar: the rocks seemed solidly wedged. Flashlight in hand, I slithered inward with my right arm forward, left arm pressed tightly to my side.

Not as bad as I had feared . . . As my whole chest entered the dimness my flashlight beam picked out the form of a spacious cavity 30 feet wide and half as high. It sloped down into shadowy vagueness. I wormed ahead, my chin passed a rocky ledge, and I could see that the sloping floor was a mere yard below.

I opened my mouth to call the glad tidings to the fatigued cavers behind me, then sputtered madly, my mouth full of tiny midges. Spitting profusely, I snaked unannounced back to daylight.

Telling the ten other eager northwestern cavers what lay ahead, I wriggled back into the hole, feet first, belly down. Soon my feet stuck out into mid-air. Another wiggle and I could bend my trunk; my toes arched down to a rubbly foothold. One good push with my knees and I was free. No time for a breath: those infernal midges weren't really tasty. Clapping my helmet on my head, I fled a few yards down the rocks, momentarily careless of little slides.

From within, one side of the rocky plug looked precarious but the center seemed well wedged. "Next!" I shouted. Accompanied by struggling noises of appropriate intensity, ten pairs of feet preceded their owners through the rocky little corkscrew. Only the petite wife of bearded Tom Hatchett showed proper nonchalance. She alone was built for holes of barely human diameter.

Happily reassembled on the slope, we skittered downward in the broad chamber. At the bottom a narrow opening was bridged by a natural floor three inches thick. To our knowledgeable eyes, it was the solidified surface of a long-gone river of molten lava which once raced through this lightless cavern with express-train speed.

We ducked beneath the span into a small keyhole-shaped opening. Beyond, the passage expanded to a comfortable ten-foot diameter. Alternately overhead were short lengths of flat ceiling and small "upper levels"—small tube segments that seemed to start nowhere and ended blindly a few feet beyond. Here and there piles of fallen rock marred the symmetry of the widening passage. In many ways it resembled the large solution caverns of the level limestones of the eastern United States. But here were other features which could never occur in limestone. Some were uncommon even in "ordinary" lava tube caverns. Four hundred feet from the entrance, we gazed upward at layers of thin-bedded lava visible edge-on in the ceiling. In a straighter section of the sinuous tube, a delicate arch spanned the upper part of the passage like an outflung wisp of lava. Beyond, an overhanging ledge dropped precipitously to the rocky floor of a cavern three times the size of the entrance corridor. Along the east wall, a dozen thick lava coatings—each heavily glazed by hot magmatic gases as if blazed in by a blowtorch—indicated that there had been a succession of flows through this abandoned conduit.

Along that wall, ledges halved the 25-foot descent. We anchored a 15-foot ladder and hastened down. Passing the rotting bones of an incautious bear, I scurried onward to watch the others descend. Looking back, I saw another large lava tube beneath our entry coridor. What processes could produce such a pattern?

Choosing first the larger, breakdown-strewn downslope passage, we progressed farther and farther from the entrance. Endless piles of col-

Clinkery lavafall in Dynamited Cave.

lapsed rock brought complaints from our muscles. Beyond were level surfaces of sharp-rippled granular lava, undisturbed since the moment of congealing.

Along the walls were long gouges and ledgelike deposits left behind by earlier flows. Locally, shiny blue-black glaze had slumped along the wall like newly pulled toffee. Here and there, small tubular stalactites had formed from the dripping glaze. To our delight, the breakdown decreased as we advanced. Fine natural gutters channeled our procession.

As we strode onward, the floor dropped out of the narrowing passage into a rounded ripple-floored room 30 feet in diameter. We skittered down a 10-foot "waterfall" of clinkery red lava. At the far end of the pleasant chamber was the fabulous pit of which we had heard—a 40-foot vertical bore which had no place in the classical theories of lava tube development. Nevertheless, there it was.

Concentrating our lights, we studied our surroundings. Beyond the wide, black maw of the pit, the main cavern level continued around a bend. The crossing would not be easy. A few feet overhead, a third tube level extended back toward the entrance—how far we could not tell. Nowhere had it intersected the tubes through which we had reached this bizarre pit.

A strong breeze swept past us and into the curving tube ahead. We looked at each other speculatively. A big room lay below the pit, promising still greater wonders. We were tired, but far from exhausted. We had plenty of ropes and ladders. Which route should we try?

The "impossible" pit was too tempting. While the ladders were being uncoiled, three of us rappelled to the bottom and studied the broad chamber. When the others joined us, we strolled along a pleasant, vaulted corridor, only occasionally plagued by mounds of breakdown. Two hundred delightful yards led us to superb gutters of pink lava. Soon they arched smoothly downward into yet another pit, funnelling downward the entire width of the passage. We could not skirt its slippery throat. We could merely scowl at a natural archway beyond, the gate to a huge dim chamber.

One volunteer with a spark of remaining energy hopefully descended the new pit on belay. It ended 45 feet down and did not connect to the room visible beyond. Decision was not difficult. The splendid chamber beyond would await many more trips. Without dissent, we turned and plodded toward the entrance.

The origin of these conduits of flaming gases and molten lava has long puzzled speleologists. In the walls and floor of Dynamited Cave and many another lava tube cavern, silent clues permit us flashes of understanding. The story they tell is fragmentary, for our comprehension is new and imperfect. The features of these heat-scarred sewers of volcanic outpourings vary enormously—perhaps even more than those of limestone caverns. He who would learn their peculiar language must study dozens, scores of far-scattered caves.

Short-lived in comparison with limestone caves, lava tube caverns are peculiar to regions of rather recent volcanic activity. None is found in the eastern United States, for that geologically stable region has had no lava flows for untold ages.

In the western United States, these tube-bearing flows are only indirectly related to the great volcanoes. For every majestic volcanic mountain, hundreds of little cinder cones dot the western landscape. Near these cones or even seemingly isolated from other vulcanism, liquid or clinkery semi-incandescent lava often welled up through rifts in the earth's crust. Sometimes only a single thin layer of lava poured out of such fissures. Elsewhere great lava plateaus built up from successive flows over periods of many million years.

Two types of basaltic lava flow from such fissures. One type known by the Hawaiian name *aa* (pronounced Ah-Ah) emerges almost reluctantly. It bulges forth in a glowing, clinkery mass that tinkles as it pushes along at a walking rate. *Aa* is found in some lava tube caves but does not form them. The tube-forming type of lava is known as *pahoehoe,* another Hawaiian term (pronounced Pah-hoey-hoey). Containing much gas dissolved in the liquid rock, it behaves rather like boiling lead. On even a gentle slope, *pahoehoe* flows rapidly in narrow tongue and broad waves until it cools and solidifies. Especially when deeply buried by succeeding waves of lava, it may retain its heat and plasticity for a considerable time.

Sometimes only a single smoking tongue flowed from a fissure. More often, wave after wave of flows rolled onward, covering those beneath more and more deeply. Some of the molten rivers flowed onward for many miles, charring and burying everything in their paths.

Molten *pahoehoe* does not long remain incandescent. Where it

Looking back across the Big Room of Dynamited Cave to cavers on and below the ledge where our first exploration turned back. The slope a few feet above the seated caver is part of an extensive middle-level complex almost invisible from this angle. The overhanging uppermost level has never been reached but appears to have been completely filled by the successive flows seen in the top of the photo.

emerges today on the island of Hawaii, its temperature is about 2,000 degrees, but the surface cools rapidly and hardens. Only along channels of on-racing torrents or where the surface cracks open can its inner red glow be seen. Many of the more recent, uneroded flows preserve encased details of trees and other objects. Rhinoceros Cave, in eastern Washington's Grand Coulee country, is the cast of a long-extinct beast, engulfed in lava several million years ago. That little cave is entered through a hole where the flank of the archaic animal once was. Some of its leg bones are still in place. It is uncanny to sit where a monstrous example of the American rhinoceros once stood. In some recent flows, engulfed trees did not even catch fire. Strange indeed were these cave-forming flows.

As the great rivers of *pahoehoe* rolled downslope, the crust cooled

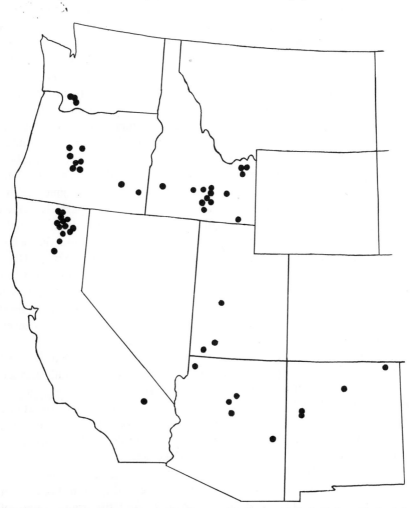

Distribution of lava tube caverns in the conterminous United States. Each dot represents one lava tube cavern or group.

and solidified—perhaps to the consistency of well-chilled tar—long before the central part of the flow. Where the flow was most active, it remained fluid long after the *pahoehoe* had hardened on each side of a well-demarcated channel.

If the volume of flow through such a channel slowed without formation of a crust, a lava trench developed as the level sank. More commonly, crusts developed and the conduit space became a lava tube cavern. Through unroofed sections, observers in Hawaii have watched *pahoehoe* flowing in the uppermost level of such newborn caves.

Some large lava tube caves like Oregon's noted Lava River Cave are little more than simple linear conduits through which passed only a single flood of *pahoehoe*. Certain branched lava tube caverns like Subway Cave near Mt. Lassen are hardly more complex. For many years such caves were generally thought to demonstrate the inner processes of the lava rivers. Now we know that formation of such a tube was only the beginning of the fiery story. New tubes occasionally broke into preformed tubes below, but not all multilevel tubes were thus formed. Dynamited Cave has suggested to some of us that newly formed lava beds are far from static. Perhaps in their still-plastic depths, tube formation is controlled by very minor variations in temperature and pressure.

At the lower end of some flows, narrow lava rivers sometimes fanned out into a deltalike pattern of branching, occasionally interlacing tubes. Particularly spectacular is the Labyrinth system of California's Lava Beds National Monument. Though fragmented by collapsed segments and subdivided by man's limited imagination into a score of separately named caves, the Labyrinth system originally comprised several miles of branching, partially interconnected tubes. In its Catacombs section, the deltalike pattern is expanded into three dimensions. Washington State's Dead Horse Cave is an even more intricate network, formed of much smaller passageways: a curious ants' nest of interlocking crawlways that seems to belong in limestone rather than in lava. What spectacles must have occurred in such caves when liquid lava cascaded from level to level, surged from corridor to corridor!

After the outlines of the tubes were stabilized, subsequent flows of varying natures often greatly modified them. Tubes occasionally formed inside lava tubes. Others are plugged with types of lava foreign to the original flow. Today the study of these strange caverns stands at the threshold where the study of limestone caves stood twenty years ago. Anyone who believes that lava tube caves are simple, uniform structures should browse through our rocky books of the ages.

Dynamited Cave is far from the largest lava tube cavern known. On the island of Hawaii, many known caves are unexplored because of their veneration as tombs of ancient royalty—and the grandparents of commoners still living. One—Kazumura Cave, under study by biologist

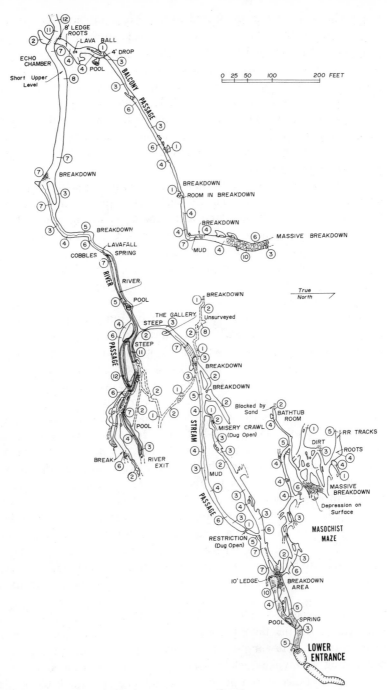

Map of lower end of Deadhorse Cave, an unusually complex lava tube cave in Washington State. Mapping by Oregon Grotto of the National Speleological Society, cartography by Jim Nieland. © 1972 by National Speleological Society, courtesy Charles Larson and Jim Nieland

Frank Howarth, now at the famed Bishop Museum—may be the world's longest, with more than 32,000 feet of passage already mapped. Already respected for his dogged role in the exploration of McFail's Cave, Frank is a Cornell "limey" turned fanatical lava tuber. Today he is noted for his discovery alike of unexpected scientific principles governing the biology of lava tubes, and of a unique Hawaiian cave hazard. Exploring what may be the world's oldest lava tube caves on the island of Maui, Frank found himself wading ever-deeper into a natural slurry of molasses from pineapple and sugar cane plantations overhead. He quit when the goo reached his umbilicus.

In the upper Snake River plateaus of eastern Idaho, a single tube has been traced for more than 20 miles. Most of it, unfortunately, is a collapsed trench. Two mainland lava tube caves are known to be more than 11,000 feet long. One of these is Duck Creek Lava Tube, a new discovery in southern Utah, concerning which the U.S. Forest Service has released no details except the length: 12,054 feet. The other is Washington State's Ape Cave, listed at 11,215 feet but with additional lengths yet unmapped.

When our first scouting party excitedly returned from Ape Cave, babbling of "a lava cave nearly three miles long," I skeptically volunteered to eat every inch of it over two miles. Since our longest listed lava tube cavern then was little more than half that length, I thought the odds were on my side. Northwestern cavers still goad me with fanciful recipes for fried, baked, or boiled 1,000-foot lengths of lava tube.

Except near the lower end, where guided tours have been proposed as part of a Mount St. Helens National Monument, Ape Cave is rough going. At one point a creek seasonally penetrates the roof: one of America's nicest natural showerbaths for those who enjoy 37-degree water. No one knows its exact length. Its lower end is an ever-lower crawlway filled with water-packed sand, at least a hundred feet longer than when we started working on it but still showing no signs of either ending or opening up into the large cave segment which should exist here. Nearby are a score of other lava tube caverns with unusually complex features. Some are small but others are measured in thousands of feet. One reaches a diameter of almost 100 feet. Another we have been able to date accurately through a series of curious chances: Lake Cave.

Three thousand feet into Lake Cave, a low, oblong opening about four feet above the floor permits a small creek to splash out into the main corridor. Crawling up the stream channel, we found ourselves penetrating a claylike rock alongside the tube. Embedded in the soft rock—or hard dirt, if you prefer—were innumerable small black specks. I dug one out. It was a small root, turned to a hard, glazed charcoal by the heat of the basalt which had flowed just overhead. And, as I squished ahead, the cavers in the lead grinned like Cheshire cats. Accidentally exposed by the action of the little underground stream were the carbonized roots of a large tree, perhaps a Douglas fir. Above them was a short remnant of a trunk extending up to the engulfing lava overhead.

Delightedly I reached into my pack for a plastic bag. Carbon-14 analysis of this charcoal would tell us how many years ago the lava had flowed over this pleasant, forested slope. The report was 2,250 years, give or take 150. And, as I reached for a sample, the outer charcoal came away, revealing the uncharred heartwood of the root stump. It looked as fresh as on the day of devastation.

Like most limestone caves, nearly every lava tube cavern has its own story. Lava Beds National Monument is as famous for its history as for its geology. Here the United States Army fought perhaps its most futile campaign.

It seems that in northern California during the early days many pioneers considered the only good Indian a dead one. This, of course, applied particularly to Indians occupying good land. As in most areas of pioneer America, however, that belief was not unanimous. Other

Rumored a sasquatch lair, Ape Cave is actually a designated geologic site a few steps from a paved U.S. Forest Service road. A rail fence surrounds the entrance, seen at the extreme right. The lava flow in which it formed came either from about halfway up Mount St. Helens, seen in the background, or from a fissure near its foot.

settlers were at least as friendly to the Modoc Indians as to certain rapacious Caucasians.

Previously peaceful, the Modocs underwent particularly harsh handling by the Army and a dominant clique of settlers. Transmountain appeals to white friends in Yreka went unanswered, and a band of Modocs fled from the Klamath Indian Reservation to their beloved

This lavaball in Ape Cave appears to have become wedged when the lava was still flowing.

ancestral homelands in the Lava Beds. When "Captain Jack," their leader, shot General Canby during a white-flag palaver, compromise became impossible. As amply chronicled elsewhere, a handful of Modoc warriors outfought and outgeneraled 1,200 troops for five months. In this rugged land they knew like each others' faces, every rock and pit was a natural stronghold. Finally forced away from the secret water holes, Captain Jack surrendered and was hanged, neither party able to comprehend the other's alien reactions. Today cavers in the remoter regions of the Lava Beds still encounter relics of this needlessly tragic struggle. We share Captain Jack's passionate love of this weirdly beautiful land and its inner secrets.

Familiar with the temperate climate of most limestone caverns, the casual visitor to Lava Beds National Monument may be startled to find sizable deposits of ice in several of its caves. Commonly ice exists only in inconspicuous frozen ponds at their lowest points. Here and there, however, is glistening glory. Occasional swirls of gleaming ice are surmounted by crystal-clear monuments of pristine natural sculpture. In hidden grottoes of these ice caves, upside-down snowbanks gleam brilliantly against a brick-red lava backdrop. And invisible coatings of glare ice cover many a treacherous rock. I usually fall flat on my coccyx as soon as I enter an ice cave.

The scientific world has known of ice caves for four hundred years. Many fantastic explanations were offered. It remained for Edwin Swift Balch, a member of Philadelphia's famed Franklin Institute to resolve the problem at the beginning of the century. Modern scientific techniques have amplified his original explanation, and recent research has added tremendously to the number of ice caves known in the United States. Nevertheless, Balch's book, *Glacieres, or Freezing Caverns,* published in 1900, still remains the classic.

Balch explained the phenomenon quite simply. Ice caves are located in areas with severe winters. Cold air is heavier than warm air. If a poorly ventilated cave is aligned in such a way that it can serve as a trap for cold air, winter air entering it is well protected. Moisture entering the cave is promptly frozen.

If much air circulates in the cave, the ice will be short-lived. Most limestone caves are well ventilated. Unless a limestone cave is in an unusually inclement area, even winter ice is rarely found far beyond the entrance. Their bedrock generally approximates the average temperature of the surface. Shallow-lying lava tubes, however, often have limited circulation of air and make excellent traps for cold air. Excluding some limestone caves in northern Alaska's remote Brooks Range which extend into permafrost, most of the major ice caves of the United States are lava tube caverns.

In these natural deep-freezes, a considerable variety of transparent

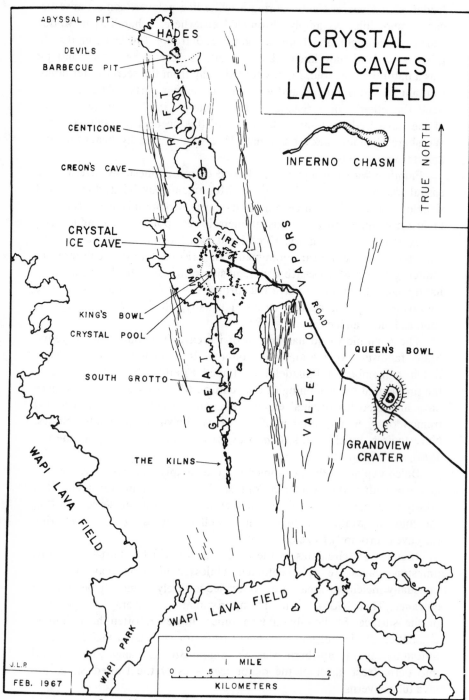

CRYSTAL ICE CAVES LAVA FIELD

ABYSSAL PIT

HADES

DEVILS
BARBECUE PIT

CENTICONE

CREON'S CAVE

CRYSTAL
ICE CAVE

KING'S BOWL

CRYSTAL POOL

SOUTH GROTTO

THE KILNS

INFERNO CHASM

TRUE NORTH

RIFT

RING OF FIRE

GREAT

VALLEY OF VAPORS

ROAD

QUEEN'S BOWL

GRANDVIEW
CRATER

WAPI LAVA FIELD

WAPI LAVA FIELD

WAPI PARK

J.L.P.

FEB. 1967

0 1 MILE
0 .5 1 2
KILOMETERS

Crystal Ice Caves Lava Field. Courtesy James Papadakis

speleothems may develop. Icicles are not so common as one might expect. Any warm air penetrating into a glaciere rises to the ceiling since it is lighter than the cold air already present. The most stagnant cave exhales and inhales slightly with changes in barometric pressure, so icicles do not last long. Ice stalagmites are much more prominent. Draperies may persist on slanting ceilings or hanging from ledges. Frost crystals sometimes reach enormous size. In at least one Idaho cave their span is almost two feet. As in grocery-store deep-freezes, they form from moisture-laden air seeping into the frozen depths. The beauty of their myriad sparkling reflections is one of the supreme rewards of caving. But one's carbide lamp or body must not be brought too close or remain too long, or the intricate crystals coalesce into tiny droplets and vanish as they melt.

Fresh cave ice usually has few bubbles, cracks, or other flaws. Older ice, undergoing constant change, varies considerably. Often it is prismatic. Sometimes it is stratified like a cake. In a few caves, the cavern ice flows, forming true subterranean glaciers. Where they encounter suitable pits, as in Idaho's Crystal Falls Cave, great jade-green columns flow slowly, silently from level to level.

Since most American glaciers are lava tube caverns, it perhaps seems strange that several of our greatest are of other types. Walled with lava but of very different nature are Crystal Ice Caves in Idaho's Great Rift, a National Landmark since 1968.

For many years sheepherders had known of this crack zone in the lava plain southeast of Craters of the Moon National Monument. From the air, it can be traced for forty-six miles. In places a barely perceptible fissure, elsewhere it widens into gaping pits and sinks as much as 200 feet in diameter. Along its course, spatter cones tell of small-scale but dramatic eruptions in the not-so-distant past: a pattern strongly reminiscent of certain of the mysterious rills of the moon. Some are known to have occurred hardly more than 2,000 years ago.

Sheepherders who picked their way to the bottom of the most prominent sink were rewarded by finding a large, icy pond—the only water hole within miles. Obsidian chips and occasional artifacts indicated that others had preceded them in the chill depths. Back beneath the south wall of the sink, a narrow cavern extended into blackness. But who cared? Usually its corridor was blocked by the pond, thinly crusted with unsafe ice.

Adventurous residents of the Magic Valley towns eventually followed the sheepherders' trails to the main crack. A few geological oriented visitors were able to read much of the area's history from the walls of the great bowl-shaped sink (actually an explosion crater caused by a steam eruption along the rift). The upper 10 to 25 feet consists of thin, recent flows. At that point, an old soil level separates them from massive, ancient flows laid down in thick layers. Occasional roofed-over sections of the fissure reveal remnants of the lava which it spewed outward. Here and there vertical flows can be seen where they solidified while arching outward to the surface of the plateau, then draining back into the depths.

At a few points explorers were able to scramble deep into the fissure. The great sink alone was 150 feet deep. In other sections, black pits or tiny vertical crevices seemed to disappear into icy black depths. Few cared to challenge the dual risk of the fissure and its redoubtable environs. Though only 20 miles from American Falls, as recently as 1950 only a handful of hardy souls had dared this sheepherders' haven in the lava plain.

In 1956 a local schoolteacher and an Idaho State College geology student set about exploring the fissure. Climbing down a steep slope with the aid of a fixed rope, young David Fortsch dropped his flashlight into a nasty hole he had not intended to investigate. While Perry Fenstermaker stood by, Fortsch slithered down a short, icy slope. Unexpectedly he burst out into a spacious netherworld sparkling with frost crystals. Great glassy pillars and glistening ice cascades stood revealed in his lantern's light. Volcanic fissures don't have this sort of cavern. Yet here it was.

The two Idahoans wandered onward amid towering columns of glare ice, slipping and sliding downward until they reached a level ice floor. Around them, pristine glassiness rose far upward toward giant icicles. The cave was at least 360 feet long, they told unbelieving friends, and 30 feet wide: not vast but magnificent.

Their friends expressed considerable doubt. Wryly, the two discoverers named their find Liars' Cave. A 1964 survey found the length of Liars' Cave to be 370 feet.

Fenstermaker was not content with this remarkable discovery. Hole after hole along the fissure remained unchecked. Some were shallow and easy. Some still defy exploration. Between his Liars' Cave and the great sink, he found a sheer 130-foot shaft that led to a spacious Great Cavern, sloping down toward Liars' Cave and beneath it. The spacious Great Cavern contained a superb ice cascade, but its glistening splendor was only a hint of the glory of Liars' Cave—now Crystal Ice Cave.

Less than a mile farther south was a cavity that seemed ten times as deep—the South Grotto. By 1957 word of "a hole out there they estimate to be 1,300 feet deep" reached Bill Echo, then director of the Idaho Speleological Survey. Bill flew over the area, then scouted it on foot. Although ground water evidently lay 750 to 850 feet below the lava plain, he wrote me that "I would tend to give considerable credence to their story." The rest of us hooted him down. A fissure cave deeper than Neff or Carlsbad? It was so ridiculous that none of us ever bothered to go to the fissure.

Because their pattern tends to trap cold winter air, the most important ice caves of the United States are lava tube caves. *Photo by Charlie and Jo Larson*

Longitudinal section of part of Crystal Ice Caves Rift, Idaho, based on sketch by James Papadakis. The King's Bowl is at the southern end of the section shown.

Except that Jim Papadakis did. Seemingly far out of place in icy, volcanic Idaho, this original developer of Texas' Cavern of Sonora talked with Bill Echo in 1961. Promptly he headed for Aberdeen. There he recruited two cave-happy high school boys, who guided him to the rift—and Liars' Cave. That name died then and there.

Dazzled by the icy splendor, Jim dreamed of opening its beauties to the public. Soon he leased the land from the Bureau of Land Management. A single trail down the side of the giant sink—The King's Bowl—accomplished much of the task. Two years of drilling, blasting, and back-breaking excavation completed a 1,200-foot tunnel alongside the Great Rift. Numerous windows opened Crystal Ice Cave for everyone to view and admire.

Even Crystal Ice Cave may not be the climax of this amazing fissure northwest of American Falls. In 1963 Jim's young guides went exploring in "impossibly deep" South Grotto. They guess a depth of 800 feet, only 50 feet above the surface of the ground water.

At first glance this hardly seems more reasonable that Fenstermaker's 1,300-foot guess. To date, Idaho cavers have been able to map only to a depth of 690 feet. But this time we don't shrug off a "1,300-foot" South Grotto. At 690 feet, their headlamps revealed no bottom, and the water table may not be continuous everywhere in these volcanic beds.

Moreover, Papadakis' surveys showed Crystal Ice Cave 160 feet deep. Fenstermaker had guessed only 90 feet.

Much as I hate to admit it, the limeys are at least partially correct, however. Someday we may indeed find that the world's largest caves are not in limestone, as I explained in the last chapter. Yet this will scarcely diminish the overwhelming importance, on a national or global scale, of limestone caves. In the United States and throughout the earth, lava tube and other volcanic caves will never achieve a place of more than secondary importance.

Yet mankind may look to these lesser caves for ultimate survival. Even cavers have been slow to comprehend this, for much of man's understanding of their significance has bypassed routine channels of speleology.

Charter members of the Southern California Grotto of the National Speleological Society in 1948 included a brilliant young astronomer at Mount Wilson Observatory: Karl Henize. An eastern limey newly transplanted to California, he eventually tired of the grim early days when we seemed to have run out of worthwhile caves within driving distance of Los Angeles. Despite the bond of joint ventures in Lilburn and lesser

A flashbulb behind a stalagmite of ice in Crystal Ice Cave reveals details of its pattern. *Photo by James Papadakis*

California caves, we lost touch. A generation later I nearly fell out of my chair when his voice came over television from NASA's Mission Control, talking matter-of-factly to teammates on the moon. Although one of the earthbound members of the team, astronomer-caver Henize had become an astronomer-astronaut.

Quite a few lava cavers had already overstrained our eyes, hopefully searching NASA photos for lava tubes and trenches amid the myriad volcanic features of the moon. Almost as soon as we realized that its huge dark maria are immense lava plains, we began wondering how many lava tubes they contained. As we obtained ever-better views of cracks in the lunar surface that look remarkably like The Great Rift, other possibilities intrigued us. Shelter from intense solar radiation, mineral sources—perhaps even vast quantities of water: the values of such caves to stranded astronauts and moon colonies were obvious.

Speculation ran wild. Under the moon's lesser gravity, lava caves might expand enormously without collapsing. On the other hand, without oxygen lava might not get hot enough to form tubes of adequate size for man's use. The eyestrain intensified. Vulcanospeleologists and astrogeologists alike found sinuous channels that looked remarkably like magnified versions of lava trenches here on earth. Most were so huge as to well-nigh rule out any possibility that they were collapsed tubes. Yet, until the Apollo astronauts demonstrated to millions of entranced viewers that Hadley Rille is *not* a collapsed lava tube 2,000 feet in diameter, we could at least hope.

In sober fact, much tinier sinuous rills—mostly at the lower limit of resolution of the best early photographs of the lunar surface—have always been much likelier candidates. Some of these are segmented by what may well be cavernous segments, just as in the western United States and Hawaii. These and segmented fissures remain uninvestigated by any of our pioneers on the moon.

And, in the vast volcanic landscapes of Mars, the likelihood of lava tube caves is even greater.

Although deliberately selecting landing sites lacking interest to vulcanospeleologists, NASA clearly remains alert to these possibilities. I wrote to Karl, twitting him about abandoning his beloved limestone caves. Responding in similar vein, he mentioned the astronauts' special training in vulcanospeleology, "just in case." The Apollo 14 crew, it seems, took an extra-good look at The Great Rift.

A few thousand millennia hence, our sun will flare in fiery death, engulfing a bubbling slagball once called Terra as the frozen outer giants flare in momentary cosmic puffs. For a generation and more, man has known that he must go to the stars long before that inexorable day. Already our most foresighted are working toward man's ultimate survival on new Terras yet undiscovered amidst a universe whose size is beyond human comprehension. Lush, uninhabited planets which man can populate, enjoy will be desperately few. Yet, amidst the 150,000 million suns of our home galaxy alone, statistically they exist.

Alas for science fiction! No easy teleportation, no quick space warping will ease our planet hunting throughout our galaxy; far less among those unimaginably beyond. Yet man even now progresses toward those new Terras—and properly so, for long will be the recurring Dark Ages which man will face and conquer along his way. To survive, man must progress whenever, however he can.

When man reaches those new Terras, limeys can take heart. Despite aliennesses inconceivable today, Earth-type planets at the proper distance from Sol-type suns of the necessary age will be found to have warm seas that grow limestone. Plus the necessary tectonics to crack it and raise it above sea level, and rain to grow caves in it that will be enjoyed by man and his brother xenospelunkers across the universe.

Along the way, however, it will be the grim volcanic caves of countless uninhabitable planets that our space castaways, our desperately leap-frogging colonists, seek for brief respites from the merciless void of space.

On a galactic scale, the limestone cavern is but a delightful special case. The future of man may depend on those molten sewers we call lava caves.

So You Want to Be a Caver?

When I wrote the first edition of this book a decade ago, I added six pages of suggestions intended to keep avid readers out of trouble—and from causing it. Today matters are different. In 1974 I expanded these six pages into a 348-page book, *American Caves and Caving,* also published by Harper & Row. Its subtitle tells its contents: *Techniques, Pleasures and Safeguards of Modern Cave Exploration.* Hopefully it will see revised editions as we continue to progress. If you want to be a caver, I suggest you study a copy. Some other pretty darn good cavers recommend it almost as much as I do.

Two suggestions are still worth mentioning here, however. Before investing in even *American Caves and Caving,* visit a commercial cave, to be sure that you don't have unsuspected claustrophobia. Then, if you didn't panic the moment you got underground, contact the National Speleological Society to obtain information on responsible local groups near your home. The address is simple: Cave Avenue, Huntsville, Alabama. The office telephone number is in the Huntsville phone book. If the office ever moves away from Huntsville, you can obtain it through the American Association for the Advancement of Science, of which the N.S.S. is an affiliate. The address of the A.A.A.S. is available in almost every public library.

Good luck and good caving!

Glossary

aa. A rough type of lava in which lava tubes do not form.

ablation. The combination of melting, evaporation, and other processes by which glaciers shrink and glacier caves form.

acetylene. The flammable gas burned by carbide lamps, produced by the chemical reaction of water with calcium carbide.

angel's hair. A somewhat fanciful name applied to clumps of unusually delicate gypsum needles which angle outward and curve slightly.

angel's wing. A gracefully folded drapery. Also applied to dripstone-hung palettes.

anthodite. A confused term originally applied to an unusual type of helictitic speleothem, but subsequently to a variety of complex speleothems.

anticline. An upward arching of rock strata.

aragonite. A mineral found in some caves, chemically composed of calcium carbonate, usually in the form of needle-like crystals but also found as stalactites, helictites, and other speleothems.

ascender box. A metal housing for two pulleys, worn high on the chest in some systems of standing rope ascents.

ascenders. Mechanical rope-gripping devices used in standing rope ascents. Sometimes also used to include knots which serve similar functions.

auto-belay. (1) Use of a Prusik knot or related technique during rappelling in such a way that it will tighten and hold the rappeller in place if released; (2) a special rockclimbers' device providing a dynamic belay in case of a fall.

bacon rind or bacon-rind drapery. A thin drapery with bands of color mimicking a huge strip of bacon.

basalt. A common type of lava. *Aa* and *pahoehoe* are forms of basalt.

bedding plane. The surface between two layers of sedimentary rock.

belay. Knowledgeable use of a safety rope. For details see my *American Caves and Caving,* Harper & Row, 1974.

block-creep cavern. A long, narrow cave parallel to the face of a cliff, resulting from cracking away and "creeping" of a block of rock.

bowline knot. A slip-proof climbers' knot.

boxwork. A complex of intricately intersecting thin blades of calcite or other mineral, projecting from the bedrock of a cave.

brake bar. A short metal bar with a hole at one end whereby it is threaded onto a carabiner or rack, and a groove at the other which fits snugly against the other arm of the rappel device.

breakdown. Any material which has fallen from the ceiling or wall of a cave, but usually applied to considerable accumulations. Also used as an adjective, describing cavern chambers or other features formed or extensively modified by the process of breakdown.

Brunton compass. A compact precision instrument used for accurate cave surveying.

calcareous. Pertaining to lime and related materials: limestone, calcite, coral, travertine, etc.

calcite. The commonest of cave minerals, forming most speleothems observed by the average American caver. Chemically it is composed of calcium carbonate.

canopy. A ledge or remnant of false floor festooned with stalactites.

carabiner. Metal "snap rings" used in many ways in caves. Some have a locking screw gate, others a simple gate which opens under direct pressure.

carbide. Speleologically, calcium carbide, a solid chemical used as fuel for miners' lamps. See **acetylene.**

carbon-14. The radioactive isotope of carbon, with a molecular weight of 14. Since it decays at a fixed rate, the age of carbon-containing materials can be calculated from the quantity of Carbon-14 present.

carbonic acid. The weak acid resulting from interaction of carbon dioxide and water.

cave. A natural cavity below the surface of the earth, large enough to enter, with some portion in essentially total darkness. The term is often used more loosely, e.g., Marmes Cave.

cave bubble, coral, cotton, hair, raft, etc. Descriptive terms for speleothems more or less resembling the specified object.

cave ice. Ice naturally formed in a cave. Sometimes used incorrectly for a delicate type of rimstone or shelfstone.

cave mile. Technically, 5,280 feet of underground passage. Humorously, any underground distance over 100 feet or so.

caver. One who explores caves.

cavern. Same as cave. Sometimes a mild connotation of grandeur.

cave system. An interrelated, basically continuous complex of caves, often separated by impassable segments.

caving. The exploration and/or recreational use of caves.

ceiling channel. A distinct channel dissolved or eroded upward into the ceiling.

chert. A very hard, flintlike rock that occurs in beds or nodules in some limestones.

chimney. (1) Any vertical opening more than about one foot in diameter, more specifically one which is rounded and lacks the characteristics of a dome pit; (2) a narrow, tubular volcanic pit; (3) to ascend or descend any narrow orifice by using both walls as climbing surfaces or by pressure against both walls.

chock. A natural or artificial chockstone serving much the same function as a piton.

collapse chamber. A cavern chamber formed or heavily modified by breakdown.

column. A compound speleothem produced by the fusion of a stalactite and stalagmite. Cf. **pillar.**

commercial cave. A cave with an admission charge. Paths and other improvements are usually present.

conduit. A roughly circular or oval subterranean passage which serves or has served to conduct large volumes of water or lava.

conglomerate. A sedimentary rock composed of fragments of other rocks, naturally cemented together.

conservation. Protection of ecological, scenic, scientific, recreational, wilderness, and other resources and values of caves.

coralloid. A small, nodular speleothem, usually of calcite or lava, often occurring in intricate complexes. Also termed **cave coral.**

corridor. A comparatively long, level portion of a cave.

crampons. Mountaineers' "climbing irons" affixed to the boot soles for better purchase in some steep snow and ice climbing.

crawl or crawlway. A cavern passage too low for stooping.

crevasse. A glacier fissure.

cupola. A domed arch of the ceilings of some lava tube caverns.

curtain. (1) A broad, wavy drapery; (2) a long row of intermingled stalactites.

dead cave. A cave in which the speleothems are no longer moist and enlarging.

derigging. Removal of the ropes and other gear used in vertical caving.

dolomite. (1) A sedimentary rock somewhat like limestone but less rapidly soluble because of a considerable proportion of magnesium carbonate; (2) a mineral composed of calcium and magnesium carbonate.

dome-pit or domepit. A roughly circular natural shaft in limestone or other soluble rock, with sheer, slightly ribbed walls. Usually several feet or yards in diameter.

drapery. A thin, pendant speleothem, often convoluted.

dripstone. Any stalactite, stalagmite, or other speleothem formed through the action of dripping water or lava. Cf. **flowstone.**

drop. A vertical or near-vertical pitch, especially one suited to standing rope descents. Also used as a verb for the latter.

dry suit. Divers' flexible waterproof garb. Cf. **wet suit.**

duckunder. A point where explorers must "duck under" a low spot to get from one place to another. Some are waterfilled; see **siphon.**

epsomite. Natural epsom salts. Chemically, hydrated magnesium sulfate.

expansion bolt. A rockclimbers' tool. After a hole has been drilled into rock, the bolt is inserted and expands, holding to the rock other hardware attached to it.

false floor. A thin layer of flowstone, lava, or other material which conceals a space below.

fault. A plane or zone on which a block of the earth's crust has been displaced.

firn cave. A cave in snow which has not compacted sufficiently to develop the density of a glacier.

fissure. A narrow crack in rock. Often used loosely for a narrow passage.

flake. A slab of ice which peels away from the wall or ceiling of a glacier or firn cave.

flow groove, ledge, line, mark, etc. Descriptive terms for features of lava tube caves which appear related to flows through the tube.

flowstone. A surface coating of mineral, usually calcite or ice, deposited from a descending film of water or lava. Cf. **dripstone.**

flute. See **scallop.**

formation. (1) A geological term referring to a specific unit of bedrock; (2) a confusing popular term for **speleothem;** (3) anything which has "formed" in a cave—a very loose and confusing usage.

free-fall. Descent without interference by the wall.

fumarole. An outlet for volcanic gases. A few are cavernous.

geothermal. Pertaining to the internal heat of the earth.

geothermal cave. A cave produced by geothermal melting of snow or ice.

Gibbs ascenders. A popular cam-type device used in standing rope ascents, most commonly used in pairs for "rope walking" but also in special systems.

glacier cave. A cave in or beneath a glacier.

glaciere. Same as **ice cave,** but also including cold-trapping sites of other kinds, such as mines.

glaciospeleology. The study of glacier caves and related phenomena.

glaze, lava tube. A shiny, relatively smooth coating of some lava tube caves.

go (verb, as in "It goes!"). In caves, to offer an opportunity of exploration.

gouffre. A French term, sometimes applied to certain American pit caves.

gour. Another French term, increasingly applied to rimstone deposits (q.v.).

grotto. (1) A small side chamber of a cave; (2) a cavernous opening which does not extend into total darkness; (3) a chapter of the National Speleological Society. Sometimes also used improperly by unaffiliated groups.

ground water. Water in cavities of porous or cavernous rock or soil. Some authorities exclude water in the subsurface zone of aeration, above the water table.

guano. Speleologically, the accumulated excreta of bats.

gulf. A steep-walled sink subject to flooding by a cave stream; characteristically floored with level stream deposits.

Gurnee can. A tapered cylindrical metal can in which gear is dragged through crawlways.

gypsum. A sedimentary rock and mineral composed of calcium sulfate, softer and more soluble than limestone and calcite.

gypsum barrel, cotton, crust, flower, grass, hair, plate, rope, sand, etc. Descriptive terms for various gypsum speleothems.

gypsum cave. A cave formed in gypsum, usually by much the same processes which form limestone caves. Occasionally misapplied to caves containing gypsum deposits.

hairy. Current caver cant for a fearsome spot.

hard hat. A caver's helmet.

helical knot. A sliding hitch with a free end, used as an ascent knot.

helictite. A speleothem which looks as if it ought to have become a stalactite but seemingly ignored the law of gravity. While most are contorted or forked, some are straight.

heligmite. A helictite directed upward, "like a stalagmite."

hydrology. Speleologically, the study of underground water and its actions.

hypothermia. Significant lowering of the temperature of the vital inner organs of the body.

ice ax. A long, somewhat hatchet-like mountaineers' tool used for controlling slides on steep snow and for chopping steps in snow and ice.

ice cave. A cave in which ice forms and persists through much or all the summer and autumn.

igneous rock. Rock of volcanic origin.

immersion suit. Heavy rubberized one-piece garb originally designed to keep the wearer dry when floating for long periods.

inchway. A crawlway so tight that explorers must force their way along, seemingly inch by inch.

joint. A crack in bedrock, caused by movement of the earth's crust or other natural processes.

Jumar ascender. A precision rope-gripping device used for standing rope ascents and other vertical caving.

karst. Topography characterized by underground drainage developed through solution of bedrock. Usually in limestone, dolomite, or gypsum.

lava ball, tongue, etc. Descriptive terms for various features of lava tube caves.

lavafall. A solidified lava cataract.

lava seal, lava siphon. A plug in a lava tube consisting of lava which hardened in place while filling it to the ceiling, leaving an open space on one or both sides.

lava trench. A long, narrow gulch that is a collapsed or never-roofed segment of a lava tube.

lava tube. A natural conduit of lava, somewhat cylindrical in shape, a few feet to a few miles in length.

lava tube cave. A cave formed as part of an abandoned conduit of *pahoehoe* lava.

lead (pronounced "leed"). An opening that appears to continue.

lily pad. A special form of shelfstone formed around a stalagmite which has been largely submerged.

limestone. A type of sedimentary rock largely or completely formed of calcium carbonate. Because it is readily dissolved by slightly acid water, most of the world's important caves are in limestone.

limestone cave. A loose term for a solutional cave in almost any kind of rock.

limey. A caver who vehemently prefers caves in limestone to those of other types.

littoral. Pertaining to the zone between high- and low-water marks on a beach or cliff. "Littoral caves" are formed in this zone.

live cave. A cave in which speleothem deposition is in progress.

marble. Limestone which has been recrystallized and often molded by heat and pressure deep in the earth.

master cave. A locally dominant throughway or trunk corridor.

meteorology. The study of the processes and contents of air and related phenomena.

moonmilk. A white, putty-like form of flowstone, formed by one of several spelean minerals.

moulin. A domepit-like structure of glaciers.

oolite. A small rounded or faceted concretion.

oulopholite. A curved, fibrous gypsum crystal or group thereof. More commonly called "gypsum flower."

pahoehoe. The relatively smooth-surfaced, once-fluid type of basaltic lava in which lava tube caves form.

palette. A broad, thin, disc-shaped speleothem. Dripstone often hangs from the margin.

permafrost. Permanent ice within alpine or arctic portions of the earth's crust.

petromorph. A cavern feature exposed by solution of surrounding limestone, i.e., boxwork.

phreatic. Pertaining to the zone of water below the water table. In the phreatic zone, all cavern passages are filled with water.

pillar. A solitary vertical or nearly vertical bedrock remnant. Cf. **column.**

pit. A natural shaft, large enough to descend.

pitch. A specific length of vertical or near-vertical wall.

piton. A thin, wedgelike blade which rockclimbers and occasionally cavers hammer into cracks for attachment of carabiners and other climbing hardware. Snow and ice pitons and other special types also exist.

pothole. (1) A round, bowl-like pocket in the floor of a cave or a surface stream; (2) a British term for pit or deep sink, with or without a cave at the bottom.

Prusik knot. A special sliding hitch using a looped sling rope, popularized as an ascent knot by Dr. Karl Prusik.

pseudokarst. Karst-like phenomena of glaciers, lava flows, and other poorly soluble rocks, resulting from processes other than solution.

purgatory cave. A cave formed by accumulations of talus at the bottom of a narrow gorge.

rappel. Controlled descent of a standing rope using friction of the rope around the body or a rappel device or both.

rappel rack. A rappel device consisting of a U-shaped holder for several brake bars.

resurgence. The point of surface appearance of a karstic stream.

ribbon. An unbanded drapery which otherwise would be termed bacon rind (q.v.).

rig. (1) To prepare anchors, belays, and other vertical caving gear; (2) a vertical caver's specialized body harness and hardware.

rigging. (1) The process of preparation of standing rope or ladder descents; (2) emplaced vertical caving gear.

rimstone. (1) Thin mineral crusts formed at the rims of some cavern pools; (2) terraced spelean deposits of calcite or other minerals, forming a complex of small or large basins.

rise. A prominent seasonal or permanent resurgence characteristic of the Lost River karst of Indiana.

rockfall. The process of breakdown.

rockshelter. An overhung cavity which does not extend into total darkness. Often erroneously termed **cave.**

saltpeter. Speleologically, cavern deposits of nitrate minerals, usually in earthy deposits.

saltpeter cave. A cave containing saltpeter, especially one formerly mined for the substance.

scallop. An unevenly rounded, shallow pocket on the surface of bedrock, glacier ice, mud, and other substances, occurring in groups. Those in limestone caves are sometimes called **stream flutes.**

sea cave. See **littoral.**

sedimentary rock. Rocks deposited in layers through the action of water or wind.

sewer passage. A comparatively small, roughly tubular passage which intermittently or continuously transmits large quantities of water.

shelfstone. Extensive mineral shelves formed at the rims of some cavern pools. Also see **rimstone.**

shield. See **palette.**

sink. A depression in cavernous country resulting from (1) collapse of an underlying cavern, or (2) solution and settling along a joint or tube, or (3) lava subsidence.

sinkhole. Essentially identical with **sink.**

siphon. Obstruction of a section of cavern by water which fills the passage to the ceiling. Also see **lava siphon.**

skylight. A comparatively small gap in the roof of a cave or underground stream, allowing entry of only a little daylight. See also **window, karst** or **pseudokarstic.**

slack. Lack of tautness of a rope.

sling. A vertical caver's short length of rope or webbing.

snow cave. A natural cave in snow formed by the same processes which create glacier caves. See also **firn cave.** Note: Mountaineers' "snow caves" are artificial bivouac holes.

soda-straw stalactite. A thin-walled, hollow, tubular stalactite approximately the diameter of a drop of water.

span. A spelean natural bridge.

spar, dogtooth, nail-head, rice-crystal, etc. Types of small calcite crystals vaguely resembling their namesakes.

spelean. Pertaining to caves.

speleogen. A cave feature resulting from natural removal of bedrock.

speleogenesis. The process of origin and development of caves. The corresponding adjective is **speleogenetic.**

speleoliferous. Containing caves (applied to certain limestones, lava flows, etc.).

speleology. The study of caves and related phenomena.

speleothem. A mineral deposit formed in a cave.

spelunker. Someone who explores caves as a hobby or for recreation. The term was coined in the middle 1930s, by Roger Johnson and the late Clay Perry, from the Latin root **spelunca** (cave).

spelunking. Sport caving.

squeezeway. A cavern passage so narrow that human passage is difficult.

stalactite. Remember the popular mnemonic: they Cling to the Ceiling.

stalagmite. *Ergo,* they Grow from the Ground.

standing rope. A rope which is securely anchored and allowed to dangle.

static belay. The type of belay which seeks to halt a fall immediately, without allowing the rope to run. This is the commonest type of belaying in caves.

steam cave. See **geothermal cave.**

swallow hole. Indiana term for the point of disappearance of a sinking stream in karst. *Swallet* is a similar British term.

swiss seat. A body harness loop of webbing, encompassing the waist and gluteal regions, often held together in front by a carabiner.

syncline. A trough-shaped or down-arched bedrock fold. Cf. **anticline.**

talus. Large and/or small rocks displaced from their original position.

tandem climbing. Simultaneous standing rope ascent by more than one vertical caver. Caution: this is an extremely dangerous technique. See *American Caves and Caving.*

tandem knots. Ascent knots consisting of a set of two or more sliding hitches.

tectonics. Movements of and in the earth's crust.

throughway. A large, near-horizontal, comparatively straight passage, uniform for hundreds or thousands of feet.

travertine. Speleologically, a coarse form of flowstone or rimstone, often of organic origin. Sometimes applied to any calcium carbonate speleothem.

travertine cave. A cave in a surface travertine deposit.

tree cast. A mold of a tree engulfed in lava.

troglodyte. A cave dweller, human or otherwise.

trunk passage or channel. See **throughway** and **master cave.**

tube-in-tube. A rudimentary lava tube formed in a secondary flow inside a lava tube cave.

vadose. Pertaining to subsurface water in the zone above the water table.

vandalism. Ignorant or deliberate damage to cave resources and values. Cf. **conservation.**

vug. A crystal-lined underground cavity, normally too small to be termed a cave.

vulcanospeleology. The study of lava tube caves and related phenomena.

wall walking. A form of standing rope ascent in which the caver obtains considerable support from footholds on the wall.

water table. The upper surface of the zone saturated with ground water. The term is poorly applicable to massive limestones.

wet suit. A diver's suit penetrated by water yet able to maintain a thin layer of warmth against the body because of the insulating effect of innumerable bubbles of water trapped in its pores.

window, karst or pseudokarstic. A type of collapse sink which has unroofed a sufficient length of cavern or underground stream to allow entry of essentially full daylight. Occasionally the term is used more broadly, including **gulfs** (q.v.). Cf. **skylight.**

yo-yo. To descend and ascend a pit or vertical pitch in rapid succession, using a standing rope. Sometimes used as a noun for this process.

xenospelunkers. Postulated extraterrestrial cave explorers.

Suggested Additional Reading

To list here every reference consulted in the preparation of this book would only confuse most readers. For those interested in delving deeper into these topics, the following should serve as a good beginning.

General References

By far the best references on American caves are the numerous and varied publications of the National Speleological Society and its units. Usually they are in a form useful only to advanced cavers, however, though included here are several which may be useful introductory references.

Celebrated American Caves, Rutgers University Press, 1955, edited by Charles Mohr and Howard N. Sloane, is a compilation of excellent older articles and stories of caves throughout the Western Hemisphere. This should not be confused with *Celebrated American Caverns* (two nineteenth-century editions) by Horace C. Hovey, the first great American cave book. In 1970 Johnson Reprint Company of New York published a new edition with a 33-page biographical and analytical introduction by me. This is one of a five-part series of reprints of classics in speleology. Others are mentioned below, and it is understood that Zephyrus Press is continuing the series.

Exploring American Caves, paperback edition, Collier Books, 1962, by Franklin Folsom is a good popular introduction to American caves. Because of errors corrected in the later paperback, the earlier hard-cover edition is not recommended.

Visiting American Caves (New York, Bonanza Books, 1966) by Howard Sloane and Russ Gurnee is a useful guide to tourist caves of the United States despite inevitable changes due to inflation and other contemporary problems.

My *American Caves and Caving* (New York, Harper & Row, 1974), from which I lifted two fun pages for this book, describes techniques of safe, scientific cave exploration throughout North America.

The Amateur's Guide to Caves and Caving, by Dave McClurg (New York, Stackpole Books, 1974), is a short, comparatively inexpensive introduction to caving. The first printing, identifiable by a drawing on page 49 instead of a photo, is marred by serious errors, most of which were corrected subsequently. Jennifer Anderson's *Cave Exploring* (New York, Association Press, 1974) is another introductory work, nicely illustrated, but contains some unsafe practices like "hasty rappelling."

The British *Manual of Caving Techniques* (London, Routledge and Kegan Paul, 1969, edited by Cecil Cullingford) is excellent except for considerations of standing rope techniques, which then were little developed in that country. Tony Waltham's *Caves* is another recent British book, well-illustrated and informative at a comparatively elementary level. An American printing is available (New York, Crown Publishers, 1974).

The best short source book on limestone speleogenesis is Joe Jennings' 1971 *Karst* (Canberra, Australian National University Press), which has a fine global orientation. *Speleology: the Study of Caves* (D. C. Heath and Co., Lexington, Mass., 1964) by George W. Moore and Brother G. Nicholas Sullivan contains much basic data but suffers from a lack of due consideration of alternate concepts of some important matters.

Bats have been the subject of several recent books. Dover Publications has reprinted (in paperback) Glover Morrill Allen's 1939 classic, simply entitled *Bats*. Donald R. Griffin's *Listening in the Dark* (New Haven, Yale University Press, 1958) is another excellent academic work. His paperback *Echoes of Bats and Men* (Anchor Books, 1959) and Russell Peterson's *Silently, by Night* are less technical. Beautifully illustrated are *Bats of America* by Roger W. Barbour and Wayne H. Davis (Lexington, University Press of Kentucky, 1969), and Alvin Novick's *The World of Bats* (New York, Holt, Rinehart and Winston, also 1969). Even more beautiful is *The Life of the Cave* (New York, McGraw-Hill, 1966), by Charles Mohr and Tom Poulson. Volume 34, no. 2 of the *Bulletin* of the National Speleological Society (April 1972) contains the proceedings of a vital symposium on the ecology, physiology, behavior, and survival of bats.

Additional reading is arranged by chapters.

1 ● The Eternal Standard

Probably more has been written about Mammoth Cave than any other American cave. The 1962 sixty-three-page bibliography prepared by Frank G. Wilkes of the University of Louisville was far from complete. Its archeology is magnificently described in Patty Jo Watson's 1974 *Archeology of the Mammoth Cave Area* (New York, Academic Press). Two of the Johnson reprints (see p. 409) are on this cave: *Rambles in the Mammoth Cave in the Year 1844,* and *The Suckers Visit to the Mammoth Cave.* The new introduction to the former, by Harold Meloy, contains the best account of the history of the cave. As for Flint Ridge, *The Longest Cave* by Roger Brucker and Red Watson, recounting the 18-year struggle of the Cave Research Foundation and its precursors, is due to appear while this book is in press. Pat Crowther's own account of the grand Kentucky junction appeared in the January 1973 issue of *National Parks Magazine*. *The Caves Beyond* (Funk and Wagnalls, 1955) by Roger Brucker and Joe Lawrence, Jr., telling the story of the C-3 Expedition, a collectors' item, was reprinted by Zephyrus Press in 1975.

2 • In a Lonely Sandstone Cave

William B. (Skeets) Miller's first-person account of the Floyd Collins fiasco appeared in the *Reader's Digest* in April 1962. Many of the published accounts of the episodes are unreliable, and some are so incorrect that the Collins family has successfully sued the writers and publishers for libel. Brucker's chapter in *Celebrated American Caves* (see p. 409) was based on extensive research. Additional advances are reflected in my present chapter. Howard W. Hartley's 1925 *Tragedy of Sand Cave* (Louisville, Standard Printing Company) gives a slanted contemporary view.

3 • Beginnings Old and New

William E. Davies' *Caves of West Virginia* (West Virginia Geological Survey, 3d edition, 1965) and H. H. Douglas' *Caves of Virginia* (Virginia Cave Survey, 1964) are the basic references. A vast literature exists; both these works contain bibliographies. For Schoolhouse Cave, by far the best references are Tom Culverwell's accounts in the *Bulletin* of the Potomac Appalachian Trail Club in January 1941, June and October 1943, October 1944, and January 1945. Hot off the press is John Holsinger's *Descriptions of Virginia Caves* (Virginia Division of Mineral Resources).

4 • The Geologist and the Showman

J Harlen Bretz's *Caves of Missouri* (Missouri Geological Survey, 1956) is the basic reference although considerably superseded by various technical publications. Another of the Johnson reprints (see above) is Luella Agnes Owen's 1898 *Cave Regions of the Ozarks and Black Hills*. Aside from Mark Twain's own writings mentioned in this book the best source on Mark Twain Cave is *Adventures at Mark Twain Cave* by Dwight Weaver and Paul A. Johnson, a 64-page booklet sold at the cave. The same authors also have published a fine 94-page booklet on *Onondaga, the Mammoth Cave of Missouri*. Little has been published on Arkansas caves, which may be one reason we are losing them so fast.

5 • Of Golden Legend

My technical report, *Caves of California* (Western Speleological Survey, 1962) is not widely available. *Adventure Is Underground* (Harper, 1959) contains much on California caves.

The famous *Playboy* Magazine interview debunking Erich von Däniken's *Gold of the Gods* cave in Ecuador—and much more—appeared in its August 1974 issue. Those interested in tracing the development and commercialization of today's hollow-earth cult may wish to compare *The Hollow Earth* by well-meaning F. T. Ives (New York, Broadway Publishing Company,

1904) with its 1969 namesake by Raymond Barnard (Secaucus, N.J., University Books, Inc.).

Although some of its conclusions are now known to have been incorrect, M. R. Harrington's most important work was *Gypsum Cave, Nevada* (Southwest Museum, Los Angeles, 1933).

6 • The Greatest Cave

Willis Lee's articles appeared in the January 1924 and September 1925 issues of the *National Geographic Magazine*. Homer Black's chapter in *Celebrated American Caves* (see p. 409) is generally very good. Best yet is the National Speleological Society's *Guidebook to Carlsbad Caverns National Park*, edited by Paul Spangle in 1960 and still in demand. The epitome of the Jim White story is Ruth Caiar's *One Man's Dream*, now in a privately printed second edition sold widely in the Carlsbad area. That of Abijah Long is in *The Big Cave*, also widely available in that area.

7 • Beneath a Thirsty Land

The National Speleological Society's *Caves of Texas* (1948, Bulletin 10) is rather out of date but still delightful reading. The Texas Speleological Survey has a very active program of technical publications. *Playboy* editor Helmer's tales of the Devil's Sinkhole appeared in the August 1974 *Texas Caver. Bats, Mosquitoes, and Dollars* (Boston, Stratford Co., 1925) by Charles A. R. Campbell, has become a much-sought classic despite its inaccuracies.

8 • Tall Tales and Icy Water

The best-known works on New England and New York caves are Clay Perry's *New England's Buried Treasure* (1946) and *Underground Empire* (1948), both published by the Stephen Daye Press. The latter has been reprinted recently. Skull and McFail's caves have not been described in detail outside the speleological literature. Bulletin 15 of the National Speleological Society was devoted to Pennsylvania caves. Leroy Foote's chapter on The Leatherman in *Celebrated American Caves* (see p. 409) reflected much research. Hypothermia is covered in detail in my *American Caves and Caving* (see p. 409), which contains additional references on that and many related topics.

9 • Southeastern Caves

Caves of Virginia (see p. 411), *Caves of Tennessee* by T. C. Barr, Jr., (Tennessee Division of Geology, 1961, 1974), Larry E. Matthews' *Descriptions of Tennessee Caves* (same, 1971), and Bill Varnedoe's privately published 1973 *Alabama Caves and Caverns* are basic references. Other pertinent publications include the 1967 N.S.S. Convention Guidebook, *The Caves of Alabama*, and the Alabama Geological Survey's *Caves of Madison County, Alabama*, published in 1968 as its Circular 52. Its Bulletin 102, *Exploring Alabama Caves*, is NOT recommended, being full of misinformation and unsafe practices.

10 • Windy Caves

Cave Regions of the Ozarks and Black Hills (see p. 411) contains much turn-of-the-century local color and data about Wind Cave, but the author's conclusions on its speleogenesis are incorrect. A bibliography of Black Hills caves and some geologic data were presented in the *Black Hills Engineer,* Volume 24, no. 4 (1938). The Conns' leading article to date appeared in 1966 in the *Bulletin* of the National Speleological Society (Volume 28, no. 2). The 1959 N.S.S. Wind Cave Expedition Report finally reached print in 1964; those of 1970, 1971, and 1972 were less tardy. Bill Plummer's article on breathing caves appeared in the May 1969 *Journal of Acoustical Engineers* (Volume 46, no. 5, part 1).

11 • Davies Didn't Crawl

Caves of Virginia and *Caves of West Virginia* (see p. 411) are basic but largely outdated references about these caves. Ike Nicholson and Fred Wefer have in press a lengthy N.S.S. *Bulletin* article giving details of the years of struggle and triumph at Butler and Breathing caves. The 1970 N.S.S. Convention Guidebook contains some information on the West Virginia caves, as does *Bulletin 36* of the West Virginia Geological and Economic Survey. This report by William K. Jones, published in 1973, is entitled *Hydrology of Limestone Karst in Greenbrier County, West Virginia* but covers only part of that area.

12 • Determination and Dark Death

Caves of Indiana (Indiana Department of Conservation, 1961) by Richard L. Powell is a basic reference but quite incomplete. George F. Jackson's *Wyandotte Cave* (Livingston Publishing Company, Narbeth, Penna., 1953) has become something of a classic, and his new *Story of Wyandotte Cave* (Albuquerque, Speleobooks, 1975) is even better. Aside from the 1973 N.S.S. Convention Guidebook, virtually all the published information on Blue Spring Cave has been in the N.S.S.'s monthly newsletter. Its Greater Indiana Grotto has recently reprinted many publications on the Lost River karst as a part of the conservation effort mentioned in Chapter 16.

13 • Of Rope and Ladder

Caves of Tennessee, Caves and Caverns of Alabama, and *American Caves and Caving* are especially pertinent here. *Mountaineering, the Freedom of the Hills* (The Mountaineers, Seattle, three editions to date) and other climbing works discuss some of the techniques involved. Many pertinent accounts have appeared in reports and newsletters of the National Speleological Society and its units. Especially notable among these is Bob Thrun's 1973 *Prusiking*. Regarding Neff Canyon Cave, the basic reference is an article by Dale Green and myself in Bulletin 20 of that society (1958). John Lyon's version published in the January 1972 *N.S.S. News* is difficult to reconcile with other accounts in the winter 1973 *Journal of Spelean History* (Volume 6, no. 1) and elsewhere.

Donald R. Myrick has published a fine booklet on Fern Cave, but aside from local press reports, the only accounts of Ellison Cave have appeared in publications of the National Speleological Society and its southeastern units.

15 ● Aqualungs in the Dark

The early story of Devil's Hole has been told in more detail in *Adventure Is Underground* (see p. 411). Much has been written about Florida cave diving. Perhaps best is the August-September 1958 *Natural History* article on Wakulla Springs Cave by Stanley J. Olsen. *Manual of Caving Techniques* (see p. 410) has an excellent section on this technique. James W. Storey's booklet *Advanced Cave Diving* and Lee Somer's *Cave Diving: Equipment and Procedures* (published by the National Oceanic and Atmospheric Administration) are worthwhile.

16 ● They Call It Progress

Vital developments in cave conservation are moving at such a pace that the best sources are current publications of the N.S.S., the Sierra Club, and other conservation organizations. The shame of the inundation of Marmes Cave or rockshelter, however, is best documented in the vertical file of the Washington State Library. Basic material on Onondaga Cave, Indiana's Lost River, and Devil's Hole has been mentioned earlier. Several recent issues of *Missouri Speleology* contain outstanding documentation. Horton Hobbs's devastating 1973 report entitled *The Lost River Karst of Indiana* (properly subtitled *A Study of Conservation and Land Management Problems in a Classic Karst Area in Indiana*) is a formal Conservation Statement of the National Speleological Society. Those concerned with the fate of Rainbow Bridge may wish to refer to my May 19, 1961, article in *Science* (Volume 133, pages 1572–1579). Much has been written about Nickajack Cave. I personally admire the short section in Edwin Way Teale's *North with the Spring*. The principal *National Geographic Magazine* articles on Russell Cave appeared in October 1956 and March 1958. Preston McGrain's key 1952 article "Some Applications of Geology to the Location of Damsites in Indiana" appeared in the *Proceedings* of the Indiana Academy of Science (Volume 61, pages 232–239). Lloyd Parris' 1973 *Caves of Colorado* (Boulder, Pruett Publishing Company) is the first American book to apply to caves the conservation education principles of the famous Sierra Club Exhibit Format Series.

17 ● Beneath the Ice

Publications on glacier caves are still very scant. The closest semblance to an overview is Garry McKenzie's 1970 Western Speleological Survey report, reprinted in 1973 in *Bulletin One* of the International Glaciospeleological Survey. Gene Kiver's initial studies of the Summit Steam Caves of Mount Rainier appeared in *Science* on July 23, 1971 (Volume 173, pages 320–322). Besides those in the speleological literature, various reports and articles on the Paradise Ice Caves have appeared in *Explorers Journal, Pacific Discovery,* and *National Parks Magazine,* mostly by Charley Anderson and myself. Floyd Schmoe's account was in the June 1926 *Nature Magazine.*

18 • The Molten Sewers

The mushrooming field of vulcanospeleology now has several basic references. Especially notable are Ron Greeley's *Geology of Selected Lava Tubes in the Bend Area, Oregon* (Bulletin 71 of the Oregon Department of Geology and Mineral Resources, 1971) and *Lava Tubes of the Cave Basalt, Mount St. Helens, Washington* (with Jack H. Hyde, NASA Technical Memorandum X-62,022, May 1971). My *Caves of Washington* (Washington State Department of Conservation, 1963) remains important. Sylvia Ross's *Introduction to Idaho Caves and Caving* (Idaho Bureau of Mines and Geology, 1969) contains information on lava tube caves of that state and the fissure caves of The Great Rift. The program, guidebook, and proceedings of the symposium on vulcanospeleology (the latter not yet in print) of the 1972 N.S.S. convention contain a great deal of information not readily available elsewhere. The story of the Modoc Indians and their war has been the subject of several recent books of varying accuracy. Despite its age, the basic reference on glacieres is Edwin Swift Balch's 1900 *Glacieres, or Freezing Caverns,* reprinted as part of the Johnson series (see p. 409) with a lengthy introduction by myself which largely brought up to date Balch's list of American glacieres.

Acknowledgments

Fifteen years ago I began my acknowledgments in *Adventure Is Underground* by noting that I was obligated to so many individuals and organizations that it was impossible to thank them all properly. With each new book and with this new edition, the situation has become even worse. I could not even begin to compile a list of all the individuals and organizations which have helped in its preparation and the research which preceded it. To keep this section of manageable length, I must thank jointly all those with photo credits. The files and library of the National Speleological Society, as well as those of many of its individual units and of the Western Speleological Survey, have been of particular importance. Editors of the *Bulletin* and *News* of the N.S.S. kindly permitted use of short excerpts. My wife Len has spent untold hours in the field and on the manuscript stages of the book, and my children, Marcia, Patricia, and Ross, have helped greatly.

Some to whom I owe special acknowledgment include Tom Aley, Charley Anderson, R. R. (Bugs) Armstrong, Bill Austin, Jim Baker, R. G. Babb, the Bancroft Library of the University of California, Thomas C. Barr, Jr., Ron Beach, George Beck, Don Black, Malcolm Black, Hugh Blanchard, Don Bloch, the Bogarts of Mark Twain Cave, former Superintendent Thomas Boles of Carlsbad Caverns National Park, John F. Bridge, Bill Brown, Roger Brucker, Bruce Bryan of the Southwest Museum, Ruth Caiar, author of *One Man's Dream*, Arch Cameron, innumerable members of the staffs of Carlsbad Caverns National Park, Jewel Cave National Monument, Mammoth Cave and Wind Cave national parks, the Cascade and other grottos of the National Speleological Society, Walter S. Chamberlin, Ed Chappell, the Chattanooga, Chicago and Cincinnati public library staffs, Badger Clark for the inspiration of his western poetry and a borrowed line, Herb and Jan Conn, Lyle Conrad, Denny Constantine, Bart Crisman, Charles Coughlin, Don Cournoyer, Bill Cuddington, Tom Culverwell, Lyman Cutliff, William E. Davies, Ray V. Davis and many other citizens of Carlsbad, New Mexico, Roy Davis, Don Davison, Jr., J. G. Day, Dwight Deal, the Denver Public Library, Lester Dill, Ray Dorr, H. H. Douglas, Arthur Doyle, Bob Dunn, Ross Eckler, the staff of the El Paso Public Library, Burton and Wilda Faust, John Fish, Franklin Folsom, Cliff Forman, Gerry Forney, Charles Fort, Jerry Frahm, Michael Furcolow, M.D., Standiford (Tank) Gorin, Ron Greeley and Karl Henize of NASA, Dale Green, Jay Gurley, Russell Gurnee, Gene Hargrove, M. R. Harrington, Robert Harnsberger, Peter M. Hauer, Oscar (Oz) Hawksley, Dr. and Mrs. William H. Hazlett, Alan M. Heller, the Herschends of Marvel Cave, John Holzinger, Frank Howarth, Clarence Hronek and other members of B.C. Speleo-Research, Carl Hubbs, the Huntington Library, George F. Jackson, Senator Henry M. Jackson, Jim Johnston, Cheryl Jones, Ernst Kastning, Gladys Kellow, Eugene Kiver, Carol Kruesi, Lewis Lamon, the Library of Congress, the Louisville and Los Angeles public libraries, Phil Lucas, Barbara MacLeod for her haunting *The Grand Kentucky Junction* and much more, C. Holt Maloney of Endless Caverns, Don Martin, Jim Martin, Larry Matthews (two of them, actually), Malcolm McCombs, Ralph McGill, Alvin McLane, Tom Meador, Harold Meloy, Cal Miller, Bill Mixon, Charles Mohr, Leonard and Barbara Munson, Donal Myrick, Bobbi Nagy, the National Geographic Society, Julia Neal and others of the staff of the Kentucky Library at Western Kentucky State College, Peter M. Neely, the New Albany and New York public libraries, Kennedy (Ike) Nicholson, Stanley Olsen, A. Y. Owen, Art Palmer, Jim Papadakis, Stuart Peck, Paul Perry for courteously making available to me the Clay Perry Collection, J. S. Petrie, Edwin P. Pister, Bill Plummer, E. R. Pohl, Alonzo Pond, Edward Post, Jim Pritchard, Richard Reardon, Jim Reddell, John Reid, Lyman Riley, Don Rimbach, Bill Russell, the St. Louis and San Antonio public libraries, Peter Sanchez, Jim Schermerhorn, Vic Schmidt, Richard Schreiber, the Seattle Public Library and especially Miss Olga Gatz of its Inter-library Loan Department, Mike Shawcross, Hugh Shell, Ken Sinkiewicz and other members of the Vancouver Island Cave Exploration Group, Stan Sides, M.D., the staff of Skyline Cavern, Carroll Slemaker, Gordon Smith, Marion Smith, Donald M. Spaulding, Don Standiford, Arthur P. Stebbins, Jack Stellmack, Bill Stephenson, Mills Tandy, the TVA Department of Information, Peter Thompson, Bill Torode, George Tracey, the United States Forest Service, Philip F. Van Cleave, Bill Varnedoe, Jerry Vineyard, the Wabash College Library staff, the University of Washington and Washington State libraries, Howard Watkins, Patty Jo and Red Watson, Dwight Weaver, Fred L. Wefer, Eb Werner, Dr. Alexander Wetmore, Mr. and Mrs. Paul Whisler, Jim White, Jr., Will White, Mike Wischmeyer, Robert Earl Woodham, Ed Yarbrough, and Hermine Zotter.

And to those who helped, but whose names do not appear here, I owe a special debt of gratitude.

Index

paleontological investigations, 115–16
   Potter Creek and Samuel caves,
    110–12
Palmer, Art, 182, 183, 188–90, 262–63
Palmer, Eric, 123
Papadakis, Jim, 394
Paradise Ice Cave(s), Wash, 364–75
Parker, Norman, 41
Peacock, Charles H., 353
Pecos River, 341–42
Penn's Cave, Pa., 172
Pennsylvania, 171–72, 175–77, 180
   See also names
Perkins, Doug, 312
Perkins, John, 117
Perry, Clay, 34–35, 56–57, 60, 182
Perryville, Mo., 85–86
Peter's Cave, Mass., 173
Peters, Don, 253
Pettibone Falls Cave, Mass., 179
Pettit, Viola V., 106
Pfeiffer, Mike, 154
Pflum, Ron, 363–64, 370
Pinnix, Cleve, 24, 25
Pister, Edward, 330
Pit River Indians, 118–19
Pittsburgh Grotto of National
   Speleological Society, 236–44
Plummer, Bill, 219
Porter, Chuck, 183
Porter Creek Cave, Calif., 110–11
Porterfield, Richard, 92–94
Port Kennedy Cave, Pa., 175–76
Potomac Appalachian Trail Club, 66
Potomac Caver, 343
Poulson, Tom, 26
Powder Mill Cave, 206, 207
Powell, Richard L., 270
Powell's Cave, Tex., 167–68
Prince, Jack, 146
Pueblo Indians, 120
pupfish, 319–21, 331–33, 345
Putnam, F. W., 115, 111
Putnam, Israel, 169–70

Queen, Mike, 183

Rainbow Bridge, 341
Rainbow Bridge National Monument,
   342
Raines, Terry, 156

Ramp, Sandy, 232
Ramsay Cave, Mo., 86
Randolph, John, 53
Ray, Bill, 328
Reader's Digest, 197
Reccius, Jack, 17
Reddell, Jim, 156
Reid, John, 359
Renault, Philippe, 79
Rew, Edward A., 184–85
Rhinoceros Cave, Wash., 383
Richards, Jim, 261–63
Ridley, Oscar, 353, 356
Riley, Lyman, 47–48
Rimbach, Don, 346
Rimstone River Cave, Mo., 86
Rio Grande River, 153
Roberts, Edward E., 125, 127
Robinson, Pete, 228–29
Rockbridge Memorial State Park, Mo.,
   350
Rocky Comfort Cave, Ark., 87–88
Rogers, John, 337
Roscoe, R. J., 184
Rosecrans, William Starke, 200
Rothrock, Andrew, 269–70
Rothrock, Charles J., 269
Rothrock, Peter, 268–69
Rothrock, Washington, 269–70
Rowland Cave, Ark., 98
Ruby Falls Cave, Tenn., 195–96
Ruffner's Cave, Va., 58
Russell, Bill, 154–56
Russell Cave, Ala., 353–58
Russell, F. Bruce, 105–6
Russell, Israel C., 360, 375

Sageser, R., 340
St. Gervais, Switzerland, 360
St. Louis *Post-Dispatch,* 346
St. Louis University Grotto of National
   Speleological Society, 98
St. Mary's University Speleological
   Society, 166
salamanders, cave dwelling, 83–84, 151,
   152, 346, 353
Salsman, Garry, 329
Salt Lake Grotto of National
   Speleological Society, 282–90, 346
Salt Lake *Tribune,* 287
Saltpeter Cave, Mo., *See* Meramec
   Cavern, Mo.